Optical Communications: Components and Systems

Optical Communications: Components and Systems

Edited by
Josephine Winston

Willford Press
www.willfordpress.com

Published by Willford Press,
118-35 Queens Blvd., Suite 400,
Forest Hills, NY 11375, USA

ISBN: 978-1-64728-459-6

Cataloging-in-publication Data

Optical communications : components and systems / edited by Josephine Winston.
p. cm.
Includes bibliographical references and index.
ISBN 978-1-64728-459-6
1. Optical communications. 2. Telecommunication. I. Winston, Josephine.
TK5103.59.H36 2023
621.382 7--dc23

For information on all Willford Press publications
visit our website at www.willfordpress.com

Willford Press

Contents

Preface

Every book is a source of knowledge and this one is no exception. The idea that led to the conceptualization of this book was the fact that the world is advancing rapidly; which makes it crucial to document the progress in every field. I am aware that a lot of data is already available, yet, there is a lot more to learn. Hence, I accepted the responsibility of editing this book and contributing my knowledge to the community.

Optical communication refers to any type of communication that uses light to carry information. An optical communication system is composed of three major components, which include a transmitter, channel and receiver. Transmitter is used to encode a message into an optical signal. Semiconductor devices such as light-emitting diodes (LEDs) and laser diodes are the most commonly used transmitters. The function of a channel is to carry the signal to its destination. Receiver is a component that reproduces the message from the received optical signal. Modern communication relies on optical networking systems which use optical fiber, optical amplifiers, lasers, switches, and routers. The most common type of channel used for optical communications is the optical fiber. The light in an optical fiber is guided by the principle of total internal reflection. The main advantages of optical communication include high bandwidth, exceptionally low loss, large transmission range, and no electromagnetic interference. This book includes some of the vital pieces of work being conducted across the world on various topics related to optical communications. It will serve as a valuable source of reference for graduate and postgraduate students.

While editing this book, I had multiple visions for it. Then I finally narrowed down to make every chapter a sole standing text explaining a particular topic, so that they can be used independently. However, the umbrella subject sinews them into a common theme. This makes the book a unique platform of knowledge.

I would like to give the major credit of this book to the experts from every corner of the world, who took the time to share their expertise with us. Also, I owe the completion of this book to the never-ending support of my family, who supported me throughout the project.

Editor

1

Silicon Avalanche Based Light Emitting Diodes and their Potential Integration into CMOS and RF Integrated Circuit Technology

Kaikai Xu, Weifeng Sun, Kingsley A. Ogudo,
Lukas W. Snyman, Jean-Luc Polleux, Qi Yu and
Guannpyng Li

1. Introduction

As a rapid growing field in worldwide science and technology, silicon nano-photonics has become one of the most promising photonics integration platforms in the last decade. This is mainly due to the combination of a very high index contrast and the availability of silicon complementary metal-oxide-semiconductor (CMOS) fabrication technology, which allows the use of electronics fabrication facilities to make photonic circuitry. Unfortunately, the indirect band-gap of silicon leads to low efficiency and slow efficiency that is unexpected. The rate of electron-hole recombination in silicon material is too low to produce emitted photons in forward biased silicon p-n junctions, but light emission observed from reverse-biased silicon p-n junctions under high electric field was already reported in 1955 by Newman [1]. The radiative transition between hot carriers emits photons larger than the energy gap. Hence the luminescence during avalanche breakdown is characterized by a broad emission spectrum. An example of the high-energy edges of avalanche-breakdown luminescence is shown in Fig. 1. The low-energy edge of the emission spectrum, on the other hand, extends to energies lower than the gap energy, due to the tunneling-assisted photon emission [2].

Since electron-hole pairs are produced during avalanche breakdown, some radiative recombination occurs. Both the electrons and the holes can be heated by the electric field. Since it is attributed to the hot-carrier population under the condition of avalanche breakdown, the light emission can be used for high-speed light-emitting devices, high-speed light amplifiers, and the analytic investigation of hot carrier distribution [3].

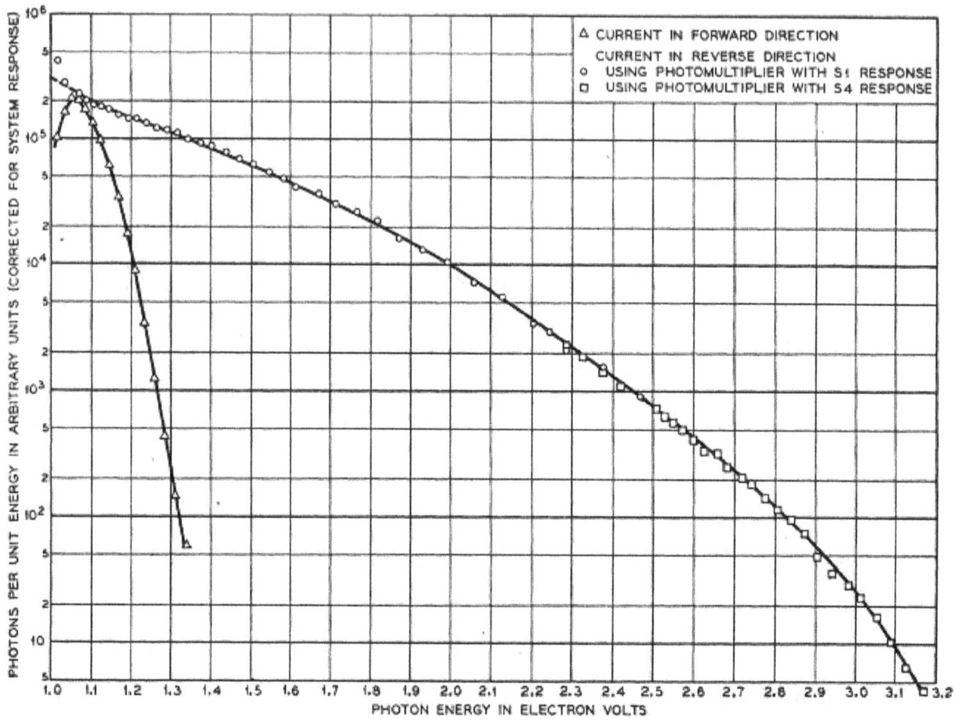

Figure 1. The emission spectra for both forward and breakdown currents corrected for the spectral response of the spectrometer-photomultiplier system (After ref. [2]).

Conventional electric interconnection in today's computers has been a bottleneck for high-speed and large-capacity data transmission. To break through this bottleneck, substituting electric interconnection by optical interconnection is a possible direction [4].

Ideally, all three components (i.e., light source, waveguide, and photo-receiver) of the optical link should be monolithically integrated with the silicon substrate chip and be compatible with complementary metal-oxide semiconductor (CMOS) technology. Silicon photo-receiver circuits can be readily embedded in silicon chips, and silicon-on-insulator (SOI) optical waveguides (i.e., SiO_2 layer) may be used as optical connections. The main difficulty lies in transmitters since light sources can not be efficiently made with silicon because it is an indirect-bandgap material. Efficient light-emitting materials, such as AlGaAs/GaAs, grown on silicon substrates by heteroepitaxy are not sufficiently reliable because of the lattice-parameter and thermal-expansion mismatch between the two materials. One practical approach to addressing the mismatch between the compound-semiconductor optoelectronic technology which is used to fabricate optical sources, and the CMOS silicon technology which is the basis of modern electronics, is the hybrid integration. This approach is based on bonding separately fabricated optoelectronic and electronic chips. A hybrid-integration process known as flip-chip bonding can integrate thousands of optoelectronic devices on a single silicon chip with lateral alignment better than 1 micron [5].

Despite the indirect band-gap, one of the most promising candidate light sources is now thought to be silicon itself because LEDs made of silicon-based materials can be integrated into the existing microelectronic and optoelectronic technologies in a highly economic way. The ideas for direct generation (i.e., electron-hole recombination) of light in silicon by the use of photonic-crystal structure remain in the domain of ongoing research, since the spatial confinement of electron-hole pair on silicon nano-crystal separated by a high-barrier oxide is able to reduce the non-radiative recombination probability and increase the luminescence. Several approaches, such as porous silicon [6, 7], silicon-silicon dioxide superlattice [8], silicon nanoparticles in silicon dioxide [9], erbium in silicon [10–12], silicon/germanium [13–17], and iron disilicide [18], have been considered to enhance its poor light emission. However, these approaches are too complicated to be perfectly compatible with the standard CMOS process technology.

In the present chapter, we will give an overview of the field-effect electroluminescence in the reverse-biased silicon p-n junctions. The following is an outline of the topics covered here. Section 2 provides the construction of a theoretical model which includes the intraband transition and the interband transition of silicon reverse-biased p-n junction electro-optical properties. Section 3 expands the analysis to the explore the silicon gate-controlled diode (a reverse-biased p-n junction with the an additional terminal) that might influence silicon LED: this will indeed show that the, in contrast to the two-terminal Si-diode LED, the three-terminal Si gate-controlled diode LED is able to control the light intensity by adjusting the gate voltage V_g. Since the N^+drain to P^+substrate junction in an N-channel MOSFET (N-MOSFET) device is also a reverse-biased p-n junction, the phenomenon of light emission from the silicon metal-oxide-semiconductor field-effect-transistor (Si-MOSFET) operating in the saturation mode is analyzed by a novel model of physical mechanism in Section 4. Section 5 and 6 present the substantial input in realizing the optical link structures. Finally, Section 7 concludes that the silicon LED will significantly play an important role in future silicon photonic-electronic-integrated-circuits (PEICs).

2. Silicon CMOS avalanche LED

Silicon the material per excellence for electronics is not used for sourcing light due to the lack of efficient light emitter and lasers. In general, high-field electroluminescence can be divided into three categories [19],

- Electroluminescence of powder phosphors (the original discovery by Destriau falls into this category– the phosphor particles were dispersed in a dielectric)
- Electroluminescence of thin films (a homogeneous thin layer of a phosphor filing up the space between capacitor plates-electrodes)
- Electroluminescence of a reverse-biased p-n junction.

In this section, photon emission from silicon was observed by using a p-n junction that operates in the reverse breakdown mode. Indeed, a PMOSFET device can works two p-n junction diodes, as shown in Fig. 2, if both the p⁺drain and p⁺source are grounded while the gate is floated. Substrate voltage V_{sub} is defined as the reverse bias of the two in-parallel connected p-n junction diodes.

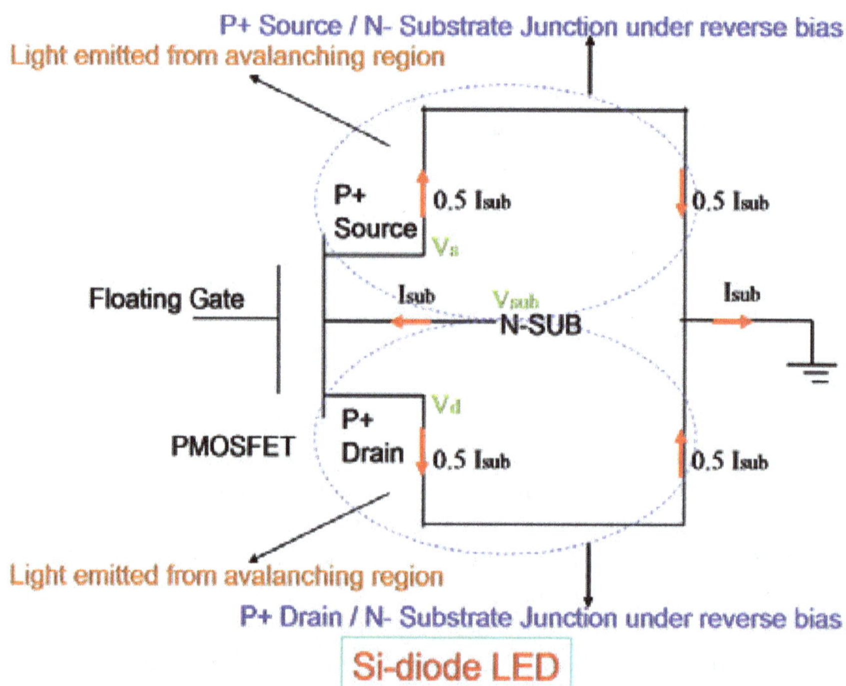

Figure 2. Schematic presentation of the Si-PMOSFET device for the case of Si-diode LED.

In Fig. 3 it is discovered that this radiation originates first in the isolated spots in the close vicinity of the junction area, and with increasing reverse bias these spots merge into a single homogeneous emitting area.

The explanation of this effect is based on the close link between electroluminescence and the electrical breakdown of a reverse-biased p-n junction, as depicted in Fig. 4. At reverse bias, the injection of electrons from n region to the p region and of holes from the p region to the n region can not be achieved. The saturation current across the junction arises from thermally generated minority carriers on both sides of the junction. Due to the high-field, an electron or hole ionizes the lattice, a free *e-h* pair is created and an avalanche like increase in the number of free carriers evolves. The current is increasing substantially, and electron-hole pairs with substantial excess energy gained by the acceleration can, naturally, recombine radiatively.

In addition to Fig. 2, Fig. 5 shows one-half of a PMOSFET-like device that can work as the reverse-biased p-n junction. The device is fabricated using the standard 3µm CMOS process with self-aligned technology. The shape of the spectra and the peaks of emission are basically

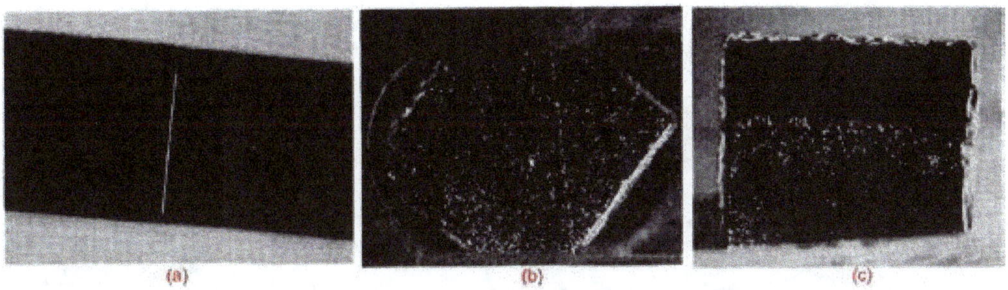

Figure 3. The dark field optical micrograph of the visible light from the reverse-biased silicon p-n junction: (a) The light emission from a grown junction interprets the surface; (b) The light emission pattern from a junction diffused to a depth of a bout 2 microns below the crystal surface, as see in the direction normal to the surface; (c) The light emission from a junction similar to that of (b) except that part of the surface has been ground off at an angle of one degree to the remainder. The sharp white line marking the boundary of the light spots indicates where the junction intercepts the sloping face. (After ref. [2]).

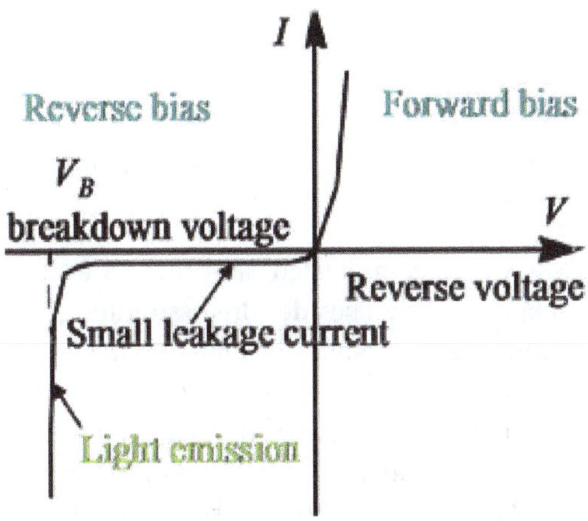

Figure 4. The current-voltage characteristic of a reverse-biased p-n junction.

unchanged under different reverse-bias conditions. Two distinguished wavelength peaks are observed to occur at 650 nm and 750 nm, as shown in Fig. 6.

Normally, in the microplasma (the prerequisite for the assumption of Braking radiation) or avalanche region of the reverse-biased p-n junction, both the holes and hot electrons are presented up to 2.4 eV (corresponding to wavelength ~550 nm) of the pair production threshold for holes and 1.8 eV (corresponding to wavelength ~650 nm) of the pair production threshold for electrons due to high accelerating field [21]. On the other hand, the accelerated carriers will lose some kinetic energies through the collision with artifical atoms (i.e., immobile charged centers) in the depletion region of p-n junction, and the lost energies will be released in the form of photons. On the origin of Bremsstrahlung radiation mentioned previously, the primary wavelength peak 650 nm and the subsidiary wavelength peak 550 nm may be explained by the two threshold values. In fact, the mechanism behind the light emission under the avalanche

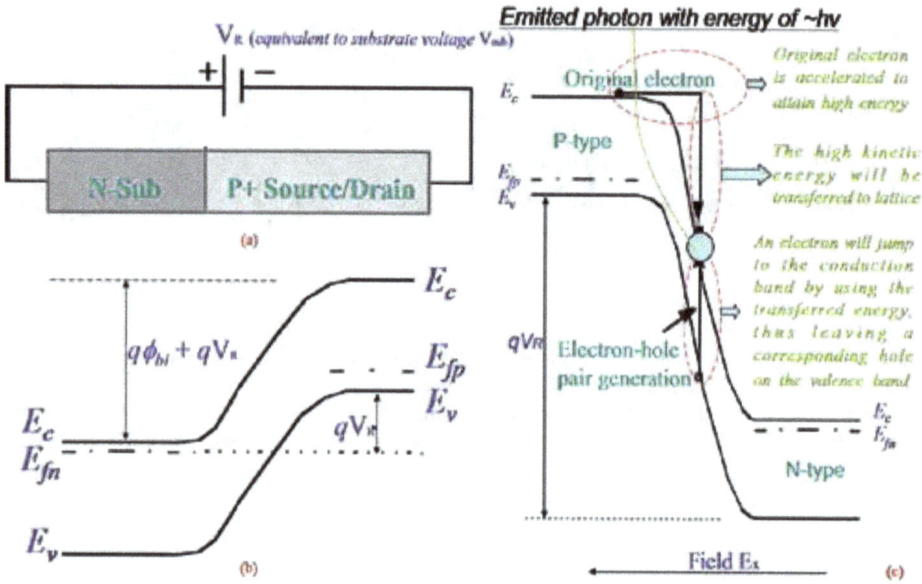

Figure 5. Energy band scheme which demonstrates current amplification due to impact ionization and depicts radiative recombination of an *e-h* pair giving.

breakdown condition is extremely complicated, and there are other possible mechanisms contributing to this type of light emission besides Bremsstrahlung radiation. In [22], a reasonable model is proposed to indicate that avalanche emission in silicon longer than ~620 nm is primarily due to indirect interband transition by high-field carrier populations, that Bremsstrahlung (i.e., indirect intraband transition) dominates the range of ~539 nm to ~620 nm, and that direct intraband transition should be the main reason for the photon emission with wavelength shorter than ~539 nm. Accordingly, it is suggested that the overall mechanism is too complicated to be explained by Bremsstrahlung only. The influence of the defects on the localized emitting spots also should be taken into consideration. In addition, the interference of the emitted light by the passivation layer on the surface of the silicon substrate may account for some of the light intensity sub-crests exhibited in Fig. 6.

The visible light emitted from silicon p-n junction reverse-biased at high voltages has been studied and analyzed. This hot-electron electroluminescence has been seen to offer a useful mean for the study of high-field effects, device integrity, transport, real-space transfer and electron energy distributions. Examination of the electroluminescence distribution reveals the electrical weak spots of the silicon device and may indicate the presence of localized breakdown. Spectral measurements suggest that a number of mechanisms contribute to the visible light emission, including indirect intraband transitions and band-band recombination. Lacaita *et al.* explained that the light emission processes in avalanching reverse-biased p-n junctions should be due to electron and hole relaxation and recombination processes occurring in the high-field avalanching region in theory [23]. In contrast, from experimental evidence and models, Bude *et al.* concluded that intraband electron relaxation processes of electrons

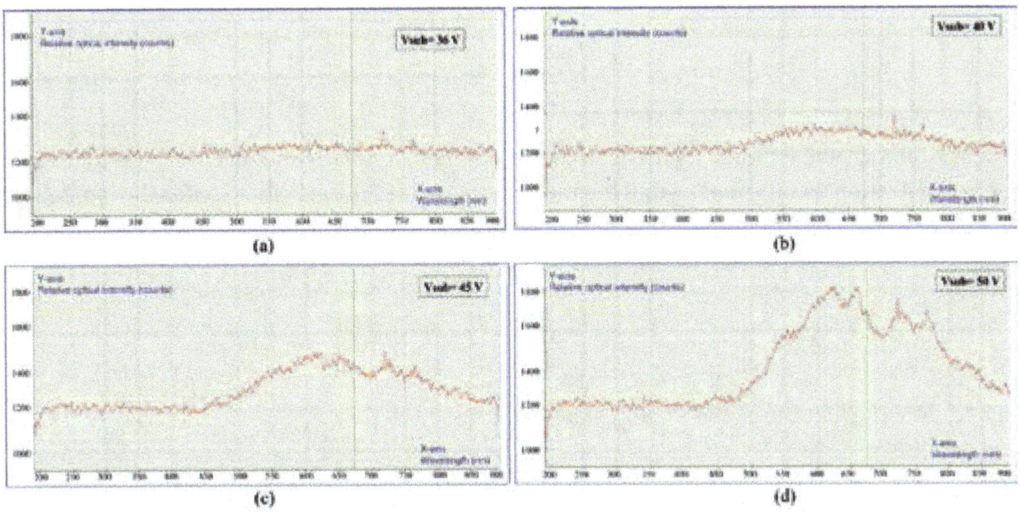

Figure 6. Output emission spectra of the Si light-emitting-device, gate is floated, V_s=V_d=0 V. Emitting wavelength ranges from 500 nm to 850 nm in the measured range of 200 nm to 900 nm (After ref. [20]).

occurring within the conduction band of the silicon itself should be the major reason [24]. It should be noted that a detailed study of the spectrum further allows analysis of the carrier distribution, scattering direction and conduction electron temperature in the space-charge region of silicon p-n junctions.

Commonly, the mechanism of this optical radiation could be explained by the classical electromagnetic theory in which an electron collides with a singly charged Coulombic center. Furthermore, the phenomenon and the rate expression of continuous x-ray spectrum (i.e., Bremsstrahlung radiation) are able to interpret the optical emission observed in the reverse-biased silicon diode if the space-charge region is approximately treated as micro-plasma. It is noted that the x-ray scattering process is distinguished from optical scattering because the energy of the incident x-ray photon with frequency ν_0 becomes large enough with respect to the rest energy of the electrons to give a significant frequency shift to the scattered radiation [25]. In fact, the incident carriers generated by impact ionization will be accelerated by the high field in order to collide with the Coulombic charged centers (i.e., artificial atoms [26]), and then the loss of kinetic energy will be released in the form of photons. The ratio of the Coulombic interaction energy to the thermal energy is given by

$$\Gamma_e = \frac{1}{4\pi\varepsilon_{si}} \frac{q^2}{rkT_e} \tag{1}$$

where k is the Boltzmann constant, T_e is the effective temperature, ε_{si} is the dielectric constant of silicon, and q is the magnitude of elementary charge. r, which is defined as the inter-particle spacing (i.e., Wigner-Seitz radius), is

$$r = \left(\frac{3}{4\pi n}\right)^{1/3} \tag{2}$$

where n denotes the ion density. In accordance with the theory of "hot carrier", the condition for full ionization usually means that the thermal energy of the particles exceeds the ionization energy of the atoms from which the plasma is formed

$$E_{thermal} = \frac{3}{2}kT_e > E_{ion} \tag{3}$$

In addition, the lifetime of the excited state is expressed as [27]

$$\tau = \frac{h}{kT_e} \tag{4}$$

where h is the Planck constant. Since Coulomb collisions serve for distribution of the reduced velocity of carrier, the variation of velocity is usually written as

$$\Delta v = \frac{hK}{m} \tag{5}$$

where K is the photons wave number and m is the ions mass.

The motion of thermal carriers resembles that of a dilute gas, and the distribution of the kinetic energy of thermal carriers can be described by the Maxwell-Boltzmann distribution, derived from the kinetic theory of gases. The probability of carriers having an energy E is expressed in the form

$$\frac{n(E)dE}{n} = \frac{2\pi\sqrt{E}}{\left(\sqrt{\pi kT_e}\right)^3}\exp\left(-\frac{E}{kT_e}\right)dE \tag{6}$$

where $n(E)dE$ is the number of carriers of energy from E to $E + dE$, n is the total number of carriers in the space-charge region which approximately is gas-microplasma, k is the Boltzmann constant, and T_e is the absolute temperature. This distribution gives an energy corresponding to the most probable velocity, or 0.025 eV at 300 K. The temperature, T_e, associated with a Maxwell's distribution of a carrier population, is also called the "carrier temperature".

Using Maxwell's equation in which the particles are the sources for current and charge distribution, the exact field at the position of the particle can be obtained from self-consistent calculation. Consequently, the kinetic equation becomes

$$\frac{\partial f_\alpha}{\partial t}+\vec{v}\bullet\nabla f_\infty+\frac{q}{m}\left(\vec{\mathrm{E}}+\vec{v}\times\vec{\mathrm{B}}\right)\bullet\nabla_v f_\alpha=\left(\frac{\partial f_\alpha}{\partial t}\right)_{coll} \tag{7}$$

where the left hand side contains only averaged quantities and the so-called collision terms on the right hand side contains all microscopic interactions [28]. In general, the existence of radiative elastic collisions (*Bremsstrahlung*) is the scattering of an electron by an external field, accompanied by the emission of a photon. Considering charge-charge or charge-neutral elastic collisions, the collision frequency is calculated as follows. The collision time $\tau=\frac{1}{v}$ is the time for a charge to experience a significant deflection (i.e., change in momentum). The equation of conversion of momentum for the species α is obtained by multiplying Eq. (7) by mv_α and then integrating over v_α. The collision term for momentum transfer can be evaluated for drifting Maxwell distribution functions, and it is found that

$$\int_{-\infty}^{+\infty} dv_\alpha m v_\alpha\left(\frac{\partial f_\alpha}{\partial t}\right)_{coll}=\sum_\beta m_\alpha n_\alpha v_{\alpha\beta}\left(\mu_\alpha-\mu_\beta\right) \tag{8}$$

where μ_α and μ_β are the drift velocities of species α and β. The *charge-charge* collision frequency is given by

$$v_{\alpha\beta}=\frac{\left(m_\alpha+m_e\right)}{3\pi^{3/2}m_\alpha^2 m_\beta}\frac{q_\alpha^2 q_\beta^2}{\varepsilon_{si}^2}n_\beta\left(\frac{2\kappa T_\alpha}{m_\alpha}+\frac{2\kappa T_\beta}{m_\beta}\right)^{-3/2}\ln\left(\frac{12\pi\varepsilon_{si}\kappa T}{q_\alpha q_\beta}\sqrt{\frac{\varepsilon_{si}\kappa T}{e^2 n}}\right) \tag{9}$$

where the symbols of α and β denote charge 1 (electron or hole) and charge 2 (hole or electron), respectively. For charge-neutral collisions,

$$v_{q\beta}=\frac{8\pi^{1/2}}{3}\frac{m_\beta}{m_\alpha+m_\beta}n_\beta\sigma^2\left(\frac{2\kappa T_\alpha}{m_\alpha}+\frac{2\kappa T_\beta}{m_\beta}\right)^{1/2} \tag{10}$$

where σ is the sum of the effective radii of the interacting particles and q denotes the quantum dot which has many properties of natural atoms and is also known as artificial atom.

The optical radiation may also occur via a number of independent competing processes, including the transfer of energy to lattice vibrations (creating one or more phonons) or to another free electron (Auger process). Different types of transition may also take place at surfaces and indirectly via traps or defect centers, which are energy levels associated with impurities or defects associated with grain boundaries, dislocations, or other lattice imperfections that lie within the forbidden band. An impurity or defect state can act as a recombination center if it is capable of trapping both an electron and a hole, thereby increasing their probability of recombining. The presence of these defects and the high electric field inside the junction result in both a reduction in free carrier density and a reduction of their mobility due to increased scattering of the remaining free carriers with the additional charge centers [29]. Hence, impurity-assisted recombination, which is radiative, should be able to partially characterize the light emission spectrums. More experimental analyses have been made to gain more fundamental insight into the physical nature and origins of the light emission process in avalanching n^+p junctions [30, 31].

Avalanche breakdown has been known to occur along the depletion region, approximated treated as "micro-plasma", and they are visible as shining points. These luminescence micro-plasma spots are connected with defects of the crystal lattice and when they are excited each "micro-plasma" site light up at its own breakdown voltage. It is noted that avalanche process is generally known as an inherently fast process. The modulation speed for the avalanche breakdown mode in the Si-diode LED is also determined by the *RC* time constant. In the PMOSFET device, with the dynamic on-resistance *R* of the "P^+Source/Drain to N-Substrate" junction in the 10^{-2} Ω•cm range and the reverse-bias junction per-unit length capacitance in the range of pF/cm, the *RC* time constant will be in the range of tens of femto-second. Although the capacitance-voltage characteristics of reverse biased p-n junction decreases slowly with reverse-bias voltage V_{sub} as the depletion width W_d increases with reverse-bias voltage V_{sub}, a strong recombination in the avalanche region will induce negative capacitance phenomena that make the capacitance $C_{dep}L_m$ decreases rapidly when the Si-diode LED is fully turned on to avalanche breakdown with hard characteristics (i.e., the multiplication factor is equal to infinity) [32]. Therefore, it is reasonably predicted that the *RC* time will be in the range of tens of pico-second, and such a time delay is capable of producing modulation in excess of 10 GHz. The transit time of the excited carriers is given by

$$\tau_0 = \frac{W_d}{v_s} \tag{11}$$

where W_d is the depletion width of the "P^+Source/Drain to N-Substrate" junction and is the saturated drift velocity of the carriers. Here v_s is also known as the thermal velocity of the carriers, i.e.,

$$v_s \equiv \sqrt{3kT/m} \cong 10^7 \, cm/s \tag{12}$$

Hence, a carrier drifting at v_s ~10^7 cm/s through the length of the depletion region of ~2 μm will set the theoretical limit of light modulation at 20 psec, thus leading to a switching speed ~10 GHz in the Si avalanche electroluminescence devices [33].

To find the relationship between the transit time τ_0 and the reverse current I_{sub}, the characteristics of the p-n junction depletion region is analyzed in detail via the rate of electron-hole generation rate *R*.

$$R+\Delta R=\frac{\left(n_0+\Delta n\right)\left(p_0+\Delta p\right)}{n_0 p_0}R=\frac{n_0 p_0+p_0\Delta n+n_0\Delta p+\Delta n\Delta p}{n_0 p_0} \tag{13}$$

On the other hand, "depletion" is equivalent to saying that most of the space-charge region will be completely depleted of carriers, and the product of excess carrier concentrations, $\Delta n\Delta p$, is a very small term that is negligible. For this case, Eq. (13) reduces to

$$\frac{\Delta R}{R}=\frac{\Delta n}{n_0}+\frac{\Delta p}{p_0} \tag{14}$$

Thus the radiative lifetime of excess carriers is

$$\tau_0=\frac{\Delta n}{\Delta R}=\frac{1}{R}\frac{n_0 p_0}{n_0+p_0} \tag{15}$$

Within the depletion region, $n_0 = p_0 = n_i$; therefore, the effective lifetime, also called transit time, within a reverse-biased depletion region is given by [34]

$$\tau_0=\frac{n_i}{2R} \tag{16}$$

and the reverse current generated in the reverse biased "P$^+$Source/Drain to N-Substrate" junction can also be expressed as

$$I_{sub}\sim\left(\tau_0\right)^{-1} \tag{17}$$

The transit time of excited carriers drifting at approximately the carrier saturation velocity (which is of ~10^7 cm/s) to reach the depletion edge of the p-n junction is defined as the time t required for a filament formation (which could set an upper limit to light modulation speed). On the other hand, a localized fall in the breakdown field and enhancement of fields in the depletion region are analyzed to further understand the transient nature of the filament formation using the 2-D device simulator Silvaco-Atlas. Fig. 7 shows the electric field is of the order of 10^5 V/cm, which implies the transit time is a few pico-seconds and the theoretical limit of light modulation speed is of the order of a few tens of gigahertz. A p-n junction diode device fabricated using a 0.18-μm n-well CMOS process was tested to obtain that light modulation as high as 20 GHz could be observed in a reverse-biased silicon p-n junction operating in avalanche breakdown mode [35]. Accordingly, these signals caused light modulation at 20 GHz frequency is given in Fig. 8.

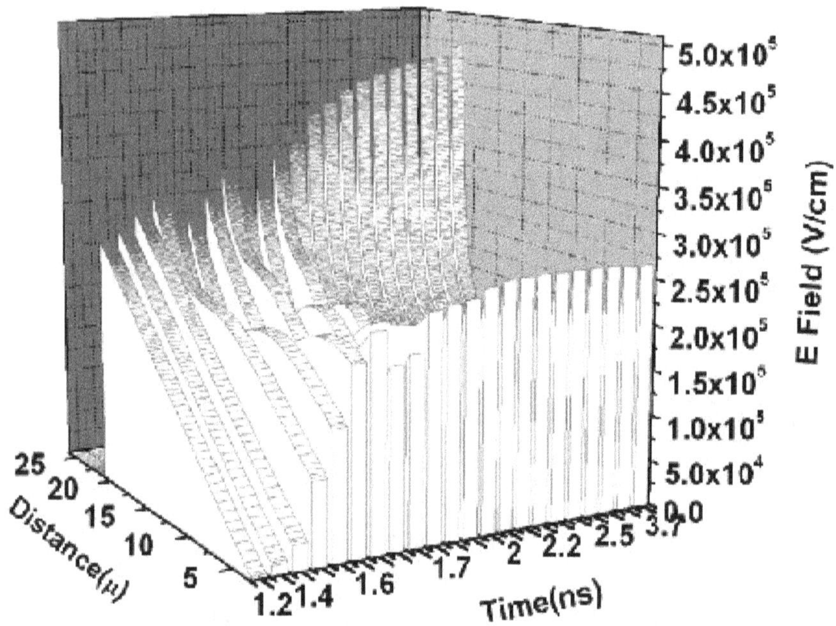

Figure 7. Formation and propagation of an avalanche-breakdown shockwave shows the variation in the electric field (in V/cm) as a function of time and position (After ref. [35]).

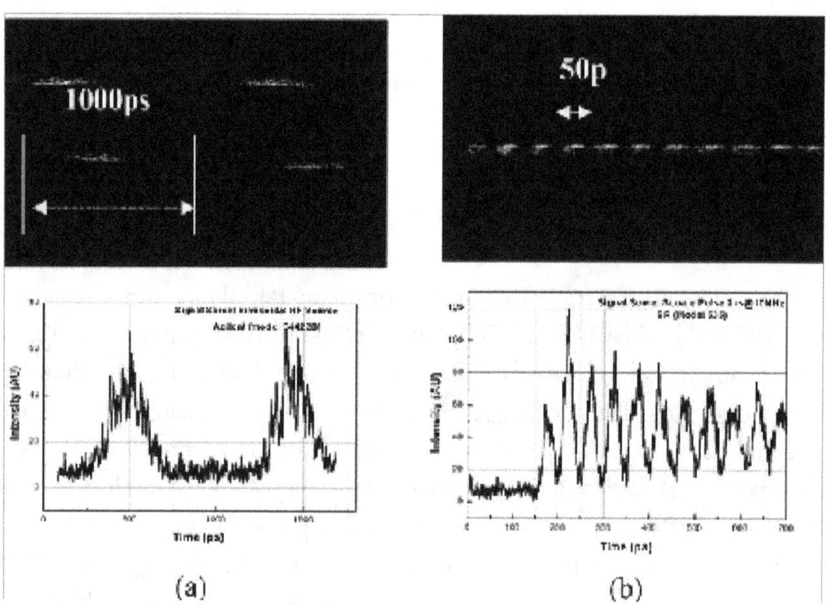

Figure 8. Light emission and intensity measurements of a test structure at: (a) 1 GHz modulation and (b) 20 GHz modulation. The spectrum of the emitted light is distributed between 430–800 nm. (After ref. [35]).

Again, Fig. 2 shows that the proposed Si-PMOSFET device which is fabricated using the standard 3-μm CMOS process can operate as the two-terminal Si-diode LED. In the diode mode, some characteristics of the Si-PMOSFET device are given in [20, 36].

3. Integrated gate-controlled silicon LED

In addition to the study of silicon diode LED in which the light emission is due to the avalanche breakdown of the silicon p-n junction, this section demonstrates the performance of a gate-controlled diode MOS-like multi-terminal device. In contrast to the silicon diode LED (i.e., a two-terminal device), the silicon gate-controlled diode MOS-like LED (i.e., a three-terminal device with an insulated gate on the surface of p-n junction) can use the gate terminal to perform both spatial and light intensity modulation [37–39]. Fig. 9 shows a cross section of the Transconductance Light Emitting Device (TRANSLED) by using a NMOSFET device with a self-aligned poly-silicon MOS gate overlapping an n^+p drain/source junction periphery with a gate oxide thickness of 160 Å working in the accumulation mode, the first gate-controlled three-terminal Si-LED ever reported. Accordingly, the light intensity modulation via the gate voltage is illustrated in Fig. 10, with V_{DD} and R_L kept constant [40].

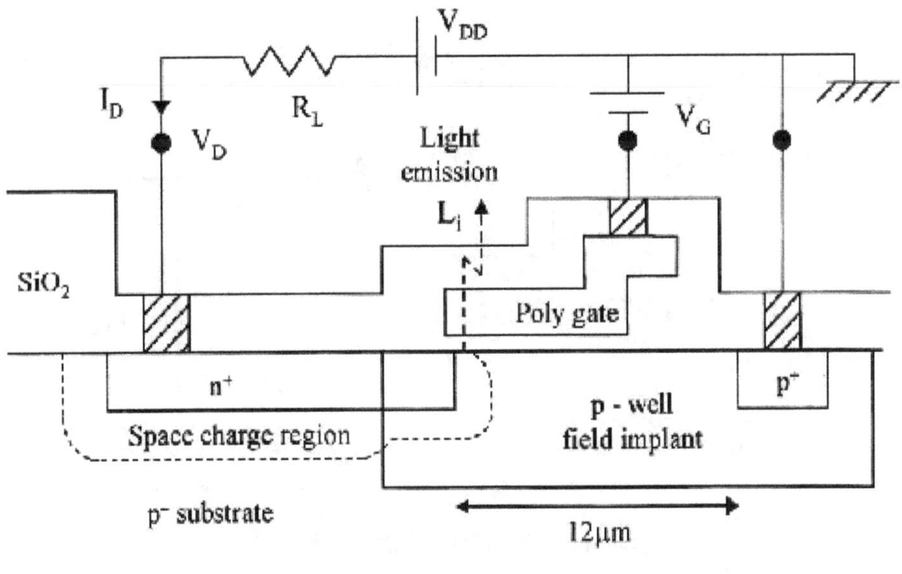

Figure 9. Schematic cross section of the three-terminal gate-controlled TRANSLED, fabricated by standard VLSI technology (After ref. [37]).

Moreover, as shown in Fig. 11, a three-terminal Silicon-PMOSFET-like LED for optical intensity modulation has been discussed in detail based on the concept of p^+n drain/source junction [38, 39].

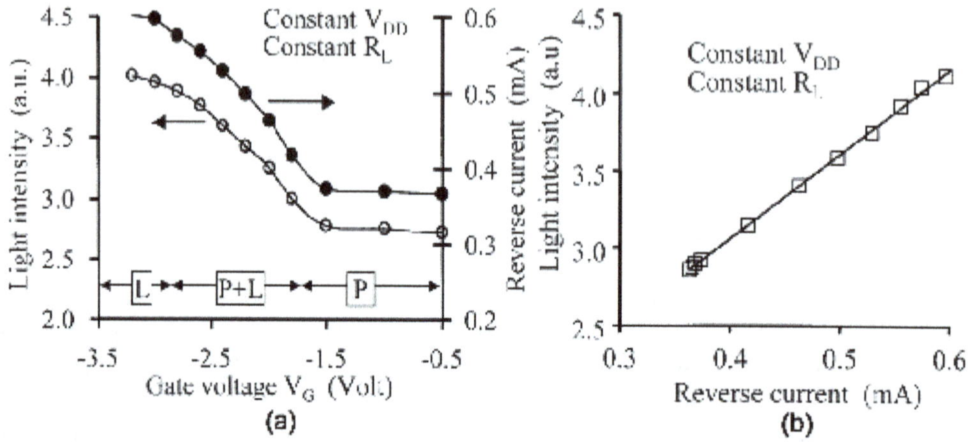

Figure 10. Light emission characteristics of the TRANSLED with constant V_{DD}. (a) Reverse current and overall light intensity, both a s function of V_G. (b) Dependence of the overall light emission on the reverse current (After ref. [40]).

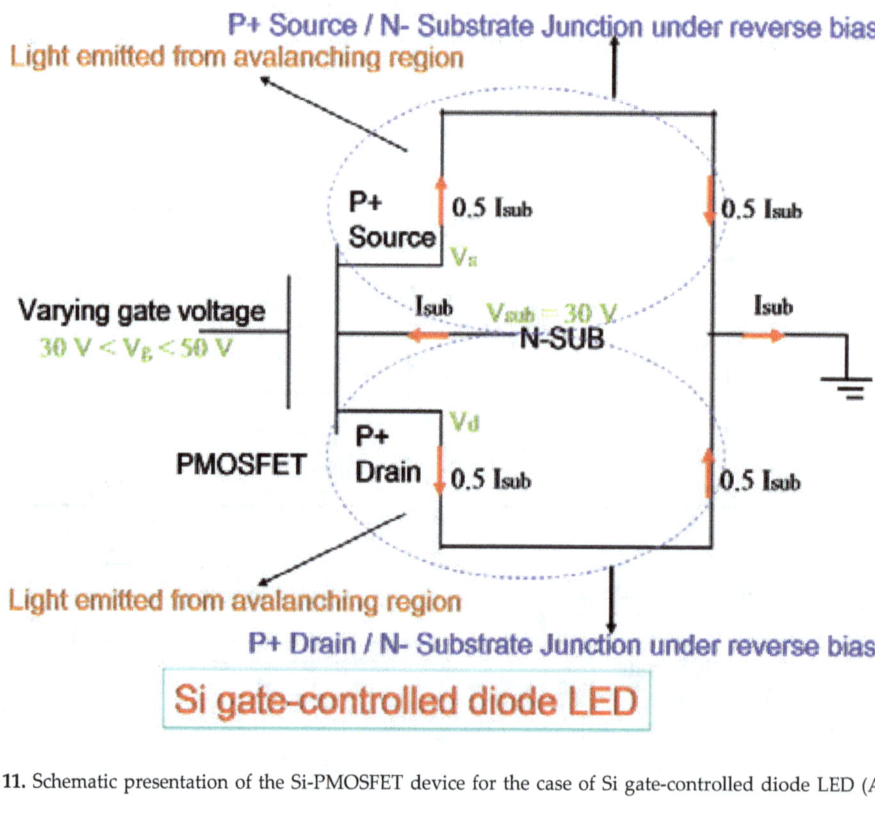

Figure 11. Schematic presentation of the Si-PMOSFET device for the case of Si gate-controlled diode LED (After ref. [41]).

Furthermore, the silicon diode LED (Si-diode LED) is an avalanche-based LED, whereas the silicon gate-controlled diode MOS-like LED (Si gate-controlled diode LED) is a field-emission-based LED. A discussion of the differences observed between avalanche (Si-diode LED) and field emission LED (Si gate-controlled diode LED) performance is presented in [42], and it is discovered that Si CMOS light sources appears with much higher efficiencies by using a gate terminal to modulate the injection-avalanche p-n junction that emit at 400–900 nm. The enhanced intensity should be mainly due to two possible reasons. One reason is that electron accumulation under the gate oxide with band bending will take place if the positive gate voltage V_g is applied, thus the rate of minority carrier injection into the n-substrate is determined jointly by the P^+S/D to N-Sub bias (i.e., V_{sub} in this measurement because V_d=V_s=0 V) and the gate to P^+S/D bias (i.e., V_g in this measurement because V_d=V_s=0 V). At a certain PN junction reverse bias, the injection rate could be significantly higher if the gate voltage Vg is applied to make the diode be a gate-diode. This phenomenon is very similar to the carrier barrier lowing effect that occurs in the lateral bipolar transistor [43]. The increased injection rate caused by the positively increased gate voltage V_g may lead to the enhanced emission power observed in the three-terminal device, gate-diode. The other reason why the gated-diode has obviously enhanced emission intensity is that the accumulation of electrons at the surface of the N-Substrate confines the photon emission to the surface of the device, thus reducing the optical absorption by the silicon material itself. In general, an improved quantum efficiency of the gate-controlled diode Si-LED can be obtained, with respect to avalanche Si-LED operating at the same reverse currents (i.e., the avalanching current I_{sub} in this study).

In gate-controlled diode, the breakdown voltage of the p-n junction will be adjustable so that the reverse current I_{sub} flowing through the p-n junction at a fixed reverse-bias voltage is changed. It is observed that the light, which is emitted from the defects located at the p-n junction, depends closely on the reverse current I_{sub}. In regard to the phenomenon of electro-luminescence, the relationship between the optical emission power and the reverse current I_{sub} is linear.

Assuming the depletion region of p-n junction to be a micro-plasma, the mechanism can be approximately treated as Bremsstrahlung radiation. Usually, the electron-ion Bremsstrahlung power emitted by the volume of plasma V_p into the solid angle $d\Omega$, and in the wavelength interval $\lambda_s \rightarrow \lambda_s + d\lambda_s$, is

$$P_B d\Omega d\lambda_s = 2.09 \times 10^{-36} g Z^2 \left(\frac{n_e n_i}{\lambda_s^2 \sqrt{T_e}} \right) \exp\left(-\frac{1.24 \times 10^{-4}}{\lambda_s T_e} \right) V_p \frac{d\Omega}{4\pi} d\lambda_s \tag{18}$$

where Z is the charge on the ion and the Gaunt factor g is temperature and density dependent [44]. After integration, Eq. (18) becomes

$$W_D = \frac{n_e q^2 v}{m_e c \varepsilon_{si} \omega_i^2} \frac{P_i}{A} \tag{19}$$

where W_D is the average power dissipated per unit volume in units of Wm^{-3}, n_e is the carrier concentration in units of m^{-3}, v is the optical frequency, m_e is the effective mass of electron, c is the speed of light, ω_i is the angular velocity, ε_{si} is the dielectric constant of silicon, and $\frac{P_i}{A} = \frac{c\varepsilon_{si}E_{io}^2}{2}$ is the incident power per unit area [45]. In accordance with the relation between hot-carrier effects and electron temperature, the energy distribution function $f(E)$ of photon intensity can be expressed as

$$f(E) = C\exp\left(-\frac{E}{kT_e}\right) \tag{20}$$

where C is a constant, E is the energy of photon, k is the Boltzmann's constant, and T_e is the effective temperature of electron. At the same time, the relative change of electron temperature is given by [46]

$$\frac{\Delta T_e}{T_e} = \frac{\int_0^{\tau} W_D dt}{1.5\kappa T_e n_e} \tag{21}$$

where τ is the incident pulse duration.

In the case of Si gate-controlled diode LED, higher emission efficiency obtained at lower electric-driving power is attributed to the surface-assisted radiative transitions [39]. Ref. describes the electrical-optical characteristics, and the theoretical calculation and the theoretical calculation and simulation modeling for the topic of modulation speed. The limiting speed of light modulation in reverse-biased silicon p-n junction is analyzed by operating the Si-PMOSFET device in the modes of Si-diode LED and Si gate-controlled diode LED. In Si-diode LED, light emission is due to the avalanche breakdown of p^+source/drain to n-substrate junction and the modulation speed of a few tens of GHz is induced by the intraband transitions of hot carriers and the transit time for sweeping the minority carriers out of the depletion region. In Si gate-controlled diode LED, the p^+source/drain to n-substrate junction breakdown process is triggered by the gate voltage V_g, leading to a localized fall in the breakdown field and enhancement of fields near the p+source/drain to n-substrate junction corner underneath the gate, and the field of the order of 10^6 V/cm is higher than the critical field for avalanche breakdown. Because of the enhancement in electric field, the light modulation speed in Si gate-controlled diode LED will be even much higher [41].

Two additional type of multi-terminal Si-LED's were developed, namely integrated gate controlled (MOS-like structure) array and integrated carrier injection from nearby junction (BJT-like structure) Si-LEDs [47].

4. Light emission from Si-MOSFET's operating in the saturation region

The photo-carrier generation mechanism is fundamentally related to the presence of hot electrons at the drain end of the scaled n-channel MOSFET (N-MOSFET) in which the drain-to-substrate is reverse-biased. Due to the increase in channel current, the drain-substrate p-n junction will reach breakdown, thereby visible light is emitted from the junction region. Furthermore, the conflict between experimentally measured result and theoretically calculated result of the dependence of drain-source current I_{ds} on substrate current I_{sub} shows that there is an secondary impact ionization induced current which is believed to be attributed to the photo-induced process of electron-hole pair generation in the n-type substrate of the N-MOSFET device. This secondarily generated hot-electron (*SGHE*) injection current is also called the minority current observed in the substrate. It is noted that the minority current is the current that is different from the substrate current I_{sub} even if the relationship between the two types of current is almost linear.

Since I_{sub} is mainly due to the mechanism of impact ionization, the relationship between I_{sub} and I_{ds} is given by

$$I_{sub} = I_{ds}\int_0^{L_D} A_i \exp\left[-B_i/\mathrm{E}(y)\right]dy \tag{22}$$

where $\alpha = A_i \exp[-B_i / E(y)]$ is the impact ionization rate, A_i and B_i are the ionization constants and are mainly a function of silicon material parameter itself, and L_D is the length of break-down region near the drain-to-substrate corner. Solving Eq. (22) using a quasi two dimensional (2-D) analysis of field $E(y)$, I_{sub} can be derived from I_{ds}

$$I_{sub} = I_{ds}\left(A_i/B_i\right)\left(V_{ds} - \kappa V_{dsat}\right)\exp\left[-L_D B_i/\left(V_{ds} - \kappa V_{dsat}\right)\right] \tag{23}$$

where the peak field E_m is located at $V_{ds} - \kappa V_{dsat} \approx L_D E_m$. $\kappa = 1$ will make sense if L_D is equal to the effective ionization length L_e that is defined as $L_e = \sqrt{(\varepsilon_{si} / \varepsilon_{ox}) t_{ox} X_j}$ where ε_{si} is the permittivity of silicon, ε_{ox} is the permittivity of SiO_2 (i.e., the insulator layer in N-MOSFET), t_{ox} is the thickness of the SiO_2 layer, and X_j is the junction depth of the drain/source region [48]. In addition to the substrate current, the *SGHE* minority current is given by $I_{SGHE} = C_1((N_C q B_i \lambda_m) / (E_g m_e A_i)) I_{sub}$ where N_c is the number of charged Coulombic centers per volume, λ_m is the mean-free-path, E_g is the band-gap energy, and m_e is the effective mass of electron. As the first two terms are both constants, I_{SGHE} is linear with I_{sub}. Consequently, the relationship between I_{sub} and the optical emission power $P_{optical}$ is linear. It should be noted that the distribution of above-mentioned hot-carrier is defined by the Boltzmann approximation $f(E) = \exp(-E / kT_e)$. Meanwhile, the indirect emission rate of photons in the high-field of is expressed as [49]

$$R_i(hv) \propto \int_0^{hv-E_g} hvf(E)\left[1-f(E-hv)\right]\sqrt{E\left(hv-E_g-E\right)}dE \qquad (24)$$

Substituting $f(E)$ into Eq. (24), it follows as

$$1-f(E-hv) \approx \exp\left[-(E-hv)/(kT_e)\right] \qquad (25)$$

which implies the Bremsstrahlung radiation because of the quasi-Maxwellian distribution [50]. For the case above, the modulation of light intensity is investigated through the measurement of substrate current in NMOSFET at various gate voltages [51].

5. Realization of first iteration on-chip optical links with Si-Av LEDs, and Si-Ge detectors

Analyses have recently been performed on the viability of integrating Si Av LEDs into standard Si CMOS integrated circuitry or nearest related technology. For this purpose a selection of suitable high speed avalanche diodes based on technology as described above and Si–Ge detector technology as developed by Polleux *et al.* for 650–850 nm detection have been implemented [52]. Several complete optical link structures comprising each of a Si Av LED, an integrated waveguide and a Si–Ge detector were all integrated on the same chip have been all integrated onto the same chip, utilizing a standard Si–Ge RF process that is routinely used for realizing high speed RF integrated circuitry for the cellular phone industry. It was decided to use the Si–Ge RF process as it relates very closely to standard CMOS processes, but it gives an added advantage of realizing Si–Ge detectors in the integrated circuitry, which could then operate at below 1 micron optical wavelength and at very low production factors.

Particularly the realization of suitable optical waveguides were extremely challenging using this technology and approach, and a subsequent thorough investigation were piloted to utilize especially standard "trench and pillar" technology as offered by standard CMOS and RF bipolar processes in order to realize such waveguides. Fig. 12 shows some of the first iteration attempts which includes some details of our first designs. A 0.35 μm RF bipolar process as generally available for the fabrication of RF cell phone components was utilized for the realization process. The process offered pillar/columnar npn epi-layer structure which could be isolated laterally by means of various oxide as TEOS layers.

The analyses show that both silicon nitride and Si oxi-nitride offer good possibilities for transmitting radiation at low loss in the wavelength regime 650–850 nm. Both SiON and Si_3N_5 offer high refractive index of 1.6–1.95 and 2.2–2.4 respectively against a background of available SiO_2 as cladding or background refractive index layers in CMOS silicon [53]. The analytical results by Gorin *et al.* [54] show that a tail-off of absorption from the shorter wavelengths towards the higher wavelengths, via high loss characteristics of 4.3 dB cm^{-1} at 530 nm versus only about 1 dB cm^{-1} at 650 nm and 0.1 dB cm^{-1} loss at 750 nm.

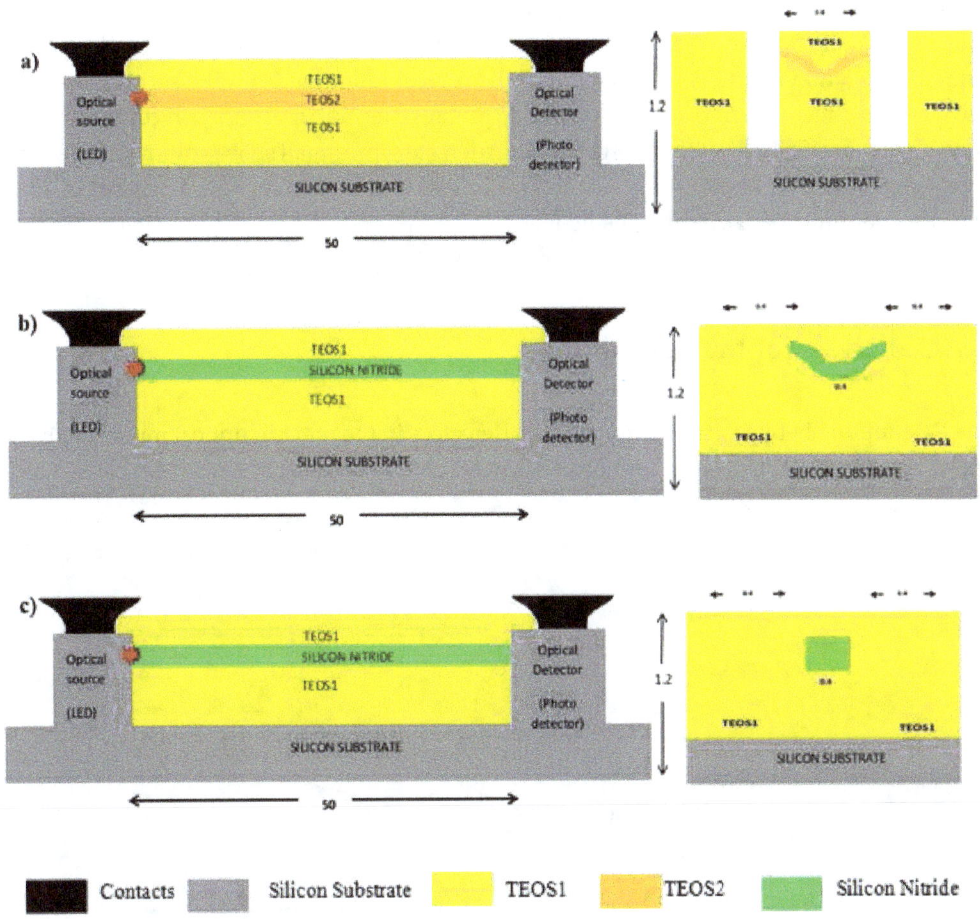

Figure 12. Schematic presentations showing the three basic designs of the waveguides of the optical links as described in this study, using Si Av LED, silicon based waveguides and Si–Ge based detectors. Cross-section layouts of the respective waveguides are given for each design. All dimensions are given in micron and not to scale.

In a first design in Fig.12 (a), the waveguide structure was placed in the outer over layers of the plasma deposit layers. A channel crevice were etched in the TEOS over layers of n=1.46 and then filled with a second TEOS layer. This layer was then densified by a thermal process, increasing its refractive index to about 1.48. A V-shaped cross-section as defined by built-in processing procedures was chosen in order to ensure the highest radiation coupling of the optical source, which was of submicron dimension and spherical in nature, and which was positioned slightly subsurface of the surface of the optical source columnar structure.

In a second design as in Fig 12 (b), an etched crevice of 0.4 micron in the first TEOS layer was filled up with a silicon nitride over layer, followed by further CVD deposited TEOS oxide overlayers. Hence a high refractive index core of "n=2.4" are formed with a surrounding index of "n=1.46".

In a third design, as in Fig 12 (c), the lateral width of the silicon nitride layer was reduced to about 0.3 micron in order to form a narrow higher index core, which would enable a less multi-mode propagation and lower dispersion of the optical radiation from the source. At shorter wavelengths, silicon nitride and silicon oxi-nitride reveals higher absorption coefficients with good detection efficiency with respect to silicon detectors in a small space volume, while longer wavelengths reveals lower loss transmission but with less efficiency in detection in silicon detectors per unit space volume of silicon.

6. AC analysis for the optical links

The RF coupling between the source and the detectors in the on-chip optical micro-links were analysed with a HP 50GHz vector network analyzer (VNA) series 8510 (50 MHz–40 GHz).

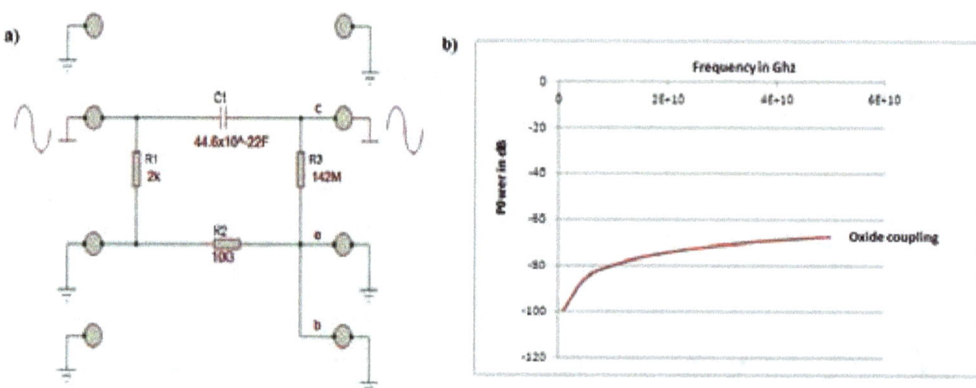

Figure 13. RF Simulation circuitry and plot used for RF coupling analyses of design waveguide-Design1as in Fig. 12 (a).

Simple AC network models of the structure were constructed for each test structure. Fig. 13(a) shows such a circuit for TS1 in simulating the optical link between source and detector. In essence, electrical-to-electrical (E/E) components are microwave two-port circuits and these contained calculated lumped elements for each component as seen by the source signal when placed at the modulation input of the optical source. The value of each lumped element was calculated using known device Design 1 structure dimensions, doping, resistivity and dielectric coefficients of the respective oxide layers. During model circuit simulations, the output power in dB was plotted against the frequency as shown in Fig 13(b).

If such lateral capacitance was indeed present between the detector and the optical source a higher coupling would be observed as modulation frequency is increased. More specifically, an increased RF coupling is observed with increase in frequency, due to the non-linear capacitive coupling.

From our calculations and simulation analyses from the simple waveguide AC network model, the oxide coupling in the structure seems to have a residual-80 dB parasitic coupling when no optical signal was administered at the Si Av LED.

Fig 14 shows some final coupling results for two of the realized optical link structures with regard to utilizing well aligned Si Av LEDs as light emitting sources, an optical waveguide of the type as in Fig. 14(b) and standard Si-Ge detectors as realized by Polleux *et al.* [52].

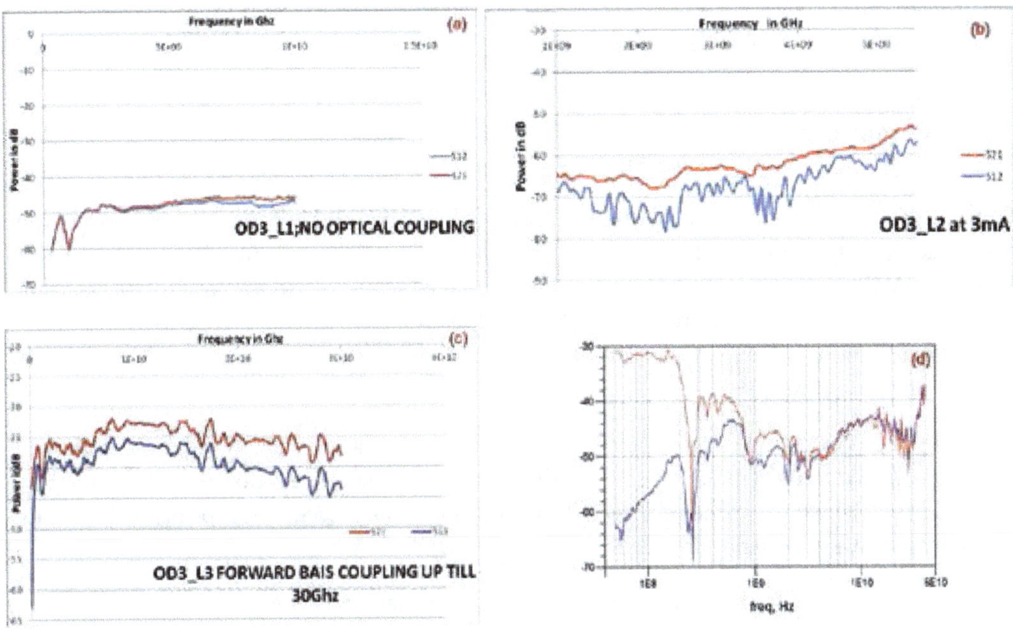

Figure 14. RF coupling results for the realized on-chip micro-optical links. S21 values are indicated as red curves, while S12 are indicated as blue curves. The RF signals were applied at Si Av LED type of devices as elucidated in this chapter (After ref. [53]).

The analyses show clearly higher S21 than S12 for most of the frequency range as investigated. This clearly indicates that some optical propagation did occur along the waveguide. The high S12 value that is observed at higher frequencies is attributed to parasitic conduction along the semi-insulating substrate, when specific capacitance values are analyzed according to the model as in Fig. 12. The capacitive coupling along the oxide layer seems to be minimal and of several orders lower. The fact that higher S21 than S12 is observed up to about 30 GHz and beyond is indicative of the high potential modulation frequencies that can be achieved with Si Av LEDs and even Si-Ge forward bias configurations. Certain design aspects of these structures could still be much improved.

Particularly, further work has been done in utilizing a graded junction concept in order to increase the emission intensity of light as emitting from Si Av LEDs and a high intensity 100 nW 5 GHz Si Av LED using the 0.35 μm micro RF bi-polar process has recently realized [55]. Latest results indicate that due to the enhanced scattering as realized for the highly energized electrons when they penetrate enhanced scattering zones as offered by higher densities of dopant atoms in traverse regions of the device, increases emission intensities substantially. Currently about 100 fold increase in emission intensities was observed as compared to normal p-n junction configurations. Some tentative results have recently been published in this regard, but substantial refinement still needs to be performed [55]. The concepts as utilized in this device design, relates largely to the carrier recombination and carrier scattering concepts as elucidated in Section 2 of this chapter. It can be anticipated that utilizing this technology in optical link structures, may enhance the preliminary optical link performances as obtained above even more.

7. Conclusion

This chapter discusses a topic of interest, namely the phenomenon of light emission from silicon devices. This topic is important since the search for light emitting devices compatible with the CMOS technology and having an efficiency which is useful for certain applications, is still drawing a lot of attention. Further, this is a potential device that can be fully integrated with any standard CMOS integrated circuitry with no adaptation of the CMOS design and processing procedures [56].

Moreover, light emission from Si NMOSFET operating in the saturation region is reviewed, and it is believed that the light emission is due to the snap-back breakdown that occurs at the high field region near the corner between drain and substrate [57, 58]. The type of light emission is also a sensitive hot-carrier degradation monitor [59].

In applications where high speed data transfer and modulation are required, it is founded that the potential high modulation bandwidths with avalanche-based Si light emitting devices (Si Av LEDs) could be of the order of a few tens of GHz in theory [36, 41]. In order to further analyze the high-speed light modulation in avalanche breakdown mode for Si diodes, Snyman *et al.* utilize the concepts of graded junction, carrier energy and momentum engineering to realize a high intensity 100 nW 5 GHZ Si Av LED using the 0.35 μm micro RF bi-polar process [55], and the silicon light source in combination with silicon-nitride (SiN) based waveguides and high speed silicon-germanium (Si-Ge) based optical detectors [53].

Acknowledgements

The College of Optoelectronic Engineering at the Chongqing University, China, is especially thanked for supporting simulated and experimental analyses as well as facilitating spectroscopic analytic work. I am deeply indebted to Prof. Hongda Chen from the Institute of Semiconductors, Chinese Academy of Sciences, and Prof. Jean-Luc Polleux from the ESIEE Paris, France, and Prof. Lukas Snyman from the TUT in South Africa who provide much-need encouragement while I am finishing the work. Mr. Siyang Liu from the Southeast University,

China, is especially thanked for supporting the earlier electrical simulation and modeling analyses.

Author details

Kaikai Xu[1*], Weifeng Sun[2], Kingsley A. Ogudo[3,4], Lukas W. Snyman[3], Jean-Luc Polleux[4], Qi Yu[1] and Guannpyng Li[5]

*Address all correspondence to: kaikaix@uestc.edu.cn

1 State Key Laboratory of Electronic Thin Films and Integrated Devices, University of Electronic Science and Technology of China, Chengdu, China

2 National ASIC System Engineering Research Center, Southeast University, Nanjing, China

3 Department of Electrical Engineering, Tshwane University of Technology (TUT), Pretoria, South Africa

4 Université Paris-Est, ESYCOM, ESIEE Paris, UPEM, Le Cnam, Cité Descartes, Noisy-Le-Grand-Cedex, France

5 Integrated Nanosystems Research Facility (INRF), University of California, Irvine, California, USA

References

[1] R. Newman, "Visible light from a silicon p-n junction," *Phy. Rev.*, vol. 100, no. 2, pp. 700-704, 1955

[2] A. Chynoweth and K. Mckay, "Photo emission from avalanche breakdown in silicon," *Phys. Rev.*, vol. 102, no. 2, pp. 369-376, 1956

[3] L. Selmi, M. Pavesi, H. Wong, A. Acovic, and E. Sangiorgi, "Monitoring hot-carrier degradation in SOI MOSFET's by hot-carrier luminescence techniques," *IEEE Trans. Electron Devices*, vol. 45, pp. 1135-1139, 1998

[4] D. Miller, "Device requirements for optical interconnects to silicon chips," *Proc. IEEE*, vol. 97, no. 7, pp. 1166-1185, 2009

[5] J. Liu, A. Tolvgard, J. Malmodin, and Z. Lai, "A reliable and environmentally friendly packaging technology-flip-chip joining using anisotropically conductive adhesive," *IEEE Trans. Comp. Packag., Manufact. Technol.*, vol. 22, no. 2, pp. 186-190, 1999

[6] K. Hirschman, L. Tsybeskov, S. Duttagupta, and P. Fauchet, "Silicon-based visible light-emitting devices integrated into microelectronic circuits," *Nature*, vol. 384, no. 6607, pp. 338-341, 1996

[7] G. Reed and A. Kewel, "Erbium-doped silicon and porous silicon for optoelectronics," *Mater. Sci. Eng. B*, vol. 40, no. 2/3, pp. 207-215, 1996

[8] Z. Lu, D. Lockwood, and J. Baribeau, "Quantum confinement and light emission in SiO_2/Si superlattices," *Nature*, vol. 378, no. 6554, pp. 258-260, 1995

[9] T. Komada, J. Kelly, F. Cristiano, A. Nejim, P. Hemment, K. Homewood, R. Gwilliam, J. Mynard, and B. Sealy, "Visible photoluminescence at room temperature from microcrystalline silicon precipitates in SiO_2 formed by ion implantation," *Nucl. Instr. Methods Phys. Res. B*, pp. 387-391, 1995

[10] B. Zheng, J. Michel, F. Ren, and L. Kimerling, D. Jacobson, and J. Poate, "Room-temperature sharp line electroluminescence at λ=1.54 μm from an erbium-doped silicon light-emitting diode," *Appl. Phys. Lett.*, vol. 64, no. 21, pp. 2842-2844, 1994

[11] M. Castagna, S. Coffa, L. Caristia, A. Messina, and C. Bongiorno, "Quantum Dot Materials and Devices for Light Emission in Silicon," in *Proc. 32 nd ESSDERC*, Sep. 24-26, 2002, pp. 439-442

[12] C. Brown, "ST lights up silicon LED for CMOS fab lines," EE Times, La Hulpe, Belgium, Rep., Oct.29, 2002

[13] L. Vescan and T. Stoica, "Room-temperature SiGe light-emitting diodes," *J. Lumin.*, vol. 80, no. 1-4, pp. 485-489, 1999

[14] J. Liu, X. Sun, D. Pan, X. Wang, L. Kimerling, T. Koch, and J. Michel, "Tensile-strained, n-type Ge as a gain medium for monolithic laser integration on Si," *Opt. Exp.*, vol. 15, no. 18, pp. 11272-11277, 2007

[15] J. Liu, X. Sun, L. Kimerling, and J. Michel, "Direct-gap optical gain of Ge on Si at room temperature," *Opt. Lett.*, vol. 34, no.11, pp. 1738-1740, 2009

[16] X. Sun, J. Liu, L. Kimerling, and J. Michel, "Room-temperature direct bandgap electroluminescence from Ge-on-Si light-emitting diodes," *Opt. Lett.*, vol.34, no.8, pp. 1198-1200, 2009

[17] X. Sun, J. Liu, L. Kimerling, and J. Michel, "Direct gap photoluminescence of n-type tensile-strained Ge-on-Si," *Appl. Phys. Lett.*, vol. 95, no. 1, pp. 011911-1–011911-3, 2009

[18] D. Leong, M. Harry, K.J. Reeson, and K.P. Homewod, "A silicon/iron-disilicide light-emitting diode operating at a wavelength of 1.5 μm," *Nature*, vol. 387, no. 6634, pp. 686-688, 1997

[19] I. Pelant and J. Valenta, *Luminescence Spectroscopy of Semiconductors*, Oxford University Press Inc., New York, 2012

[20] K. Xu and G. Li, "Light-emitting device with monolithic integration on bulk silicon in standard complementary metal oxide semiconductor technology," *J. Nanophoton.*, vol. 7, no. 1, 073082, 2013

[21] J. Moll and R. Overstraeten "Charge multiplication in silicon p-n junctions," *Solid State Electron.*, vol. 6, no. 2, pp. 147-157, 1963

[22] N. Akil, S. Kerns, D. Kerns, A. Hoffmann, and J. Charles, "A multimechanism model for photon generation by silicon junctions in avalanche breakdown," *IEEE Trans. Electron, Dev.*, vol. 46, no. 5, pp. 1022-1028, 1999

[23] A. Lacaita, F. Zappa, S. Bigliasrdi, and M. Manfredi, "On the bremsstrahlung origin of hot-carrier-induced photons in silicon devices," *IEEE Trans. Electron. Dev.*, vol. 40, no. 3, pp. 577-582, 1993

[24] J. Bude, N. Sano, and A. Yoshi, "Hot-carrier luminescence in silicon," *Phys. Rev. B*, vol. 45, no. 11, pp. 5848-5856, 1992

[25] S. Glenzer, G. Gregori, F. Rogers, D. Froula, S. Pollaine, R. Wallace, and O. Landen, "X-ray scattering from solid density plasmas," *Phys. Plasmas*, vol. 10, no. 6, pp. 2433-2441, 2003

[26] T. Brandes, "Coherent and collective quantum optical effects in mesoscopic systems," *Phys. Rep.*, vol. 408, iss. 5-6, pp. 315-474, 2005

[27] W. Itano and D. Wineland, "Laser cooling of ions stored in harmonic and penning traps," *Phys. Rev. A*, vol. 25, no. 1, pp. 35-54, 1982

[28] H. Zohm, "Physics of "hot" plasma," *Let. Notes Phys.*, 670, pp. 75-93, 2005

[29] F. McLean, J. McGarrity, C. Scozzie, C. Tipton, and W. Delancey, "Analysis of neutron damage in high-temperature silicon carbide JFETs," *IEEE Trans. Nucl. Sci.*, vol. 41, no. 6, pp. 1884-1894, 1994

[30] L. Snyman, H. Aharoni, M. du Plessis, and R. Gouws, "Increased efficiency of silicon light-emitting diodes in a standard 1.2-μm silicon complementary metal oxide semiconductor technology," *Opt. Eng.*, vol. 37, no. 7, pp. 2133-2141, 1998

[31] L. Snyman, M. du Plessis, E. Seevinck, and H. Aharoni, "An efficient low voltage, high frequency silicon CMOS light emitting device and electro-optical interface," *IEEE Electron Device Lett..*, vol. 20, no. 12, pp. 614-617, 1999

[32] C. Wang, C. Zhu, G. Zhang, J. Shen, and L. Li, "Accurate electrical characterization of forward AC behavior of real semiconductor diode: giant negative capacitance and nonlinear interfacial layer," *IEEE Trans. Electron Devices*, vol. 50, no. 4, pp. 1145-1148, 2003

[33] F. Stellari, F. Zappa, S. Cova, C. Porta, and J. Tsang, "High-speed CMOS circuit testing by 50 ps time-resolved luminescence measurements," *IEEE Trans. Electron Devices*, vol. 48, no. 12, pp. 2830-2835, 2003

[34] D. Sandiford, "Carrier lifetime in semiconductors for transient conditions," *Phys. Rev.*, vol. 105, no. 2, p. 524, 1957

[35] A. Chatterjee, B. Bhuva, and R. Schrimpf, "High-speed light modulation in avalanche breakdown mode for Si diodes," *IEEE Electron Device Lett.*, vol. 25, no. 9, pp. 628-630, 2004

[36] K. Xu, "On the design and optimization of three-terminal light-emitting device in silicon CMOS technology," *IEEE J. Sel. Topics Quantum Electron.*, vol. 20, no. 4, 8201208, July/Aug, 2014

[37] M. du Plessis, H. Aharoni, and L. Snyman, "A silicon transconductance light emitting device (TRANSLED)," *Sens. Actuators, A.*, vol. 80, no. 3, pp. 242-248, 2000

[38] K. Xu and G. Li, "A three terminal silicon-PMOSFET like light emitting device (LED) for optical intensity modulation," *IEEE Photonics J.*, vol. 4, no. 6, pp. 2159-2168, 2012

[39] K. Xu, "Current-voltage characteristics and increase in the quantum efficiency of three-terminal gate and avalanche-based silicon LEDs," *Appl. Opt.*, vol. 52, no. 27, pp. 6669-6675, 2013

[40] M. du Plessis, H. Aharoni, and L. Snyman, "Silicon LED's fabricated in standard VLSI technology as components for all silicon monolithic integrated optoelectronic systems," *IEEE J. Sel. Top. Quantum Electron.*, vol. 8, no. 6, pp. 1412-1419, 2002

[41] K. Xu, "Electro-optical modulation processes in Si-PMOSFET LEDs operating in the avalanche light emission mode," *IEEE Trans. Electron Devices*, vol. 61, no. 6, pp. 2085-2092, 2014

[42] K. Xu and G. Li, "A novel way to improve the quantum efficiency of silicon light-emitting diode in a standard silicon complementary metal-oxide-semiconductor technology," *J. Appl. Phys.*, vol. 113, no. 10, 103106, 2013

[43] Z. Yan, M. Deen, and D. Malhi, "Gate-controlled lateral PNP BJT: Characteristics, Modeling and Circuit Applications," *IEEE Trans. Electron. Devices*, vol.44, no.1, pp. 118-128, 1997

[44] C. Fortmann, A. Wierling, and G. Ropke, "Influence of local-field corrections on Thomson scattering in collision-dominated two-component plasma," *Phys. Rev. E*, 81, 026405, 2010

[45] E. Gusakov, A. Gurchenko, A. Altukhov, A. Stepanov, L. Esipov, M. Kantor, and D. Kouprienko, "Investigation of ETG mode-scale component of tokamak plasma turbulence by correlative enhanced scattering diagnostics," *Plasma Phys. Control. Fusion* 48, 371-376 (2006)

[46] B. Kronast, H. Rohr, E. Glock, H. Zwicker, and E. Funfer, "Measurements of the ion and electron temperature in a theta-pinch plasma by forward scattering," *Phys. Rev. Lett.* 16, 1082-1085 (1966)

[47] M. du Plessis, H. Aharoni, and L. Snyman, "Two-and multi-terminal silicon light emitting devices in standard CMOS/BiCMOS IC technology," *Physics Status Solidi A*, vol. 201, no. 10, pp. 2225-2233, 2004

[48] N. Akil, S. Kerns, D. Kerns, A. Hoffmann, and J. Charles, "A multi-mechanism model for photon generation by silicon junction in avalanche breakdown," *IEEE Trans. Electron Devices*, vol. 46, no. 5, pp. 1022-1028, 1999

[49] S. Yamada and M. Kitao, "Recombination radiation as possible mechanism of light emission from reverse-biased p-n junctions under breakdown condition," *Jpn. J. Appl. Phys.*, vol. 32, no. 10, pp. 4555-4559, 1993

[50] H. Elghazi, A. Jorio, and I. Zorkani, "Analysis of silicon light emission under breakdown condition using an indirect intraband model," *Opt. Commun.*, vol. 281, iss. 12, pp. 3320-3323, 2008

[51] M. Gurfinkel, M. Borenshtein, A. Margulis, S. Sade, Y. Fefer, Y. Weizman, and Y. Shapira, "Study of hot-carrier-induced photon emission from 90 nm Si MOSFETs," *Appl. Surf. Sci.*, vol.248, issu. 62-65, 2005

[52] J. Polleux, F. Moutier, A. Billabert, C. Rumelhard, E. Sönmez, H. Schumacher, "A Strained SiGe layer Heterojunction Bipolar Phototransistor for Short-Range Opto-Microwave Applications", in *IEEE International Topical Meeting on Microwave Photonics*, MWP2003, Budapest, Hungary, Sep. 2003

[53] K. Ogudo, L. Snyman, J. Polleux, C. Viana, Z. Tegegne, and D. Schmieder, "Toward 10-40 GHz on-chip micro-optical links with all integrated Si Av LED optical sources, Si N based waveguides and Si-Ge detector technology," *Proc. SPIE*, vol. 8991, 899108-1–16, 2014

[54] A. Gorin, A. Jaouad, E. Grondin, V. Aimez, and P. Charette, "Fabrication of silicon nitride waveguides for visible-light using PECVD: a study of the effect of plasma frequency on optical properties," *Opt. Exp.*, vol. 16, iss. 18, pp. 13509-13516, 2008

[55] L. Snyman, J. Polleux, K. Ogudo, C. Viana, and S. Wahl, "High intensity 100 nW 5 GHz silicon avalanche LED utilizing carrier energy and momentum engineering," *Proc. SPIE*, vol. 8990, 89900L-1–12, 2014

[56] K. Xu, "Electric field modulation of light emission in silicon gated diodes," Ph.D. dissertation, University of California, Irvine, CA, July 2014

[57] G. Teh, W. Chim, Y. Swee, and Y. Co, "Spectroscopic photon emission measurements of n-channel MOSFETs biased into snapback breakdown using a continuous-pulsing transmission line technique," *Semicond. Sci. Technol.*, vol. 12, no. 6, pp. 662-671, 1997

[58] L. Selmi, M. Pavesi, H. Wong, and E. Sangiorgi, "Monitoring hot-carrier degradation in SOI MOSFET's by hot-carrier luminescence techniques," *IEEE Trans. Electron Devices*, vol. 45, no. 5, pp. 1135-1239, 1998

[59] K. Xu, "Hot carrier luminescence in silicon metal oxide semiconductor field effect transistor," in Advances in Optoelectronics Research, Nova Science Publisher Inc., New York, pp. 1-28, 2014

2

Light-Emitting Devices – Luminescence from Low-Dimensional Nanostructures

S.H. Mousavi, S.A. Jafari Mohammdi,
H. Haratizadeh and P.W. de Oliveira

1. Introduction

Luminescence is the emission of light that does not cause energy loss from the temperature of the emitting material. The emission of light occurs because of an excitation in the material and has different forms like photoluminescence, electroluminescence, thermoluminescence, etc. Reasonable efficiency, high stability, and easy and economical fabrication methods make light-emitting devices a good choice for mass production.

This chapter focuses on the luminescent properties of low-dimensional nanostructures and reviews the principles of luminescence. Different materials for this application and some of the best-known electroluminescent devices are reviewed. In addition, low-dimensional nanostructures, a simple method for preparing them, and the development of these structures for application in light-emitting devices are briefly described.

2. Fundamentals of luminescence

In this part of the chapter, the basis of radiation from solids that produce visible light is discussed. Luminescence is the emission of light by an excited substance. In order to create an emission, an electron needs to be excited from the ground state (E_1). During transition of the electron from the excited (E_2) to the ground state, a photon is released. In order to start this transition, we need to stimulate the electron in the excited state. This process is shown in Figure 1:

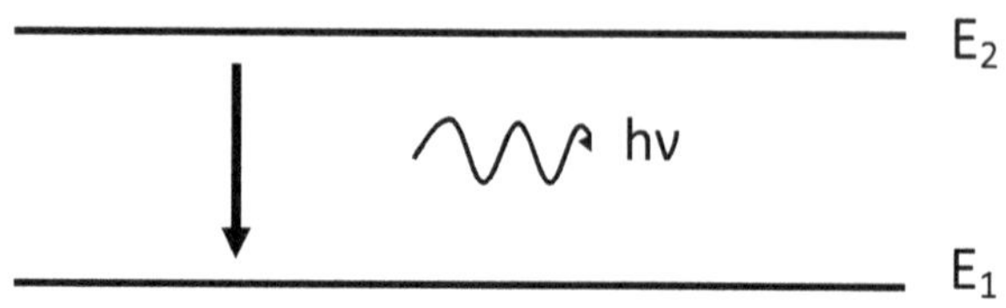

Figure 1. Transition of an electron from excited (E2) to ground (E1) state in a double-state system that results in the release of a photon.

In semiconductors, the ground state is usually referred to as electrons in the valance band while excited state electrons are known as the conduction band. Unlike the metals in semiconductors, these two states are separated by an energy gap called the bandgap (E_g). Therefore, a minimum energy of the bandgap is necessary to excite an electron from the ground to the excited state. Luminescence from semiconductors can be observed by exciting the electrons to higher states like the conduction band and subsequent decay to the ground state. There are different methods of providing the excitation that cause luminescence from a material. Depending on the excitation origin, there are several types of luminescence such as photoluminescence, electroluminescence, cathodoluminescence, chemiluminescence, thermoluminescence, etc. When an electron is promoted from the valance band to the conduction band, a hole will remain in the valance band [1].

If a semiconductor absorbs the electromagnetic radiation of a photon, an electron in the valance band can be excited to the conduction band. When the excited electron returns to a lower state, it causes the radiation of a photon in a process called photoluminescence (PL). When electrical potential is applied, the conversion of energy from electrical energy into light emission is called electroluminescence and the device that produces it is called a light-emitting diode (LED). Due to the narrow nature of the bandgap in elemental semiconductors, they are not suitable candidates for LEDs. Instead, other semiconductors with wide bandgaps are frequently used for this application. The term 'wide bandgap semiconductors' describes those that exhibit light in the visible part of the light spectrum and in the shorter wavelengths of ultraviolet.

Compound semiconductors from group III-V (such as III-nitrides), group II-VI (such as oxides and chalcogenides) are the most important types of wide bandgap semiconductors, although other semiconductors like SiC, Si, and Ge are also studied for light-emitting devices. Figure 2 shows some of the semiconductors that have been used for light-emitting devices. The materials with small lattice constants have stronger interatomic forces and their outer electrons are strongly bound to the lattice, which means that such materials have a large bandgap energy. By changing the structure and doping, many colors can be provided for LED applications. While the infra-red and red LEDs have been studied for 40 years, bright blue, violet, and UV LEDs have been the subjects of research studies in more recent years.

Even more recently, light-emitting devices based on polymer materials have been investigated. In the next section, polymer and organic light-emitting devices are discussed briefly.

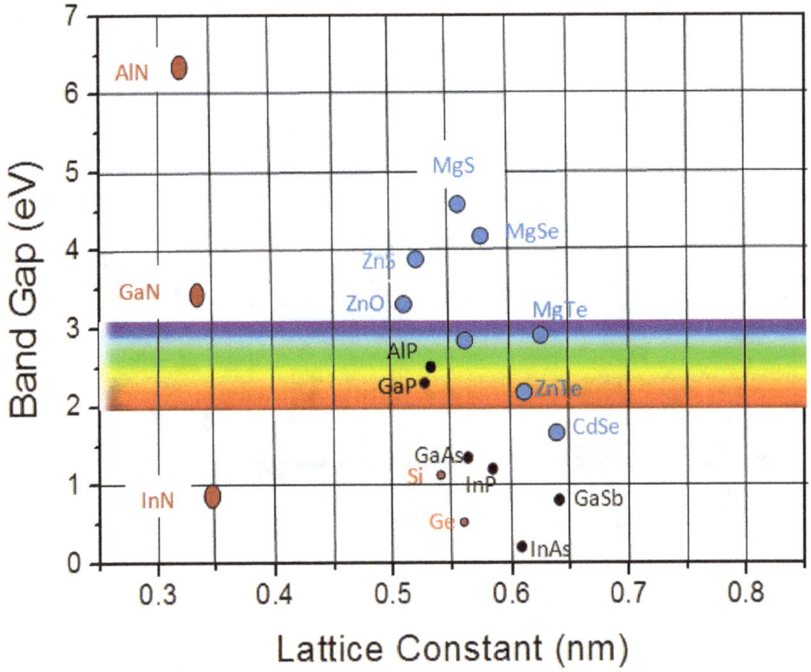

Figure 2. A selection of semiconductors for light-emitting devices. The area between 2 and 3eV are distinguished for visible wavelengths.

3. Light-emitting devices

In this section, different types of light-emitting devices and their working principles are summarized; they are grouped based on the type of material that is applied to create the light emission.

3.1. Emission from inorganic semiconductors

To provide LEDs with different colors, III-V and II-VI materials like GaAlAs, AlinGaP, InGaN, AlN, ZnO, and ZnSe are typically used. The p-n junction is situated between two electrodes, at least one of which should be transparent. Industrial LED applications are highly focused on white color as a light source and as backlighting for electronic devices. New types of LEDs with white color have been developed by mixing luminescent materials in three red-green-blue (RGB) colors. There are two common methods for the generation of white color in LED devices.

The first method uses an individual combination of red, blue, and green LEDs to mix the provided colors and produce the white color shown in Figure 3a, while the second is a combination of different phosphorous materials to generate the white color shown in Figure 3b.

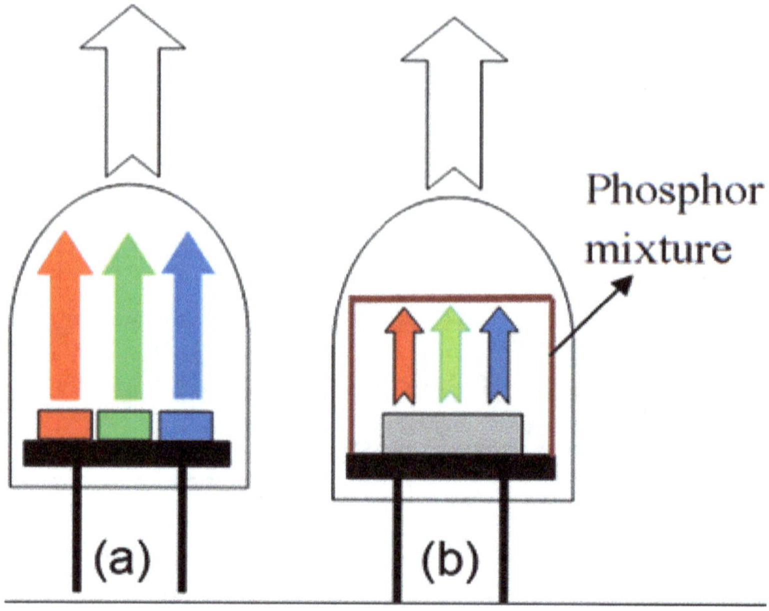

Figure 3. Mechanism of providing light-emitting diode with white color (a) Combination of three red-green-blue (RGB) LEDs (b) Mixing of different luminescent materials to generate the white color.

3.2. Emission from small organic molecules

The first emitting organic material was observed by Pope et. al [2] and led researchers to investigate organic materials as light-emitting devices [3-6]. The working principles of these devices are similar to normal LEDs with the difference that an organic compound is used as an emissive layer. Such devices are called OLEDs (organic light-emitting diodes). A typical OLED consists of at least one emissive layer and one conductive layer which are deposited between two electrodes (anode and cathode). Like semiconductors, these organic materials show other conductivity characteristics under an electrical field between two electrodes. Analogous to p-n junction semiconductors, the highest occupied molecular orbital (HOMO) and lowest unoccupied molecular orbital (LUMO) are regarded as the valance and conduction bands, respectively, of inorganic semiconductors [7].

Figure 4 shows the basic schematic structure of an OLED, which can also consist of an electron transport layer and hole transport layer. At the first stages of the fabrication of OLEDs by Philips, the devices consisted of simple structures of poly(dialkoxy-p-phenylenvinylene) (dialkoxy–PPV) that were situated between a metallic electrode and tin-doped indium oxide (ITO) as the transparent conductive layer [8]. Glass is extensively used as a substrate to fabricate this type of LED, while polyethylene terephthalate (PET) can also be used for fabrication of flexible devices [9, 10].

There has been extensive effort to fabricate OLEDs with primary colors that involved studying different organic materials. H. Fukagawa et. al. [11] reported highly efficient red phosphores-

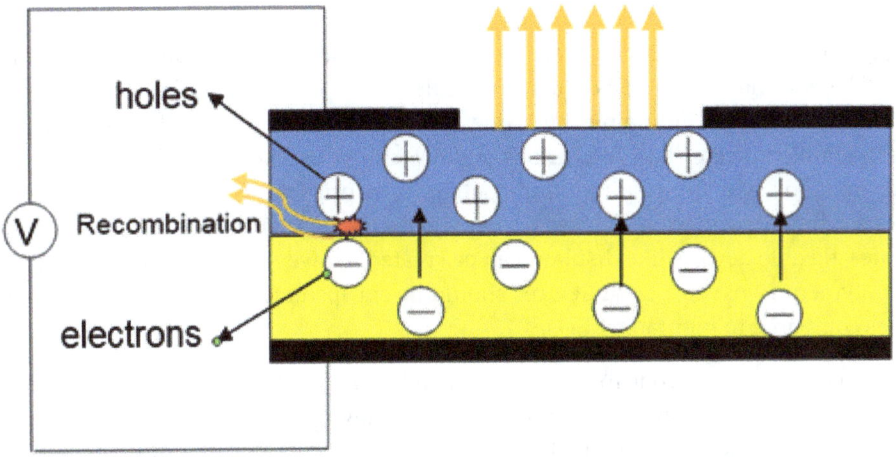

Figure 4. The basic schematic structure of an OLED consist of electrodes (cathode and anode), organic luminescence layer, and transparent conductive material.

cent OLEDs fabricated using platinum complexes as dopants, but most of the studies were performed using iridium complexes and proposed novel complexes of platinum and a useful host/dopant combination. Many host/dopant combinations have been investigated in an effort to provide red OLEDs that show high stability. The optimized device exhibited good color saturation and high efficiency as well as working with low driving voltage. The lifetime of the fabricated devices was estimated to be approximately 10,000 hours.

Green OLEDs with high efficiency, using copper complexes, were reported by S. Igawa et. al. [12]. The copper complexes are an inexpensive emissive choice and exhibit substantial thermal stability compared to other rare earth metals. They exhibit strong green emission at room temperature with a wavelength in the range of 523-544 nm. A conventional OLED structure containing electron and hole transform layers, polymer complex, and electrodes shows the bright green emission with quantum efficiencies between 11.9% and 17.7%.

The complexes that can provide blue colors are more attractive than the red and green complexes for application in OLEDs because of their short wavelength emission. Several organic emissive materials have been used in blue OLEDs. A novel benzofuropyridine complex was synthesized by Chil Won Lee et. al. [13]. These OLEDs were fabricated with two different methods of spin-coating and a vacuum-based route. Both devices exhibited an emission peak at 486 nm, with a strong shoulder peak at 493 nm also observed due to the thickness of prepared thin films. They achieved high quantum efficiencies between 18.0% and 25.0%, but they did not report the lifetime of their fabricated devices.

By mixing these three main colors, white color can be created with OLEDs. The pioneer research study for these LEDs was initiated by Kido et al. [14] by mixing three dyes (blue, green and orange) into the emissive layer to obtain the white color.

3.3. Emission from large polymer molecules

In addition to small organic molecules, polymers can also be used as an electroluminescent material. Due to the variety of polymer molecules, there is great variability in the emission wavelength of polymer light-emitting diodes (PLEDs) [15-17]. Most of the conjugated polymers are soluble in water-based solutions and therefore can be deposited with wet chemical coating methods like dip-coating, spin-coating, and printing, which makes them highly suitable for the fabrication of devices on big substrates [18]. Additionally, by using flexible substrates, flexible devices like displays can be created. However, because of the degradation of the polymers by oxygen, such devices should be encapsulated and therefore have shorter lifetimes compared to inorganic LEDs.

A lot of research has been done to increase the stability of PLEDs [19-20]. A. Berntsen et al. [19] improved the stability of polymer LEDs operating at daylight under ambient conditions. The lifetime about 5000 hours is obtained and under higher temperatures (70ºC), they have one tenth of this time. Y. Cao et al. [21] applied an ultra-thin alkaline earth metal as electrodes for PLEDs and increased their stability and invented this method to increase operating life of PLEDs in which the decay of efficiency and light output is substantially postponed.

Due to the diversity of organic materials, PLEDs can also create different colors from blue to red in visible wavelengths. Several polymer emitters can provide the three primary colors. The common approaches to fabricate the white light OLEDs include doping with small fluorescent or phosphorescent combinations, polymer blending, fabrication of devices with multiple emissive layers, and synthesize a single polymer that can provide the white color [22]. In [22], the authors reported new design techniques for high performance white PLEDs. So far the external quantum efficiency more than 20% has been obtained for these devices and the total power efficiency more than 80 lm/W has been achieved. By improving and increasing the lifetime of these structures, they can be used as energy-saving light sources (for general light, because of low energy consumption), back light for electronic devices (because of low thickness), and flat or flexible displays.

3.4. Emission from phosphor particles

Theoretically, the electroluminescence (EL) from phosphor materials is classified into two groups: injection EL and high field EL. The high-field EL can be further grouped into different types in terms of the phosphors (powders vs. thin film) and the voltage (DC vs. AC). Due to the requirements for various applications, injection EL (high intensity lighting-emitting diodes (LEDs), for example,) have a wide application in liquid crystal displays (LCDs) backlighting and are entering the lighting market.

Luminescence is the emission of lights that does not derive energy from the heating of emitting material. Reasonable efficiency, high stability, and an easy and economical fabrication method make these devices a good choice for mass production. Electroluminescent lamps can be made by sandwiching a phosphor material between two conductive electrodes. At least one of these electrodes should be transparent (TCOs like ITO, FTO or AZO layer). An EL lamp consists of phosphor material; Dielectric layer (e.g. barium titanate with a high dielectric constant to increase the electrical field between the electrodes); Conductive layers (e.g. ITO as transparent layer and silver coating as back contact); Barrier layer to keep phosphor from moisture, and dust. A schematic structure on an EL lamp is shown in Figure (5).

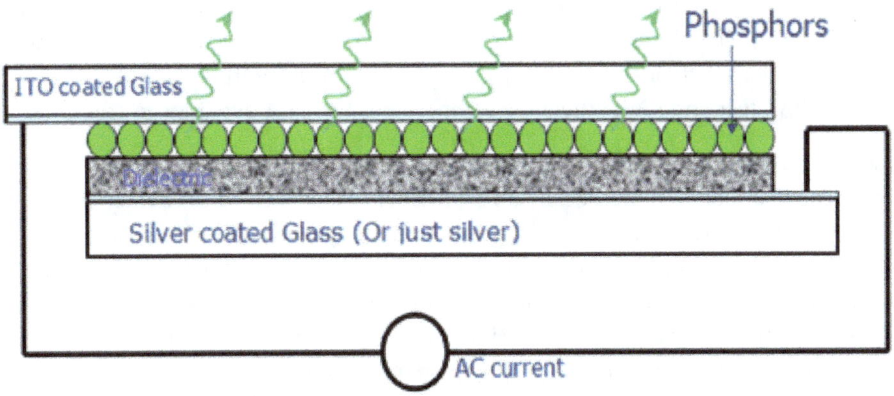

Figure 5. Structure of EL lamp.

By applying a high electrical field (provided with an AC voltage through the conductive electrodes) over the phosphor particles, the electrons will be trapped in the interface layer and will be tunneled to the conduction band of the phosphor layer. The accelerated electrons by electrical field (with enough kinetic energies) excite the impurities from the ground state to excited state. When the electrons come back to their ground state, light is emitted. For this purpose, the electrons must be able to localize in the ground state; otherwise, having a high electric field, they do not emit light. The phosphor material for application in EL devices should be an insulator and contain impurities. It also has to be transparent for the emitted wavelength.

Zinc sulfide is one the most important candidates for phosphor material that have been doped with different chemical elements like magnesium, manganese, copper, or vanadium to provide a different range of colors. Due to the simple structure, they can be used for applications in back-lighting but are still not useful for general light applications.

4. Low dimensional nanostructures

Many physical and chemical properties of materials can be changed effectively by controlling the size of materials between 1 and 100 nm. In such nanostructures, the finite size can generate novel optical, electrical, or magnetic properties that cannot be observed in bulk materials. In other words, the atoms of material on the surface show different properties and therefore by decreasing the dimensions of a material, the high surface-to-volume ratio enables them to exhibit new properties.

Nowadays, semiconducting nanostructures have been the subject of great interest due to their unique optical properties. With more investigation and development into such structures, many optoelectronics devices have been produced. Materials with a large surface area-to-volume ratio can be a good candidate in optoelectronics applications. Applications of devices based on optical properties can include lasers, solar cells, optical detectors and sensors, displays, photo-catalysis, imaging and phosphor devices [23-30].

Various methods have been used for the fabrication of nanostructures, like molecular beam epitaxy (MBE) [31,32], sol-gel [33,34], chemical vapor deposition (CVD) [35,36], lithography [37,38], physical vapor deposition (PVD) [39], and chemical methods [40-41]. Among these methods, the low cost and high efficiency technologies have attracted more attention.

In this section, we present a low-cost and industrial method to obtain such nanostructures with a chemical vapor deposition method by an electrical furnace. In addition, their structural and optical properties are studied. So far, a large number of methods have been used to synthesize wide bandgap semiconductors. Some of the currently used methods have problems regarding high temperature, high vacuum conditions, the necessity of using expensive equipment, and difficulties with quality or commercialization. The focus of this chapter is on development of a low-cost method for producing high quality nanostructures with regard to facilities and experimental conditions.

This method is based on the evaporation of source materials in the hot-zone of a horizontal electrical furnace, reacting with an active gas and finally transporting on the substrates in cold-zone of the furnace using a carrier gas. Despite the simplicity of this method, there are a lot of parameters that can affect a growth mechanism. Among these parameters, the effects of growth conditions such as synthesis temperature, temperature rate, amount of carrier and active gases, source chemical materials, and impurities on the structural properties of these nanostructures are studied.

The variation of these parameters leads us to the systematic modification of one and two dimensional (1D & 2D) nanostructures such as nanowires, nanorods, nanosaws, nanodendrites, nanobelts, and nanosheets. Reasonable emissions of these structures in the blue-green, violet, and ultra-violet (UV) wavelengths can lead us to produce light emitting devices. Also the effect of structural properties and impurities are investigated.

A horizontal furnace is used in this experiment, which is divided into two independent zones: zones (a) and (b), called the hot and cold zones, respectively. Active and carrier gases are introduced through the tube on one side in order to produce the active gas system and carry them to the cold zone of the furnace. The other side of the chamber is evacuated using a vacuum pump. Source materials with the desired molar ratio are inserted into the center of the hot zone. The substrates can be placed at different distances from the source material in the cold zone. Before increasing the temperature, the chamber is flashed with inactive gases (like argon) to remove the residual oxygen from the air. The temperature of the hot zone is increased to, and kept at, different temperatures for various durations. Finally, a thin layer of synthesized material is deposited on the substrates in the cold zone. The distance between the source material and the substrates and also the temperature of the substrates are two important parameters that should be controlled during the deposition. The schematic setup of this furnace with a quartz tube is shown in Figure 6.

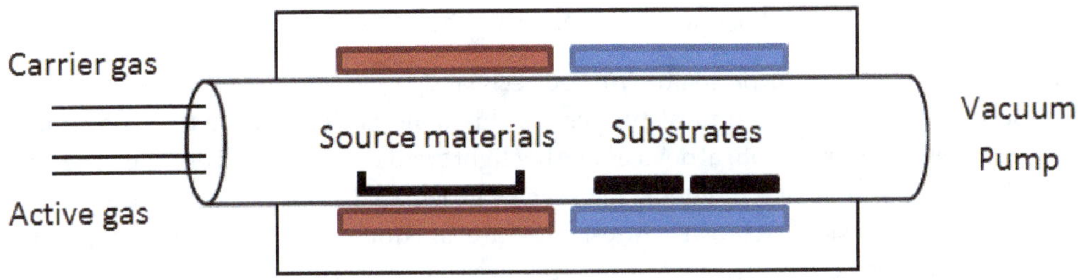

Figure 6. The schematic setup for production of low dimensional nanostructures as light emission material.

5. Luminescence from low dimensional materials

Low dimensional materials attract more interest these days for applications as opto-electronic devices. They have a high surface-to-volume ratio compared to bulk materials and it causes them to show unique properties. When we talk about the low dimensional materials, they can be categorized as: two-dimensional (2-D) like thin films and quantum wells, one-dimensional (1-D) like nanowires, nanorods or quantum wires, and zero dimensional nanostructures (0-D) like quantum dots. Each category shows interesting optical properties [42]. The emission behavior of such nanomaterials is much more complicated than the bulk materials.

The optical properties of nanomaterials depend on many parameters like dimensions, size, shape, temperature, and other variables like dopants and impurities. Even surface characteristics and the surrounding atmosphere have a strong effect on optical properties of these materials. So by controlling these properties, one can generate optical devices covering the whole visible and UV regions of the optical spectrum. The red-shift or blue-shift of photoluminescence spectra of semiconductor nanoparticles is one the most famous examples of changes in optical properties of the materials. In semiconductors, size is the most important parameter to affect the optical properties.

A. Wolcott et. al [43] studied the optical properties of CdTe quantum dots and the results showed different emission colors from CdTe quantum dots. The emission of different colors changes for the particles with different sizes. Increasing the size of the nanoparticles (that were obtained by changing the refluxing time) causes the emitted light to change from blue to red. Also A.M. Schwartzberg showed different colors emitted by hollow gold nanospheres. By varying the diameters and wall thickness, the samples produce different colors from red to violet [44].

Two kinds of wide bandgap semiconductors, such as zinc oxide (ZnO) and aluminum nitride (AlN), are the main subject of this chapter. These nanostructured wide bandgap semiconductors show good light emitting properties in blue-green, violet and ultraviolet (UV) wavelengths. As we discussed in the last section, by controlling the growth condition in this method, we can have different nanostructures with various sizes, dimensions, defects, shapes, dopants and crystallinities. Changes of the surface-to-volume ratio with size and quantum confinement effects cause dependency of the optical properties on the size and dimension.

ZnO is a wide bandgap semiconductor with a direct bandgap of about 3.2 eV at room temperature. Because of the unique optical properties, this material has been used as a suitable semiconductor for different optical devices like UV light emitters, lasers, and detectors. Because of the wide range of synthesis methods, ZnO can be grown in different shapes and structures like one or two-dimensional nanostructures. Structure and doping are two important parameters that have a strong effect on optical properties of this material. One of the major aspects of growth of ZnO is doping of it with different metals to approach an n-type semiconductor.

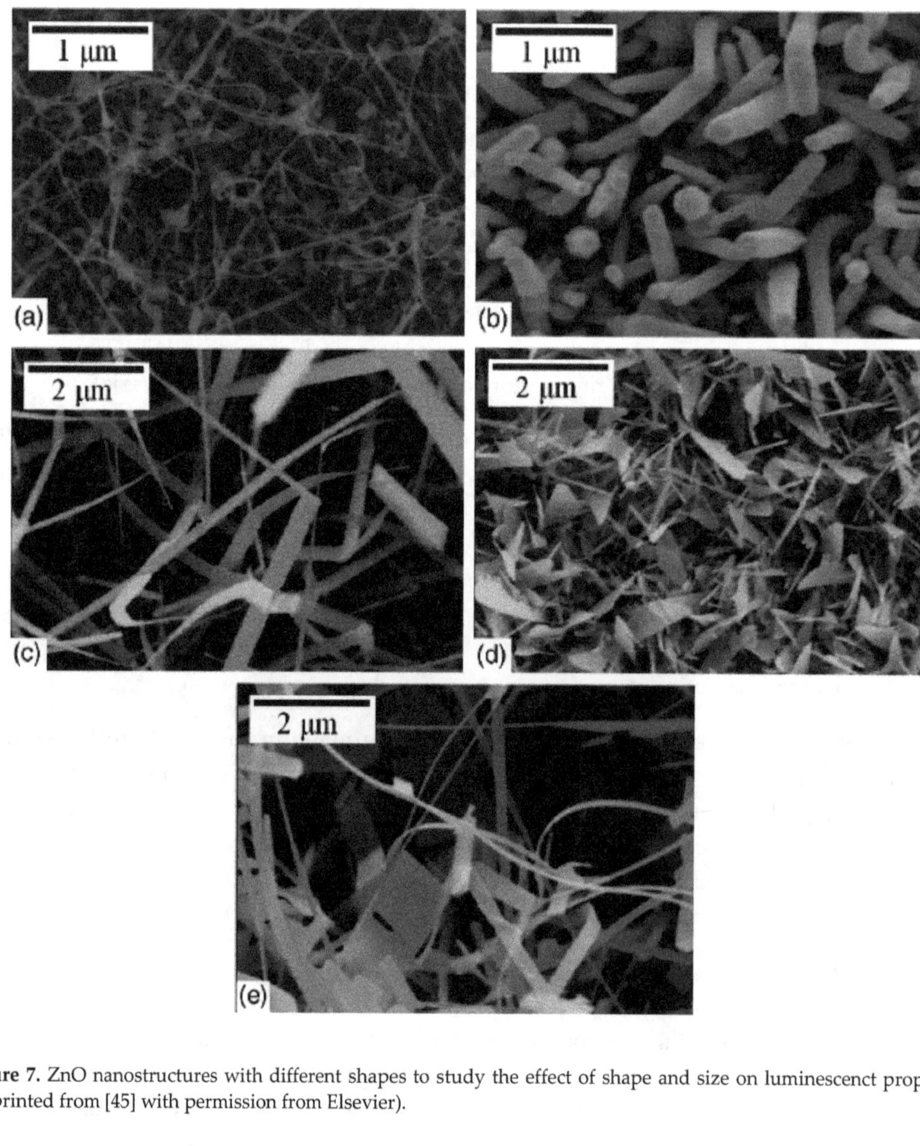

Figure 7. ZnO nanostructures with different shapes to study the effect of shape and size on luminescenct properties (Reprinted from [45] with permission from Elsevier).

In the previous work [45] by changing the growth condition, we prepared different ZnO nanostructures. Nanowires with diameters of 20±5 nm are observed in Figure 7(a). In addition, as the amount of oxygen in the carrier gas mixture increases, the nanostructures change from wire structures to nanorods as shown in Figure 7(b). Figures 7(c–e) show the 2D ZnO nanobelts (with a typical width in the range of 100 to 200 nm), nanodendrites, and nanosheets (typically 10–20 nm in thickness and 1–2 μm in width), respectively, grown on the Si substrate.

After the excitation of an electron with an external energy, electrons and holes possess higher energies and they will recombine together and form an exciton which has a lower energy state. The released energy from this recombination can result in the emission of photons or Auger electrons [7]. In photoluminescence, the recombination of electron and hole cause the radiation of a photon. In semiconductors, these radiations may result from the near band emission transitions and from the defects in the structures.

To explain this fact, we will return to the PL emission from the nanostructures in Figure 7. Figure (8) shows the photoluminescence spectra of the crystalline structure at the related spots. Three main peaks are observed. In addition to an emission in the UV region (λ=380 nm), broad visible emission is observed with two main peaks at ~485 and ~530 nm. The origin of UV emission is radiation because of the recombination of electrons near or in the conduction band with a hole near or in the valance band, which is called near band emission. Due to the bandgap of ZnO (3.37 eV at room temperature), this emission is in the ranges of UV wavelengths. The visible emission from the ZnO nanostructures has been investigated in many studies [46] and this green luminescence in ZnO was attributed to oxygen vacancy, zinc vacancies, interstitial oxygen and zinc atoms, substitution of oxygen at zinc atom positions and donor-acceptor complexes.

The intensity of emissions is dependent on the nanostructure's size. The UV emission was stronger than the visible emission for samples with larger ones. Below a certain size, the luminescent properties of ZnO nanostructures should be dominated by the properties of the surface. An enhanced deep-level emission for thinner nanostructures has been observed and attributed to their larger surface area. As we discussed, the green light emission intensity progressively increases, in proportion to the UV emission, as the nanostructure dimensions decreases. This suggests that there are a large number of oxygen vacancies in the nanowires. Generally, the defects are present at the surface of the ZnO structure. ZnO nanowires with a small diameter have a higher surface area-to-volume ratio, which results in a high level of surface defects. Therefore the visible emission intensity increases.

The effect of doping was also investigated on luminescence properties of ZnO nanostrustures. The ZnO nanostructures grown by this method are doped by only a few percent of copper and iron as the details are discussed in [45]. The photoluminescence from ZnO:Cu can generally be similar to pure ZnO nanostructures underlying physical mechanisms as free and defect-bound excitons, deep and shallow donor–acceptor pair recombinations, and deep-defect associated emissions. A low-intensity shoulder peak at about 455 nm has also been observed as shown in Fig. 9(a).

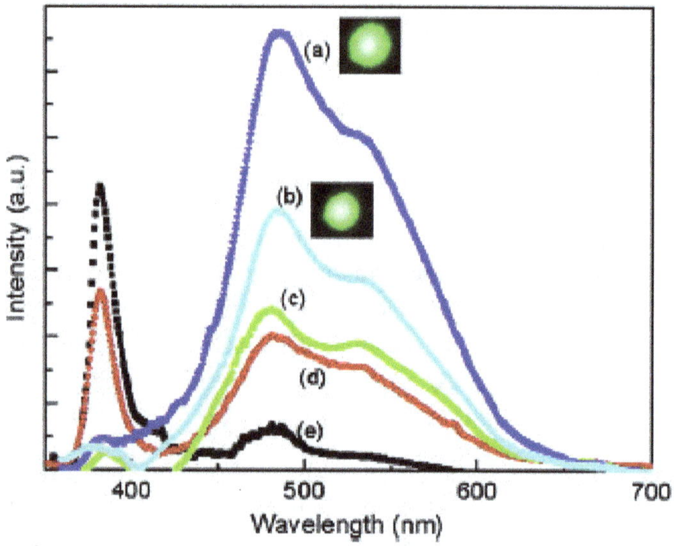

Figure 8. Photoluminescence spectra of ZnO nanostructure with different shapes and sizes (Reprinted from [45] with permission from Elsevier).

The recombination of donor–acceptor pairs involving $Zn^{2+}$and Cu^{+}states is known to yield a blue-green emission. This recombination is responsible for the increased defect luminescence in the ZnO nanostructure. The peak centered at 455 nm can be attributed to Cu^{2+}–Cu^{+}transitions where the hole remains localized on the Cu^{+}center [47,48]. In the case of the introduction of iron as a dopant, the UV emission band of ZnO has a red shift (from 385 to 485 nm) as well as the green emission band showing a red shift (from 530 to 542 nm) as shown in Fig. 9(b).

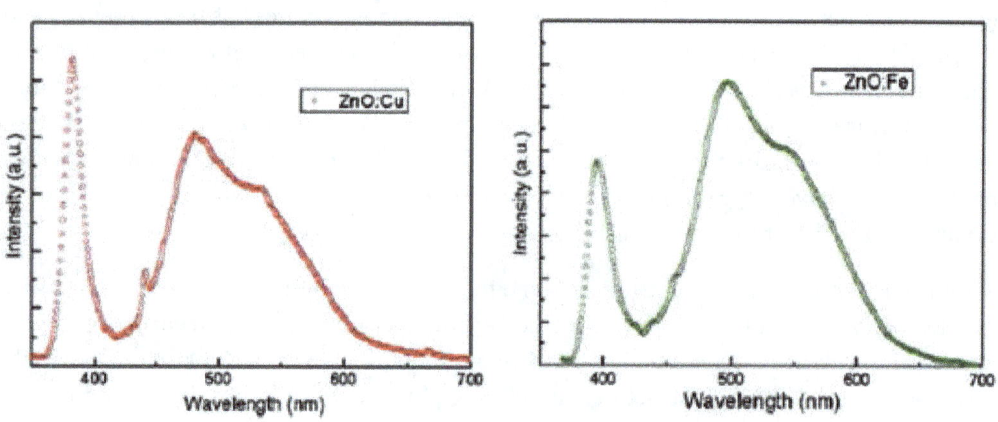

Figure 9. Photoluminescence spectra of ZnO nanostructure with a few percent of (a) copper (b) iron as dopant (Reprinted from [45] with permission from Elsevier).

III-nitrides are also interesting materials for application as light emitting materials. As another wide bandgap semiconductor, AlN, with a bandgap of about 6.1 eV at room temperature, will be introduced. AlN is a direct bandgap semiconductor with a bandgap of about 6.1 eV at room temperature, the largest among semiconductors. Thus, AlN emitters are expected to be good candidates for ultraviolet and blue-violet optoelectronics devices [49]. This work emphasizes novel results on the growth and optical properties of an aluminium nitride (AlN) nanostructure by direct nitridation. The nitridation process was done in a horizontal tube furnace at different temperatures by introducing an N_2 gas flow. Most of the last efforts for synthesizing AlN nanostructures [50, 51] are performed using ammonia gas as this gas is toxic and corrosive.

Among the synthesis methods for AlN nanowires, the direct nitridation of metallic aluminium powder has been attractive because of the low cost of raw materials and the simple nitridation setup. At the beginning of growth, the reaction temperature is higher than the aluminium melting point, so the Al is in liquid form. These liquid droplets cause the aggregation of obtaining nanostructures into spherical islands. However, we have overcome this problem by using ammonium chloride (NH_4Cl). The addition of NH_4Cl to the starting Al powder produced many pores because of the decomposition of NH_4Cl [52-54]. Therefore, the introduction of porosity during the synthesis can enhance the nitridation rate because N_2 gas easily has more access to the pores, which causes the formation of nanostructures AlN powders with low agglomeration. Different AlN nanostructures (Figure (10)) were synthesized and analyzed by scanning electron microscope.

Figure 10. SEM images of AlN nanostructures grown by direct nitridation of aluminium powder. (a, b) AlN nanowires with diameters less than 50 nm (c) AlN nanotips, and (d) long AlN nanowires with diameters about 1 micrometer.

The photoluminescence (PL) spectra were obtained by a 325nm He-Cd laser as the excitation source measured. The results show high intensity light emitting emissions in blue and violet-blue wavelengths for these structures at room temperature. As shown in Fig (11), a typical PL spectrum of these nanostructures shows a peak at 450 nm. Photoluminescence properties of these structures are investigated aiming at the potential application in the field of blue and violet-blue light emitting diodes (LEDs) [55].

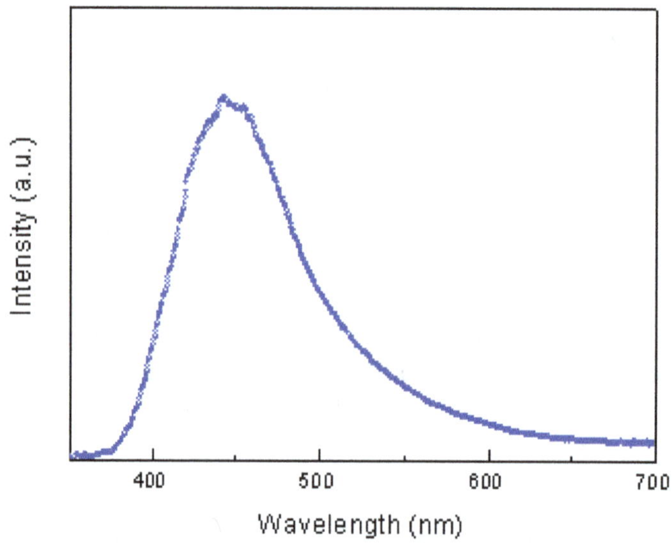

Figure 11. Photoluminescence spectra of AlN nanostructure which shows a broad peak at 450 nm.

6. Conclusions

This chapter was focused on light emitting materials and devices. The working principles of light emitting devices are discussed and different materials for this application are introduced. Different types of light emitting devices, including inorganic light emitting diodes (LEDs), organic light emitting diodes (OLEDs), polymer light emitting diodes (PLEDs), and electro-luminescence (EL) lamps are compared.

Different types of the inorganic semiconductors from group III-V (such as GaN, AlN), group II-VI (such as ZnO, ZnS) are the most important wide bandgap semiconductors, for light-emitting devices. In addition, several organic and polymer complexes that can provide the main three colors have been used in OLED and PLED deives. Due to the wide applications of white color LEDs, different methods and different studies for fabrication of these devices are discussed. Stability of light emitting devices is one the most concerns of the providers. A lot of research has been done to increase the stability of these structures under ambient conditions. The lifetime about several thousand hours is obtained. Electroluminescent lamps are other applications of light emissive materials. Luminescence is the emission of lights that does not derive energy from the heating of emitting material. Therefore due to their efficiency and high stability, these devices are good choices for mass production.

A simple chemical method based on vapor transport is applied for fabrication of semiconductor nanostructures. Nanostructures with different shapes and structures offer several advantages for light emitting devices and are receiving increasing attention as a light emissive material to improve the efficiency of the optical devices. These structures could be developed to overcome the problems for fabrication of UV and blue-violet light emitting devices.

ZnO is a wide bandgap semiconductor which because of the unique optical properties, has been used as a useful semiconductor for the optical devices like light emitters, lasers, and optical detectors. Because of the wide range of synthesis methods, it can be grown in different shapes and structures from one-dimensional structures like nanowires and nanorods to two dimensional nanostructures like nanosheets and nanodendrites. AlN with a direct bandgap semiconductor of 6.1 eV at room temperature, the largest among semiconductors, is one the most applicable semiconductors for application in light emitting device and the AlN nanostructures with different shapes and sizes are expected to be good candidates for optoelectronics devices. ZnO and AlN nanostructures are synthesized using a chemical vapor condensation method and their luminescent properties are investigated at roon temperature.

Author details

S.H. Mousavi[1,2*], S.A. Jafari Mohammdi[3], H. Haratizadeh[2] and P.W. de Oliveira[1]

*Address all correspondence to: sayed.mousavi@inm-gmbh.de, hadi_mousavi@yahoo.com

1 INM –Leibniz Institute for New Materials, Germany

2 Physics Department, Shahrood University of Technology, Shahrood, Iran

3 Aryasolar Chemical Company, Tehran, Iran

References

[1] Shionoya S, Yen WM. Phosphor Handbook. New York: CRC Press; 1998.

[2] Pope M, Kallmann HP, Magnante P. Electroluminescence in organic crystals. Journal of Chemical Physics 1963;38 2042-3.

[3] Meng H, Luo J, Wang W, Shi Z, Niu Q, Dai L, Qin G. Top-Emission Organic Light-Emitting Diode with a Novel Copper/Graphene Composite Anode. Advanced Functional Materials 2013,23(26) 3324-3328.

[4] Jankus V, Chiang CH, Dias F, Monkman AP. Deep Blue Exciplex Organic Light-Emitting Diodes with Enhanced Efficiency; P-type or E-type Triplet Conversion to Singlet Excitons. Advanced Materials 2013;25(10) 1455-1459.

[5] Lee CW, Lee JY. High Quantum Efficiency in Solution and Vacuum Processed Blue Phosphorescent Organic Light Emitting Diodes Using a Novel Benzofuropyridine-Based Bipolar Host Material. Advanced Materials 2013;25(4) 596-600.

[6] Togashi K, Yasuda T, Adachi Ch. Triphenylene-based Host Materials for Low-voltage, Highly Efficient Red Phosphorescent Organic Light-emitting Diodes. Scientific Reports. 2013;42(4) 383-385.

[7] Kitai A. Luminescent Materials and Applications. West Sussex, England: Wiley Series in Materials for Electronic & Optoelectronic; 2008.

[8] Blom PWM, Vissenberg MCJM. Charge transport in poly(p-phenylene vinylene) light-emitting diodes, Materials Science and Engineering 2000;27 53-94.

[9] Kang JW, Jeong WI, Kim JJ, Kim HK, Kim DG, Lee GH. High-Performance Flexible Organic Light-Emitting Diodes Using Amorphous Indium Zinc Oxide Anode. Electrochemical Solid-State Letterts 2007;10(6) 75-78.

[10] Wang ZB, Helander MG, Qiu J, Puzzo DP, Greiner MT, Hudson ZM, Wang S, Liu ZW, Lu ZH. Unlocking the full potential of organic light-emitting diodes on flexible plastic. Nature Photonics 2011;5 753-757.

[11] Fukagawa H, Shimizu T, Hanashima H, Osada Y, Suzuki M, and Fujikake H. Highly Efficient and Stable Red Phosphorescent Organic Light-Emitting Diodes Using Platinum Complexes. Advanced Materials 2012;24:5099-5103.

[12] Igawa S, Hashimoto M, Kawata I, Yashima M, Hoshinoa M and Osawa M. Highly efficient green organic light-emitting diodes containing luminescent tetrahedral copper(I) complexes. Journal of Materials Chemistry C 2013;1:542-551.

[13] Lee CW, Lee JY. High Quantum Effi ciency in Solution and Vacuum Processed Blue Phosphorescent Organic Light Emitting Diodes Using a Novel Benzofuropyridine-Based Bipolar Host Material. Advanced Materials 2013;25 596-600.

[14] Kido J, Kimura M, Nagai K. Multilayer white light-emitting organic electroluminescent device. Science 1995;267 1332-1334.

[15] Rozanski LJ, Castaldelli E, Sam FLM, Mills CA, Demets GJF, Silva SRP. Solution processed naphthalene diimide derivative as electron transport layers for enhanced brightness and efficient polymer light emitting diodes. Journal of Materials Chemistry C 2013;1 3347-3352.

[16] Yang X, Zhou G, Wong W. Recent design tactics for high performance white polymer light-emitting diodes. Journal of Materials Chemistry C 2014;2 1760-1778.

[17] Steckler TT, Fenwick O, Lockwood T, Andersson MR, Cacialli F. Near-Infrared Polymer Light-Emitting Diodes Based on Low-Energy Gap Oligomers Copolymerized in-

to a High-Gap Polymer Host. Macromolecular Rapid Communications 2013;34(12) 990-996

[18] Chen CY et al. Continuous blade coating for multi-layer large-area organic light-emitting diode and solar cell. Journal of Applied Physics 2011;110 094501-094518.

[19] Berntsen A, Croonen Y, Liedenbaum C, Schoo H, Visser RJ, Vleggaar J, van de Weijer P, Stability of polymer LEDs. Optical Materials 1998;9 125-133.

[20] Woo JY, Lee J and Han CS, Alumina/polymer-coated nanocrystals with extremely high stability used as a color conversion material in LEDs. Nanotechnology 2013;24 505714-505723.

[21] Cao Y. Ultra-thin alkaline earth metals as stable electron-injecting electrodes for polymer light emitting diodes, US patent 6452218B1.

[22] Yang X, Zhou G, Wong WY. Recent design tactics for high performance white polymer light-emitting diodes. Journal of Materials Chemistry C 2014;2 1760-1778.

[23] Guo P, Zhuang X, Xu J, Zhang Q, Hu W, Zhu X, Wang X, Wan Q, He P, Zhou H, Pan A. Low-Threshold Nanowire Laser Based on Composition-Symmetric Semiconductor Nanowires. Nano Letters 2013;13(3) 1251-1256.

[24] Tanner MG, Makarov V, Hadfield RH. Optimised quantum hacking of superconducting nanowire single-photon detectors. Optics Express 2014;22(6) 6734-6748.

[25] Imura K, Ueno K, Misawa H, Okamoto H. Optical Field Imaging of Elongated Rectangular Nanovoids in Gold Thin Film. Journal of Materials Chemistry C 2013;117(6) 2449-2454.

[26] Kim S, Song K, Kempa T, Day R, Lieber C, Park H. Design of Nanowire Optical Cavities as Efficient Photon Absorbers. ACS Nano 2014;8(4) 3707-3714.

[27] Li L, Yang S, Han F, Wang L, Zhang X, Jiang X, Pan A. Optical Sensor Based on a Single CdS Nanobelt. Sensors 2014,14(4) 7332-7341.

[28] Chang H. A Study on Intelligent System Design of U-Business Service by Using UbiquitousComponent Technologies. International Journal of Advanced Robotic Systems 2011;12(1) 123-127.

[29] Gourbilleau F, Dufour C, Rezgui B, Bremond G. Silicon nanostructures for solar cell applications. Materials Science and Engineering 2009;159 70-73.

[30] TaiChen J, Hsu CS. Conjugated polymer nanostructures for organic solar cell applications. Polymer Chemistry 2011;2 2707-2722.

[31] Gong Z, Niu ZC, Huang SS, Fang ZD, Sun BQ, Xia JB. Formation of GaAs/AlGaAs and InGaAs/GaAs nanorings by droplet. Applied Physics Letters 2005;87 093116-093118.

[32] Ivanov SV, Toropov AA, Sorokin SV, Shubina T, Sedova VIV, Sitnikova AA, Kop'ev PS, Alferov ZhI, Lugauer HJ, Reuscher M, Keim G, Fischer F, Waag A, Landwehr G.

CdSe fractional-monolayer active region of molecular beam epitaxy grown green ZnSe-based lasers. Applied Physics Letters 1999;74 539-544.

[33] Sui R, Charpentier P. Synthesis of Metal Oxide Nanostructures by Direct Sol–Gel Chemistry in Supercritical Fluids. Chemical Reviews 2012;112 (6) 3057-3082.

[34] Giordano C, Antonietti M. Synthesis of crystalline metal nitride and metal carbide nanostructures by sol–gel chemistry. Nano Today 2011;6(4) 366-380.

[35] Cai X, Wang F, Yan D, Zhu Z, Gu X. Luminescence characteristics and growth mechanism of awl-like ZnO Nanostructures fabricated on Ni-coated silicon substrate via chemical vapor deposition method. Ceramics International 2014;40(8) 12293-12298.

[36] Navamathavan R, Yoo H. Lee Ch. Single Nanowire Light-Emitting Diodes Using Uniaxial and Coaxial InGaN/GaN Multiple Quantum Wells Synthesized by Metalorganic Chemical Vapor Deposition. Nano Letter 2014;14(3) 1537-1545.

[37] Dev A, Dev Choudhury B, Abedin A, Anand S. Fabrication of Periodic Nanostructure Assemblies by Interfacial Energy Driven Colloidal Lithography. Advanced Functional Materials 2014;67 4577-4583.

[38] Wu H, Xu X, He M, Zhang M, Ma K, Bao M. Fabrication of size-tunable antireflective nanopillar array using hybrid nano-patterning lithography. Surface and Coatings Technology 2014;20 240-413.

[39] Kumar V, Wang H. Plasmonic Au nanoparticles for enhanced broadband light absorption in inverted organic photovoltaic devices by plasma assisted physical vapour deposition. Organic Electronics 2013;14(2) 560-568.

[40] Israr-Qadir M, Jamil-Rana S, Nur O, Willander M, Larsson LA, Holtz PO. Fabrication of ZnO nanodisks from structural transformation of ZnO nanorods through natural oxidation and their emission characteristics. Ceramics International 2014;40(1) 2435–2439.

[41] Sheng Q, Mei H, Wu H, Zhang X, Wang S. PtxNi/C nanostructured composites fabricated by chemical reduction and their application in non-enzymatic glucose sensors. Sensors and Actuators B: Chemical 2014;203 588-595.

[42] Mousavi SH, Haratizadeh H, Minaee H. Comparison of structural and photoluminescence properties of zinc oxide nanowires grown by vapor-solid and vapor-liquid-solid methods. Thin Solid Films 2012;520 4642-4645.

[43] Wolcott A, Gerion D, Visconte M, Sun J, Schwartzberg A, Chen SW, Zhang JZ. Silica-Coated CdTe Quantum Dots Functionalized with Thiols for Bioconjugation to IgG Proteins Journal of Physical Chemistry B 2006;110 5779-5789.

[44] Schwartzberg AM, Olson TY, Talley CE and Zhang JZ. Synthesis, characterization, and tunable optical properties of hollow gold nanospheres. Journal of Physical Chemistry B 2006;110 19935-19944.

[45] Mousavi SH, Haratizadeh H, Minaee H. The effect of morphology and doping on photoluminescence of ZnO nanostructures. Optics Communications 2011;284 3558-3561.

[46] Hurle DTJ. Handbook of Crystal Growth, Amsterdam: North-Holland; 1993.

[47] Hausmann A, Schallenberger B. Electronic structure of copper in zinc oxide. R. Roll, Zeitschrift Physik B Condensed Matterials 1979;34 129-134.

[48] Robbins D, Herbert D, Dean P. The origin of the α, β, γ blue no-phonon transitions in ZnO:Cu-A deep-level problem. J. Phys. C. Solid State Phys 1981;14 2859-2870.

[49] Yin LW, Bando Y, Zhu YC, Li MS, Li YB, Golberg D. Growth and field emission of hierarchiral singe-crystalline wurtzite AlN nanoarchitectures. Advanced Materials 2005;17 110-114.

[50] Shi SC, Chen CF, Chattopadhyay S, Chen KH, Chen L. Field emission from quasi-aligned aluminum nitride nanotips. Applied Physics Letters 2005;87 073109-073111.

[51] Zheng J, Yang Y, Yu B, Song X, Li X. Oriented Aluminum Nitride One-Dimensional Nanostructures: Synthesis, Structure Evolution, and Electrical Properties. ACS Nano 2008;2 134-142.

[52] Wu Q, Hu Z, Wang ZX, Hu YM, Chen Y. A simple route to aligned AlN nanowires. Diamond and Related Materials 2004;13 38-41.

[53] Paul RK, Lee KH, Lee BT, Song HY. Formation of AlN nanowires using Al powder. Materials Chemistry and Physics 2008;112:562-565.

[54] Aghdaie A, Haratizadeh H, Mousavi SH, Jafari Mohammadi SA, de Oliveira PW. Structural and optical properties of transition metal doped AlN nanowires. Semiconductors (Submitted).

[55] Mousavi SH, Gharavi MA, Haratizadeh H, Kitai A, de Oliveira PW. One dimensional aluminum nitride nanostructures: Synthesis, structural, and luminescence properties. Journal of Nanoscience and Nanotechnology 2011;11 8284-8288.

3

Wired/Wireless Photonic Communication Systems using Optical Heterodyning

Alejandro García Juárez, Ignacio Enrique Zaldívar Huerta, Antonio Baylón Fuentes, María del Rocío Gómez Colín, Luis Arturo García Delgado, Ana Lilia Leal Cruz and Alicia Vera Marquina

1. Introduction

Over the past few years, there has been an increasing effort in researching new design of indoor wireless communications systems, due to connectivity that show in a room or in a building. Currently, several companies of telecommunications use purely omnidirectional antennas in their wireless routers to transmit data to laptops in close vicinity [1]. The properties of microstrip patch antennas and arrays with their planar configuration exhibit an attractive option for indoor communications where the gain is considerably enhanced. On the other hand, the generation of microwave and millimetre-wave (mm-wave) signals by using photonic techniques are being used in radio-over-fiber (RoF) systems, distribution antenna systems, broadband wireless access networks, and radar systems etc. In all these applications the microwave signals are generated at a remote central station and distributed transparently to several simplified antenna stations via optical fiber [1]. The main goal of these systems is to reduce infrastructure cost and to overcome the capacity bottleneck in wireless access networks, allowing, at the same time, flexible merging with conventional optical access networks. Thus, in order to design a reliable RoF-based access network infrastructure, RoF techniques must be capable of generating the microwave signals and allow a reliable microwave signals transmission over the optical link. For broadband wireless systems and distribution antenna systems operating at microwave and millimeter-wave carriers, several photonic techniques for generating microwave signals have been proposed. Among the most common used techniques are: optical heterodyning [2], optical injection locking [3], optical frequency/phase locked loops (OFLL/OPLL) [4], microwave generation using external modulation [5]. Optical injection locking and optical phase-locked loops (OPLL) are expensive in practice. The use of external intensity modulation generates frequency doubling or quadrupling of the driven radiofrequency (RF) sinusoid signal [6]. This method requires an external modulator which increases both loss and cost, and is more susceptible to bias drifting of the modulators, which can affect the output spectrum. The key advantage for generating microwave or millimeter-wave signals by optical means is that very high-frequency signals with very low phase noise and high purity can be generated. By using optical heterodyne technique it is very easy to tune frequencies with a spectral linewidth of a few ten MHz and over a wide range by simply tuning the wavelength of the two optical input signals; the obtained frequencies are limited only by the photodetector bandwidth [2]. Besides, the generated signals by using this technique can be generally used as both information carriers, and as a local oscillator for transmitting and receiving both analog and digital information signals by using not only RF schemes but also through an optical fiber. On the other hand, microwave photonics, which brings together radiofrequency engineering and optoelectronics, has attracted great interest in the field of telecommunications since it is an excellent alternative for the transmission of services such as high quality audio and video, e-mail, and Internet among others [7]. Furthermore, there has been an increase effort in researching new microwave photonics techniques for different interesting application that attracts interest in research is the filtering of microwave signals by using photonic techniques. The main feature of a photonic filter is that microwave signals are directly processed in the optical domain exploiting advantages inherent to photonics such as low loss, high bandwidth, immunity to electromagnetic interference, and tunability [8]. On the other hand, network architectures such as FTTx, where x can stand for home (H), building (B), neighborhood (N), or curb (C), are a communication architecture in which the final connection to the subscribers is optical fiber. Another important application of photonic telecommunications systems, which is very closely related to the FTTx systems, is the distribution of signals by integrating optical and wireless networks and passive optical networks (PONs). This particular type of scheme is referred as fiber-radio system [9]. Along with wavelength division multiplexing (WDM) technique, it would be more advantageous if RoF is integrated with a conventional PON where a base station (BS) plays the role of an optical network unit (ONU) to support both wired and wireless services. This integrated optical access system is capable not only to reduce the cost of multifunction BSs and the whole system but also meet the demands for bandwidth, mobility and connection options of users [10]. In this sense, the purpose of this chapter is to describe two an alternative optical communications systems. The first proposed system uses a couple microstrip antennas for distributing point to point analog TV with coherent demodulation based on optical heterodyne. In the proposed experimental setup, two optical waves at different wavelengths are mixed and applied to a photodetector. Then a beat signal with a frequency equivalent to the spacing of the two wavelengths is obtained at the output of the photodetector. This signal corresponds to a microwave signal located at 2.8 GHz, which it is used as a microwave carrier in the transmitter and as a local oscillator in the receiver of our optical communication system. The feasibility of this technique is to demonstrate the transmission of a TV signal located at 66-72 MHz. The second system deals with the experimental transmission of analog TV signal in a fiber-radio scheme using a microwave photonic filter (MPF). For that purpose, filtering of a microwave band-pass window located at 2.8 GHz is obtained by the interaction of an externally modulated multimode laser diode emitting at 1.5 μm associated to the chromatic dispersion parameter of an optical fiber. Transmission of TV signal coded on the microwave band-pass window is achieved over an optical link of 20.70 km. Demodulated signal is transmitted via radiofrequency using printed antennas.

2. Optical heterodyne technique

The basic principle for generating microwave carriers is based on optical heterodyne technique, it represents a physical process called optical beating or frequency beating, where two phase-locked optical sources with angular frequencies ω_1 and ω_2 are superimposed and injected into a high frequency photodetector that permits to obtain a photocurrent at a frequency $\omega_2-\omega_1$. To explain this in more detail, let us consider the relation between the generated electrical output signal and the two superimposed optical input waves from a more physical point of view. For simplicity, we assume that the two optical input waves are linearly polarized monochromatic plane waves in the infrared which propagate in the +z direction. Let

$$\mathrm{E}_1 = \hat{\mathrm{E}}_1 \exp\left[i\left(\omega_1 t - k_1 z + \varphi_1\right)\right]\mathrm{e}_1, \tag{1}$$

and

$$\mathrm{E}_2 = \hat{\mathrm{E}}_2 \exp\left[i\left(\omega_2 t - k_2 z + \varphi_2\right)\right]\mathrm{e}_2, \tag{2}$$

be the complex electrical field vectors of the two optical waves, with field amplitudes $\hat{\mathrm{E}}_1$ and $\hat{\mathrm{E}}_2$, angular frequencies ω_1 and ω_2 and wave numbers k_1 and k_1 . The phase of each optical input wave is considered by φ_1 and φ_1 and e_1 and e_2 are the unit vectors determining the orientation of the electrical field vector of the linearly polarized optical input waves. The intensities of the constituent waves are given by the magnitude of their Poynting vectors and are therefore given by [11]

$$I_1 = \frac{1}{2}\left(\frac{\varepsilon_r \varepsilon_o}{\mu_o}\right)^{1/2} |E_1|^2 . \tag{3}$$

$$I_2 = \frac{1}{2}\left(\frac{\varepsilon_r \varepsilon_o}{\mu_o}\right)^{1/2} |E_2|^2 . \tag{4}$$

If the two incident optical waves are perfect plane waves and have precisely the same polarization ($e_1 = e_2$), the resulting electrical field E_o of the optical interference signal is the sum of the two constituent input fields and hence we can write $E_o = E_1 + E_2$. Taking the squared absolute value of the optical interference signal we obtain

$$\begin{aligned}|E_o|^2 = |E_1 + E_2|^2 = |E_1|^2 + |E_2|^2 + E_1 E_2^* + E_1^* E_2 \\ = |E_1|^2 + |E_2|^2 + 2|E_1||E_2|\cos\left((\omega_2 - \omega_1)t - (\varphi_2 - \varphi_1)\right).\end{aligned} \tag{5}$$

From equation (5) and by using equations (3) and (4), it follows that the intensity of the interference signal I_o is given by [11]

$$I_o = I_1 + I_2 + 2\left(I_1 I_2\right)^{1/2} \cos\left((\omega_2 - \omega_1)t - (\varphi_2 - \varphi_1)\right). \tag{6}$$

By launching this optical interference signal into a photodetector, a photocurrent i is generated which can be expressed as [11]

$$i = \frac{\eta_0 q}{hf_1} P_1 + \frac{\eta_0 q}{hf_2} P_2 + 2\frac{\eta_{f_c} q}{h}\left(\frac{P_1 P_2}{f_1 f_2}\right)^{1/2} \cos\left((\omega_2 - \omega_1)t - (\varphi_2 - \varphi_1)\right), \tag{7}$$

where q is the electron charge and P_1 and P_2 denote the optical power levels of the two constituent optical input waves. The photodetector's DC and high-frequency quantum efficiencies are represented by η_o and η_{f_c}. It is of course important to consider that the detector's quantum efficiency is not independent of the frequency. Several intrinsic and extrinsic effects such as transit time limitations or microwave losses will eventually limit the high-frequency performance of the detector and thus the detector's DC responsivity η_o is typically much larger than its high-frequency responsivity η_{f_c}. In our case, we can further simplify the photocurrent equation (Eq. (7)) by considering the fact that the two optical input waves are close in frequency ($f_1 \approx f_2$) whereas the difference frequency f_c is by far smaller ($f_c = |f_2 - f_1| << f_1, f_2$). If we further assume for simplicity that the power levels of the two optical input waves are equal ($P_{opt} \approx P_1 \approx P_2$), Eq. (7) becomes [11]

$$i = 2s_o P_{opt} + 2s_{f_c} P_{opt} \cos\left(2\pi f_c t + \Delta\varphi\right). \tag{8}$$

Where $\Delta\varphi = \varphi_2 - \varphi_1$. Here $s_o = \frac{\eta_0 q}{hf}$ and $s_{f_c} = \frac{\eta_{f_c} q}{hf}$ are the photodetector's DC and high frequency responsivities given in A/W. Eq. (8) is the fundamental equation describing optical heterodyning in a photodetector. The first term is the DC photocurrent generated by the constituent optical input waves and the second term is the desired high-frequency signal oscillating at the difference frequency f_c (down-converter) or intermediate frequency (IF) [1]. In our case it represents the microwave signal that we will use as both information carriers, and as a local oscillator for transmitting and receiving TV signals in a wireless communication system.

3. Experimental scheme for generating microwave signals

The heterodyne technique for generating microwave signals has been done using the experimental setup shown in Figure 1. In this experiment, two laser diodes emitting at different wavelengths are used. One of them is a tunable laser (New Focus, model TLB-3902) which can be tuned over the C band with a channel spacing of 25 GHz, and the other one is a fiber coupled distributed feedback (DFB) laser source (Thorlabs, model S3FC1550) with a central wavelength at 1550 nm. For the generation of the microwave signals, the outputs of both lasers are coupled to optical isolators to avoid a feedback into the lasers and consequently instabilities to the system. A pair of polarization controllers is used to minimize the angle between the polarization directions of both optical sources. Thus, the polarization of the light issued from each optical source is matched and therefore, there is not degradation of the power levels in the microwave signals generated from the photodetector. The output of each controller is launched to a 3 dB coupler to combine both optical spectrums. After that, an optical output signal is received by a fast photodetector (MITEQ model SCMR-50K6G-10-20-10) with a typical gain of 25 dB, and –3 dB bandwidth of 6 GHz, The resulting photocurrent from the photodetector corresponds to the microwave beat signal which is analyzed with an Electrical Spectrum Analyzer (ESA), (Agilent model E4407B). The other optical output resulting from optical coupler is applied to an Optical Spectrum Analyzer (OSA) (Anritsu model MS9710C), for monitoring the wavelength of the two beams.

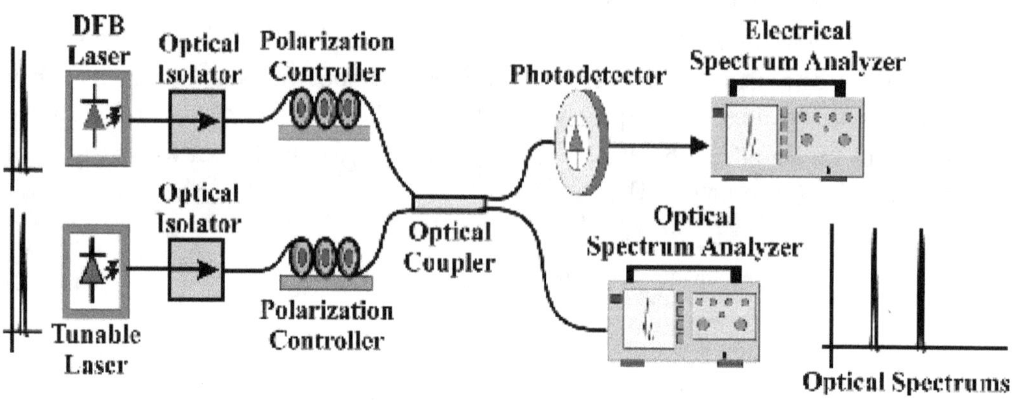

Figure 1. Experimental setup for generating microwave signals by using optical heterodyne technique.

DFB laser can be used to control not only the output power of the fiber coupled laser diode, but also the precise control of the temperature at which the laser is operating. Both controls can be used to tune the fiber coupled laser diode to an optimum operating point, providing a stable output. In this way, it is possible to observe that the wavelength of the DFB laser is shifting, by varying its temperature with a scale of 1 °C. Consequently, the beat signal frequency is continuously over the band of photodetector. On the other hand, the frequency difference from both lasers can be expressed by [1]

$$\Delta f = \frac{c}{\lambda_1} - \frac{c}{\lambda_2} = \frac{c(\lambda_2 - \lambda_1)}{\lambda_1 \lambda_2} \approx \frac{c}{\lambda^2}|\Delta\lambda|, \tag{9}$$

where λ_1 and λ_2 are the wavelengths of the two beams, respectively, and $\Delta\lambda$ is the difference between the two wavelengths. To obtain a microwave signal, in a first step, the tunable laser is biased and its optical spectrum is displayed on the OSA screen. In a second step, the DFB laser is also biased, fixing an optical power of 2.2 mW and its central wavelength is settled as near as possible to the central wavelength of the tunable laser. As can be seen from Figure 2, the value of $\Delta\lambda = 0.023739$ nm is the wavelength difference between both lasers and it corresponds to the beat signal frequency of 2.8 GHz.

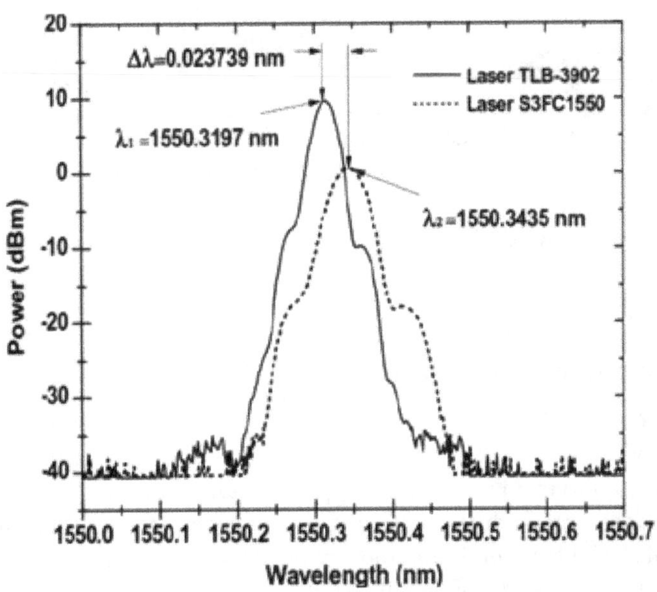

Figure 2. Optical spectrum corresponding to the mixed optical sources. The peaks located at λ_1=1550.3197 nm and λ_2=1550.3435 nm corresponds to the tunable and DFB lasers, respectively.

A precise control of the difference, between the two central wavelengths and by consequence over the frequency difference, is obtained by tuning the DFB. The wavelength variation of the laser source is obtained by changing the junction temperature between 22.8 °C, 23.2 °C and 23.7 °C corresponding to the frequency range of 0 to 5.0 GHz. Figure 3 illustrates the electrical spectrums of four generated microwave signals by using optical heterodyne. These signals are located at f_1=1.0 , f_2=2.0 , f_3=2.8 and f_4=4.0 GHz respectively. It can be seen, that the microwave signals are in good agreement with theoretical value given by Eq. (9). Therefore, when one laser source is operating at a fixed wavelength and the other is being continuously tuned, the beat frequency will shift correspondingly. In particular, the frequency of the microwave drive signal is set at 2.8 GHz.

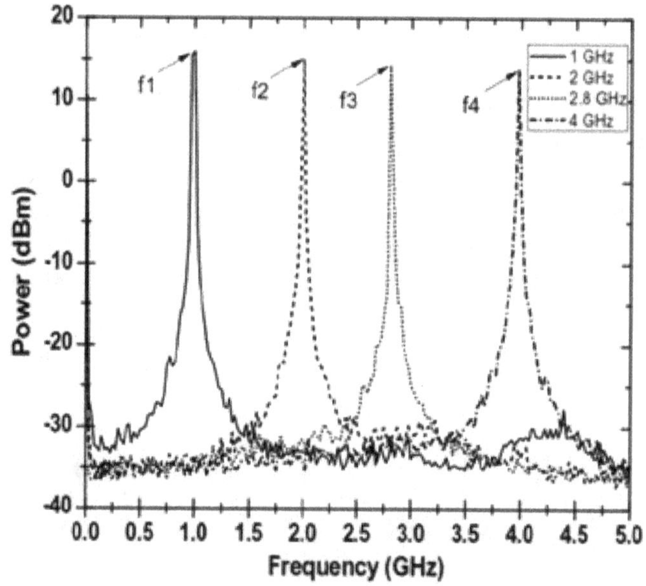

Figure 3. Spectrum for the microwave signal generated by using optical heterodyne.

4. Modulation and demodulation

Some form of modulation is always needed in an RF system to translate a baseband signal (e.g., audio, video, data) from its original frequency bandwidth to a specified RF frequency spectrum. There are many modulation techniques, for example, amplitude modulation (AM), frequency modulation (FM), amplitude shift keying (ASK), frequency shift keying (FSK), phase shift keying (PSK), biphase shift keying (BPSK), quadriphase shift keying (QPSK), 8-phase shift keying (8-PSK), 16-phase shift keying (16-PSK), minimum shift keying (MSK), and quadrature amplitude modulation (QAM). AM and FM are classified as analog modulation techniques, and the others are digital modulation techniques [12]. In this section we describe the AM modulation and demodulation due to it was used in our proposed wireless communication system.

4.1. Amplitude modulation

Analog modulation uses the baseband signal (modulating signal) to vary one of three variables: amplitude A_c , electrical frequency $(\omega_1-\omega_2)=\omega_c=2\pi f_c$; or phase $(\phi_1-\phi_2)=\Delta\phi$. According to Eq. (8), the obtained carrier signal by using optical heterodyne technique can be written by

$$p(t)=A_c\cos\left((\omega_1-\omega_2)t+\phi_1-\phi_2\right)=A_c\cos\left(2\pi f_c t+\Delta\phi\right). \tag{10}$$

Where $A_c=2s_{f_c}P_{opt}$. In amplitude modulation, if we assume that $s(t)$ is the information signal, and considering $A_c=1$, $\Delta\phi=0$, then a modulated signal can be written by

$$g(t)=s(t)\cos 2\pi f_c t. \tag{11}$$

Applying the modulation property of the Fourier transform to Eq. (11), we can find the density spectral of $g(t)$ is

$$G(f)=\frac{1}{2}S(f-f_c)+\frac{1}{2}S(f+f_c). \tag{12}$$

Amplitude modulation therefore translates the frequency spectrum of a signal by $\pm f_c$ hertz, but leaves the spectral shape unaltered. This type of amplitude modulation is called suppressed-carrier because the spectral density of $g(t)$ has no identifiable carrier in it, although the spectrum is centered at the frequency f_c .

4.2. Amplitude demodulation

Recovery the signal information $s(t)$ from the signal $p(t)$ requires another translation in frequency to shift the spectrum to its original position. This process is called demodulation or detection. Because the modulation property of the Fourier transform proved useful in translating spectra for modulation, we try it again for demodulation. Assuming that $g(t)=s(t)\cos 2\pi f_c t$ is the transmitted signal, we have

$$g(t)\cos 2\pi f_c t=s(t)\cos^2 2\pi f_c t=\frac{1}{2}s(t)+\frac{1}{2}\cos 4\pi f_c t. \tag{13}$$

Taking the Fourier transform of both sides of Eq. (13) and using the modulation property, we get

$$\Im\left[g(t)\cos 2\pi f_c t\right]=\frac{1}{2}S(f)+\frac{1}{4}S(f+2f_c)+\frac{1}{4}S(f-2f_c). \tag{14}$$

The mathematical process described in this section can be obtained by convolving the spectrum of the received signal $g(t)$ with that of $\cos 2\pi f_c t$ (i.e., with impulses at $\pm f_c$). A low-pass filter is required to separate out the double frequency terms from the original spectral components. Obviously we need a filter with a cut frequency $f_{cut} > 2f_m$ for proper signal recovery. In this case f_m represents the information frequency.

4.3. Effects in frequency and phase variations

When the local oscillator at the receiver, has a small frequency error Δf and a phase error $\Delta\theta$, then this signal can be written as

$$p_L(t) = \cos\left[2\pi\left(f_c + \Delta f\right)t + \Delta\theta\right]. \tag{15}$$

Assuming again that $g(t)=s(t)\cos 2\pi f_c t$ is the transmitted signal; then we have that at the receiver, the recovered signal can be written by

$$\begin{aligned} g(t)\cos\left[2\pi(f_c + \Delta f)t + \Delta\theta\right] &= s(t)\cos(2\pi f_c t)\cos\left[2\pi(f_c + \Delta f)t + \Delta\theta\right] \\ &= s(t)\left(\frac{\cos(2\pi\Delta f t + \Delta\theta)}{2} + \frac{\cos\left[2\pi(2f_c + \Delta f)t + \Delta\theta\right]}{2}\right). \end{aligned} \tag{16}$$

The second term on the right hand side of Eq. (16) is centered at $\pm 2f_c + \Delta f$ and can be filtered out by using a low pass filter. The output of this filter $s_F(t)$ will then be given by the remaining term in Eq. (16).

$$s_F(t) = \left[\frac{s(t)}{2}\left(\cos 2\pi(\Delta f)t\cos(\Delta\theta) - sen 2\pi(\Delta f)t sen(\Delta\theta)\right)\right]. \tag{17}$$

As can been from equation (17), the output signal is not $\frac{s(t)}{2}$, unless both Δf and $\Delta\theta$ are zero. The effects of both frequency errors and random phase errors render this demodulation of the signal unsatisfactory. It is necessary, therefore, to have synchronization in both frequency and phase between the transmitter and the receiver when amplitude modulation is used. The synchronization of the carrier signals presents no major problem when the transmitter and the receiver are in close proximity. Recovering the original signal $s(t)$ from the modulated signal $g(t)$ using a synchronized oscillator is called coherent demodulation. In our case we take advantage of proposed optical heterodyne technique permits to obtain microwave carrier and local oscillator simultaneously in the transmitter and receiver respectively.

5. Design of a patch antenna at 2.8 GHz

The microstrip patch antenna is a popular printed resonant antenna for narrow-band microwave wireless links that require semihemispherical coverage. Due to its planar configuration and ease of integration with microstrip technology, the microstrip patch antenna has been studied heavily and is often used as an element for an array. Common microstrip antenna shapes are square, rectangular, circular, ring, equilateral triangular, and elliptical, but any continuous shape is possible [13]. Furthermore, a patch antenna is an excellent device due to its small size, low cost, and good performance [14-16]. In this chapter, a rectangular printed patch antenna is proposed. Simulation results have been obtained by using Advanced Design System (ADS) that is a computer-aided-engineering software tool. The radiating structure consists of a patch and a microstrip inset-feed line, allowing that the characteristic impedance (Zo) to be improved. Figure 4 shows the geometry and configuration of the top layer. The proposed antenna in this work was designed to operate in the band S of telecommunications (2.8 GHz). FR4 is used as a dielectric substrate exhibiting a thickness $h = 1.524mm$, and relative dielectric constant $\varepsilon_r = 4.2$. In a first step, the width (W) of the patch is computed by using [17]:

$$W = \frac{c_o}{2f_c}\sqrt{\frac{2}{\varepsilon_r + 1}}, \tag{18}$$

where c_o is the light velocity in the free space, and f_o is the operation frequency. Next, the value of the effective dielectric constant ε_{eff} is evaluated considering $W/h > 1$.

$$\varepsilon_{eff} = \frac{\varepsilon_r + 1}{2} + \frac{\varepsilon_r - 1}{2}\left(1 + 12\frac{h}{W}\right)^{-1/2} \tag{19}$$

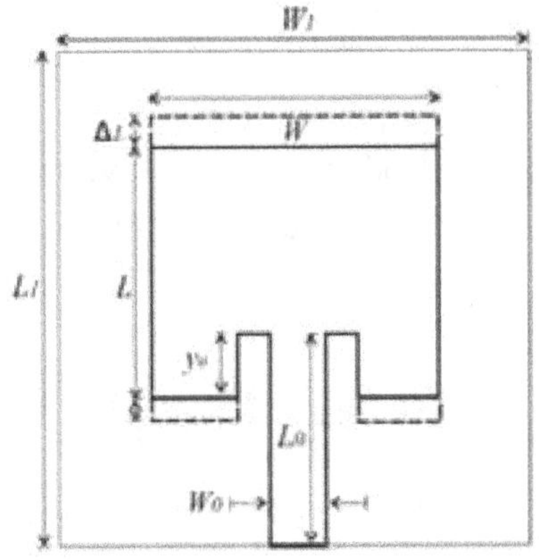

Figure 4. Layout of the patch antenna.

Border effects [17] must to be considered in the design of the antenna. For this reason, ΔL from Figure 4 can be evaluated as:

$$\frac{\Delta L}{h} = 0.412 \frac{\left(\varepsilon_{eff} + 0.3\right)\left(\frac{W}{h} + 0.264\right)}{\left(\varepsilon_{eff-0.258}\right)\left(\frac{W}{h} + 0.8\right)} \tag{20}$$

This allows that the length (L) of the patch to be evaluated as:

$$L = \frac{c_o}{2f_o\sqrt{\varepsilon_r}} - 2\Delta L \tag{21}$$

Considering the values previously obtained, the effective dimensions (L_{eff} and W_{eff}) can be calculated, respectively as:

$$L_{eff} = L + 2\Delta L \tag{22}$$

$$W_{eff} = W + \frac{t}{\pi}\left(1 + \ln\left(\frac{2h}{t}\right)\right) \tag{23}$$

From Eq. (23), t is the conductor thickness and $W/h > 1/2\pi$ must to be considered. The ground plane dimensions are computed as:

$$\begin{aligned} L_1 &= 6h + L_{eff} \\ W_1 &= 6h + W_{eff} \end{aligned} \tag{24}$$

The best dimensions which assure a good matching between the impedances ($R_{in} = Z_o = 50\Omega$) of the antenna and generator can be calculated by the use of LineCalc tool from ADS and by the next expression:

$$R_{in}\left(y = y_o\right) = \frac{1}{2\left(G_1 \pm G_{12}\right)}\cos^2\left(\frac{\pi}{L}y_o\right) \tag{25}$$

where G_1 and G_{12} are the conductance values obtained by the cavity method. Finally, Table 1 shows a summary of the dimensions for the patch and the ground plane.

Operation Frequency (Antenna)	Dimensions (cm)						
	W_o	L_o	W	L	W_1	L_1	y_o
2.8 GHz	0.13	3.08	3.32	2.56	10	10	0.93

Table 1. Dimensions of the fabricated antenna.

Figure 5(a) shows a picture of the fabricated patch antenna where a SubMiniature version A (SMA) connector is added. Figure 5(b) illustrates simulation and experimental results corresponding to the S_{11} parameter. Electrical measurements are obtained by using a Vector Network Analyzer (VNA) (Agilent Technologies model: E8361A). It is clearly observable that experimental result is in good agreement with the simulation.

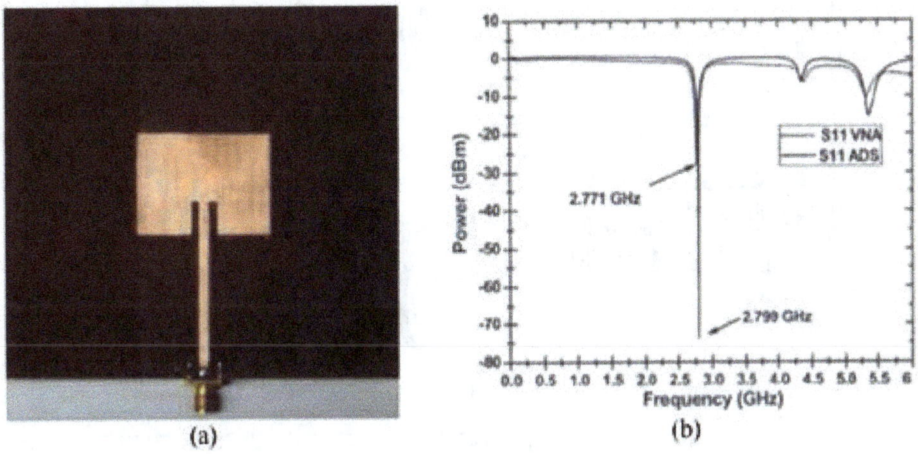

Figure 5. Fabricated antenna (a), Experimental and simulation return loss curve for the antenna (b).

6. Transmission of TV signals by using heterodyne technique

In order to show a potential application of optical heterodyne technique in the field of the wireless communications, we have proposed a coherent wired/wireless photonic communication system as shown in Figure 6. This system is not a truly wireless communication system, since an optical fiber is required to deliver both microwave carrier and local oscillator for transmitting and receiving information of TV signals as an approximation to point to point indoor wireless communications systems. From the photodetector 1 in the transmitter, a microwave signal located at 2.8 GHz is obtained and mixed with an analog TV signal located at 62.25 MHz. Then the resulting signal is amplified before being applied to our fabricated microstrip antenna. After that, the obtained modulated signal as shown in Figure 7, is transmitted through a point to point wireless link by using the microstrip antenna. Finally in the receiver, another microstrip antenna is used to receive the transmitted information, which it is processed using optical heterodyne technique again to recover in this case the TV signal (66-72 MHz). From the photodetector 2 in the receiver, a local oscillator that is synchronized, in frequency as well as in phase with to that obtained from the photodetector 1, is mixed with the received signal. Then the resulting signal is filtered and the power spectral density obtained is displayed on an electrical spectrum analyzer, where it is analyzed to measure the power level of recovered information.

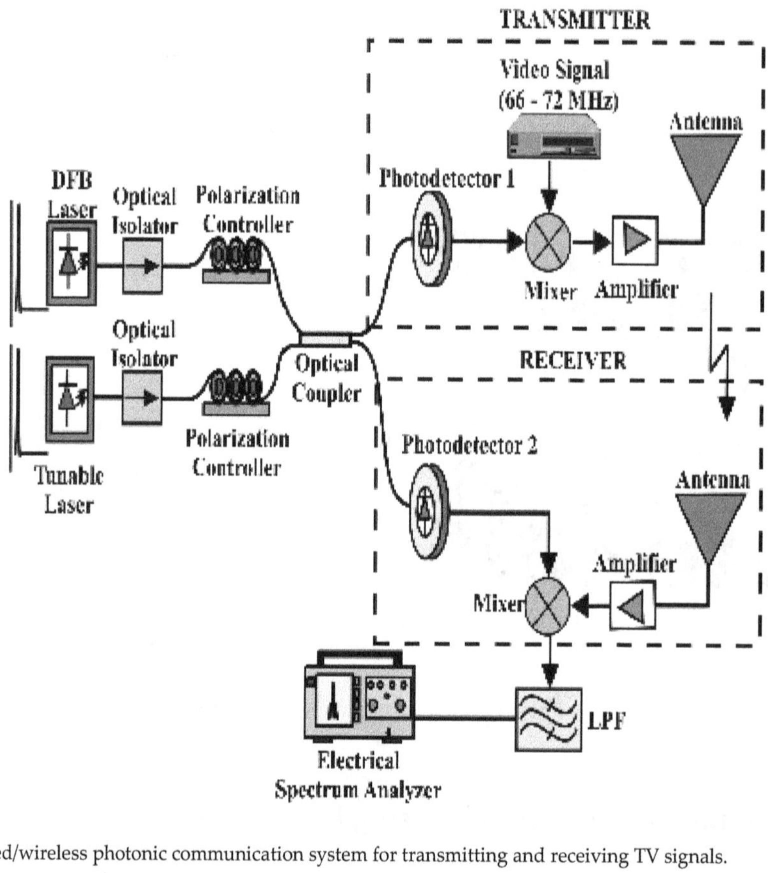

Figure 6. Wired/wireless photonic communication system for transmitting and receiving TV signals.

Figure 8 shows the frequency spectrum of an analog National Television System Committee (NTSC) TV signal at the input of the transmitter located at 67.25 MHz (before being applied to frequency mixer). In the same figure we can see the obtained analog NTSC TV signal at the output of the receiver. In order to measure the quality of the received signal, it is necessary to quantify the parameter of signal-to-noise ratio (SNR), in this case it is approximately 45 dB. The analog information is successfully transmitted from the transmitter to the receiver, and the received signal is satisfactorily reproduced on TV monitor. The differential gain and differential phase were not measured Nevertheless we demonstrated that the generated microwave signal by using optical heterodyning can be used as carrier information in a traditional communication system and we have used a TV signal of test to verify it.

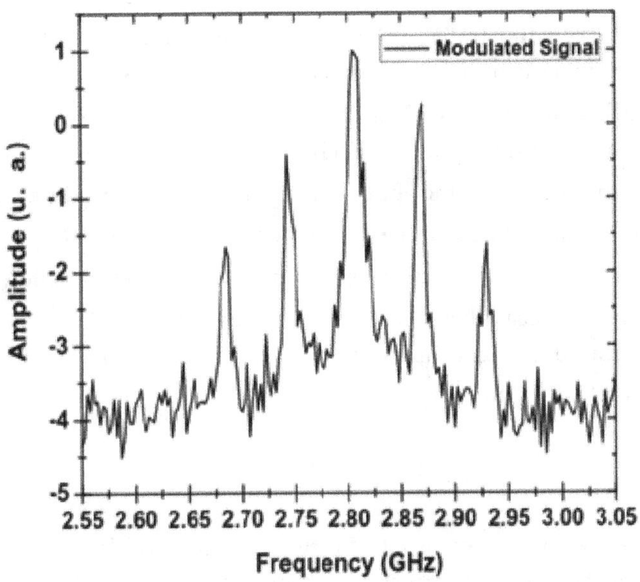

Figure 7. Electrical spectrum of the modulated signal.

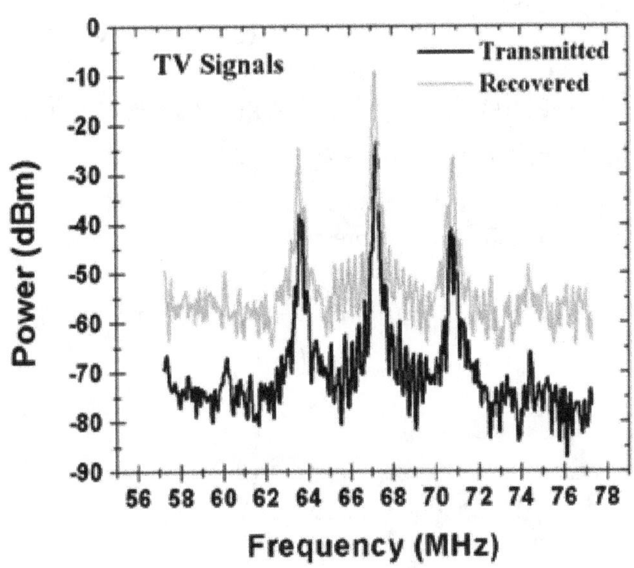

Figure 8. TV signals at 67.25 MHz, transmitted and recovered.

7. Analytical model of the microwave photonic filter

The scheme of the MPF is illustrated in Figure 9. Consider that the optical signal of a polychromatic source with spectrum $P(\omega)$, centered at an optical frequency ω_n, is launched into the input of the Mach-Zehnder intensity modulator (MZ-IM). A single spectral component of such an optical signal can be modeled by a stochastic process $e(t)=E_o(t)\exp(j\omega_n t)$, where $E_o(t)$ is the complex amplitude and ω_n is the optical angular frequency. If the intensity of such optical signal is externally modulated by an electrical signal $V_m=1+2m\cos(\omega_m t)$, where m is the modulation index and ω_m is the angular frequency of external modulation, then the optical field at the input of the optical fiber can be expressed by Eq. (26). The modulation index m is related to the electrical input signal amplitude, V_m, as: $2m=\pi(V_m/V_\pi)$, where V_π is the half wave voltage of the MZ-IM [18].

$$e_i(t)=e(t)s(t) \tag{26}$$

The optical fiber can be considered as a linear time invariant (LTI) system. If, for simplicity, the attenuation is ignored, then the transfer function of the optical link, for a given length L, is $H(j\omega)=\exp(-j\beta L)$, where β is the propagation constant. Thus, the optical field at the end of the link is given by

$$e_L(t)=e_i(t)\exp(-j\beta L) \tag{27}$$

Substituting $e(t)$ and $s(t)$ in Eq. (26), and then replacing this in Eq. (27), it becomes:

$$\begin{aligned} e_L(t) = E_o(t)\exp\left(j(\omega_m t-\beta L)\right) + E_o(t)m\exp\left(j\left[(\omega_m-\omega_n)t-\beta L\right]\right) \\ + E_o(t)m\exp\left(j\left[(\omega_m+\omega_n)t-\beta L\right]\right) \end{aligned} \tag{28}$$

In the frequency domain Eq. (28) can be expressed as:

$$\begin{aligned} E_L(\omega) = E_o(\omega-\omega_n)\exp(-j\beta L) + E_o\left(\omega-(\omega_n-\omega_m)\right)\exp(-j\beta L) \\ + E_o\left(\omega-(\omega_n+\omega_m)\right)\exp(-j\beta L) \end{aligned} \tag{29}$$

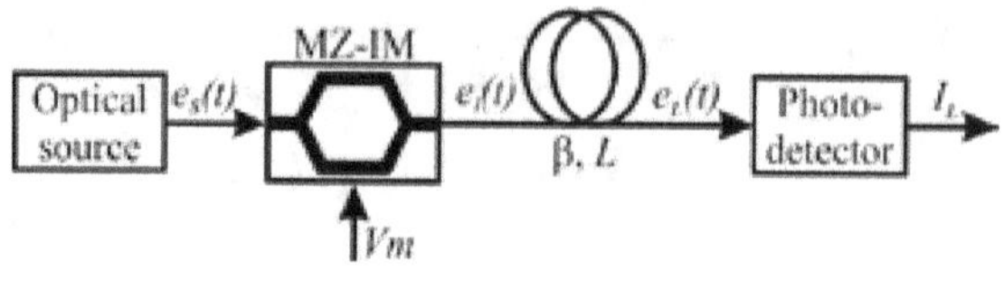

Figure 9. Scheme of the microwave photonic filter.

There are three spectral components. In the presence of chromatic dispersion, there is a propagation constant associated to each one of them, i.e. $\beta(\omega-\omega_n)$, $\beta(\omega-(\omega_n+\omega_m))$ and $\beta(\omega-(\omega_n-\omega_m))$. By denoting $W=\omega-\omega_n$, Eq. (29) then becomes:

$$\begin{aligned}E_L(\omega) = E_o(W)\exp(-j\beta(W)L) + E_o(W+\omega_m)\exp(-j\beta(W+\omega_m)L) \\ + E_o(W-\omega_m)\exp(-j\beta(W-\omega_m)L)\end{aligned} \quad (30)$$

Assuming that within the frequency range $\omega_n-\omega_m$ to $\omega_n+\omega_m$, centered at ω_n the propagation constant varies only slightly and gradually with ω , it can be approximated by the first three terms of a Taylor series expansion, and it can be shown that

$$\beta(W \pm \omega_m) = \beta(W) \pm \beta_1\omega_m + \beta_2\left[\frac{1}{2}\omega_m^2 \pm \omega_m(\omega-\omega_n)\right] \quad (31)$$

where $\beta_i=[d^i\beta(\omega)/d\omega^i]_{(\omega=\omega_n)}$.

The optical intensity, I , is obtained by integrating the power spectral density over all the frequency range, i.e.

$$I = \int_{-\infty}^{\infty} |E_L(\omega)|^2 d\omega \quad (32)$$

Considering that the MZ-IM is operating on its linear region, it is valid to note that $m^2 \approx 0$. On the other hand, if $\omega_n >> \omega_m$ then $E_o(W) \approx E_o(W+\omega_m) \approx E_o(W-\omega_m)$. Furthermore, in the frequency domain, the spectrum of the source is defined as $P(\omega)=E_o(\omega)E_o^*(\omega)$. Thus, developing the product $|E_L(\omega)|^2$ in Eq. (32) and replacing Eq. (31), it is possible to demonstrate that the intensity at the end of the optical fiber is given by:

$$I = \int_{-\infty}^{\infty} P(W)dW + 4m\cos\left(\beta_2\frac{\omega_m^2}{2}L\right)\cos(\beta_1\omega_m L)\Re\left\{\int_{-\infty}^{\infty} P(W)\exp(-j2\pi ZW)dW\right\} \quad (33)$$

where $Z=\beta_2\omega_m L/2\pi$, $W=\omega-\omega_n$ and its derivative, $dW=d\omega$. The total average intensity is $I_o=\int_{-\infty}^{\infty} P(W)dW$, and the integral $\Re\left\{\int_{-\infty}^{\infty} P(W)\exp(-j2\pi ZW)dW\right\}$ corresponds to the real part of the Fourier transform of the spectrum of the optical source. This means that the optical intensity which reaches the surface of the photodetector is proportional to:

$$F(W) = \Re\{FT\{P(W)\}\} \tag{34}$$

A spectrum with Gaussian shape can be modeled by an analytical expression as:

$$P(\omega) = \frac{2P_o}{\Delta\omega\sqrt{\pi}} \exp\left(-\frac{4(\omega - \omega_m)^2}{\Delta\omega^2}\right) \tag{35}$$

where ω is the angular frequency, ω is the central angular frequency, ω is the maximum power emission and $\Delta\omega$ is the full width at half maximum (FWHM) of the optical source. If the emission spectrum of the optical source has a Gaussian shape, as defined in Eq. (35), then the Eq. (34) becomes:

$$F(\omega) = \exp\left(-\left(\frac{\beta_2 \omega_m L \Delta\omega}{4}\right)^2\right) \tag{36}$$

In such case the FWHM of the frequency response can be determined equating $F(\omega) = 0.5$, which implies:

$$\left(\frac{\beta_2 \omega_m L \Delta\omega}{4}\right)^2 = \ln(2) \tag{37}$$

For finding the value of the frequency f_m that yields that condition, it is necessary to express ω_m in terms of f_m, i.e. $\omega_m = 2\pi f_m$. But this, in turn, yields an expression that can be reduced by expressing $\Delta\omega$ in terms of $\Delta\lambda$ and β_2 in terms of dispersion D. For $\Delta\omega$ this is done as follows: given $d\omega / d\lambda = -(2\pi c / \lambda^2)$, where c is the speed of light in the free space and λ is the wavelength of the optical signal, it is possible to establish the following correspondence:

$$d\omega = -\frac{2\pi c}{\lambda^2} d\lambda \Leftrightarrow \Delta\omega = -\frac{2\pi c}{\lambda^2} \Delta\lambda \tag{38}$$

Now, for the factor β_2, given that the group velocity, $v_g = L / \tau_g$ where τ_g is the group delay, is related to $\beta(\omega_n)$ as $\tau_g / L = d\beta(\omega_n) / d\omega$, and its derivative is $(d\tau_g / d\omega) / L = d^2\beta(\omega_n) / d\omega^2 = \beta_2$, then $(1/L)(d\tau_g) = d\omega\beta_2$. Thus, the derivative of this expression by $d\lambda$ is $(1/L)(d\tau_g / d\lambda) = (d\omega / d\lambda)\beta_2$. Furthermore, the dispersion, as a function of the wavelength is defined as $D = (1/L)(d\tau_g / d\lambda)$. This means that $\beta_2 = -D(\lambda^2 / 2\pi c)$. Finally, by substituting $\omega_m = 2\pi f_m$, $\Delta\omega$, in Eq. (38), and the expression for β_2 in Eq. (37), the frequency f_m, which corresponds to the low-pass bandwidth Δf_{lp}, can be expressed as:

$$\Delta f_{lp} = \frac{2\sqrt{\ln(2)}}{\pi DL\Delta\lambda} \tag{39}$$

where the dispersion D has units of ps nm^{-1} km^{-1}, length L is given in km, and the FWHM of the optical source, $\Delta\lambda$, in nm. This means that in the presence of an optical source, like a super luminescent light-emitting diode (LED), the frequency response of the system is low-pass, and its bandwidth is given by Eq. (39). In the context of this chapter, the optical source is an multimode laser diode (MLD). The emission spectrum of this type of optical sources can be modeled by means of an analytical expression as expressed in Eq. (40):

$$P(\omega) = \frac{2P_o}{\Delta\omega\sqrt{\pi}} \exp\left(-\frac{4(\omega-\omega_n)^2}{\Delta\omega^2}\right)\left[\frac{2P_o}{\sigma\omega\sqrt{\pi}} \exp\left(-\frac{4(\omega-\omega_n)^2}{\sigma\omega^2}\right) * \sum_{n=-\infty}^{\infty} \delta(\omega - n\delta\omega)\right] \tag{40}$$

where ω is the angular frequency, ω_n is the central angular frequency, P_o is the maximum power emission, $\Delta\omega$ is the FWHM of the optical source, $\sigma\omega$ is the FWHM of each emission mode and $\delta\omega$ is the free spectral range (FSR) between the emission modes. By using variables Z and W, as defined earlier, and substituting Eq. (40) in Eq. (34), it can be expressed as:

$$F(\omega) = \exp\left(-\left(\frac{\beta_2\omega_m L\Delta\omega}{4}\right)^2\right) * \left[\exp\left(-\left(\frac{\beta_2\omega_m L\sigma\omega}{4}\right)^2\right)\frac{1}{\delta\omega}\sum_{n=-\infty}^{\infty} \delta\left(\frac{\beta_2\omega_m L}{2\pi} - \frac{n}{\delta\omega}\right)\right] \tag{41}$$

The term between crochets indicates the presence of a periodic pattern. The frequency of the first maximum can be determined by equating:

$$\frac{\beta_2\omega_1 L}{2\pi} = \frac{1}{\delta\omega} \tag{42}$$

For finding the value of the frequency f_1 that yields that condition, it is necessary to express $\delta\omega$ in terms of f_1. In a similar way as in Eq. (38), it is possible to establish the following correspondence:

$$d\omega = -\frac{2\pi c}{\lambda^2} d\lambda \Leftrightarrow \delta\omega = -\frac{2\pi c}{\lambda^2} \delta\lambda \tag{43}$$

thus, substituting $\delta\omega$ in Eq. (42), expressing ω_1 in terms of f_1 and using $\beta_2 = -D(\lambda^2/2\pi c)$ then the frequency f_1 can be expressed as:

$$f_1 = \frac{1}{DL\delta\lambda} \tag{44}$$

and, in general, the central frequency of the n -th band-pass lobe is given by

$$f_n = \frac{n}{DL\delta\lambda} \tag{45}$$

where n is a positive integer, dispersion D is given in ps nm^{-1} km^{-1}, length L in km, and the FSR $\delta\lambda$ in nm. The bandwidth of each of these band-pass lobes is equal to:

$$\Delta f_{bp} = \frac{4\sqrt{\ln(2)}}{\pi DL\Delta\lambda} \tag{46}$$

which is twice Eq. (39). The periodic pattern in the frequency response of the system will appear only when an MLD is used in the system. This behavior will allow that microwave signals to be filtered and transmitted over a wide range of frequencies.

8. Experimental setup of optical and wireless transmission

In a first step, the MLD used in this experiment (OKI OL5200N-5) is optically characterized by means of an optical spectrum analyzer (Agilent, model 86143B). Figure 10 corresponds to the measured optical spectrum obtaining λ_o=1553.53 nm , $\Delta\lambda$=5.65 nm , and $\delta\lambda$=1.00 nm for a driver current of 25 mA. The use of a laser diode temperature-controller (Thorlabs, model LTC100-C) allows us to guarantee the stability of the optical parameters to thermal fluctuations.

In a second step, considering a length L=20.70 km of single-mode-standard-fiber (SM-SF) exhibiting a chromatic fiber-dispersion parameter of D=16.67 ps/nm km. Eq. (45) allows us to determine the value of the central frequency corresponding to the first filtered microwave or first band-pass as

$$f_1 = \frac{1}{DL\delta\lambda} = \frac{1}{(16.67\text{x}10^{-12}\text{seg/nm}\cdot\text{km})\cdot(20.70\text{ km})\cdot(1.0\text{ nm})} = 2.8\text{ GHz}$$

Eq. (39) permits us to determine the value of the low-pass band as

$$\Delta f_{lp} = \frac{2\sqrt{\ln(2)}}{\pi DL\Delta\lambda} = \frac{2\sqrt{\ln 2}}{(\pi)\cdot(16.67\text{x}10^{-12}\text{seg/nm}\cdot\text{km})\cdot(20.70\text{ km})\cdot(5.65\text{ nm})} = 271.85\text{ MHz}$$

Finally, according to Eq. (46), the corresponding bandwidth of the band-pass window is Δf_{bp}=543.70 MHz .

At this point, it is well worth highlighting the advantageous use of the chromatic dispersion parameter to obtain the filtered microwave signal. Once the main parameters are known, the topology illustrated in Figure 11 is assembled in order to evaluate the frequency response of the MPF.

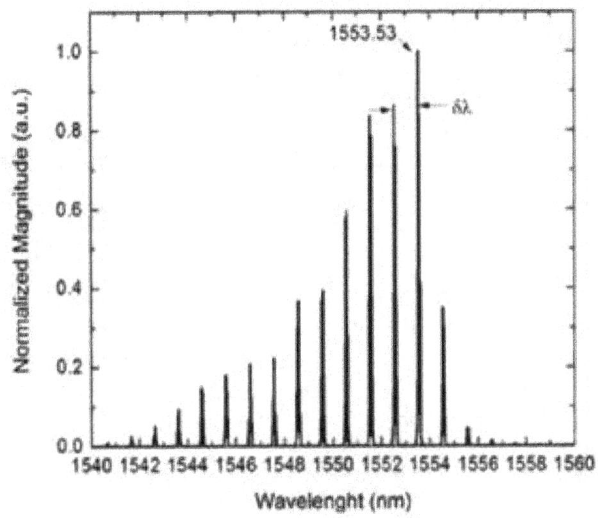

Figure 10. Optical spectrum for the MLD used in the experiment.

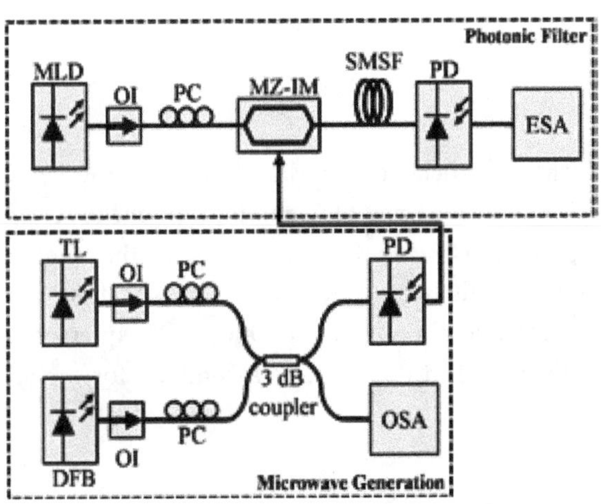

Figure 11. Experimental microwave photonic filter.

At the output of the MLD, an optical isolator (OI) is placed in order to avoid reflections to the optical source. Since the MZ-IM (Photline MX-LN-10) is polarization-sensitive, a polarization controller (PC) is used to maximize the modulator output power. The optical signal is launched into the MZ-IM. The microwave electrical signal (RF) for modulating the optical intensity is supplied by using optical heterodyne as described in Figure 1. The registered frequency response is located from 0.01 to 4 GHz at 0 dBm. The intensity-modulated optical signal is then coupled into a 20.70 km of SM-SF coil. The length of the optical fiber is corroborated by using an optical time domain reflectometer, OTDR (EXFO, model FTB-7300E). At the end of the link, the optical signal is applied to a fast Photo-Detector (PD, Miteq DR-125G-A), and its output connected to an electrical spectrum analyzer (Anritsu, model MS2830A-044), in order to measure the frequency response of the MPF. Figure 12 corresponds to the measured experimental frequency response where a low-pass band centered at zero frequency and the presence of a band-pass band centered at 2.8 GHz are clearly appreciable.

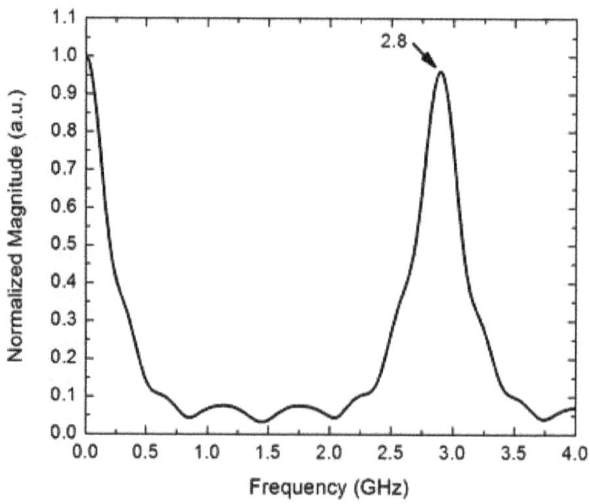

Figure 12. Experimental frequency response of the filter.

The bandwidth of 543.70 MHz associated to the band-pass window centered at 2.8 GHz allows us to guarantee enough bandwidth in case of fluctuations (in the order of nanometers) between mode spacing. On the other hand, a considerable increase on the length of the optical fiber due to thermal expansion is practically impossible. These considerations permit us to guarantee a good stability for the microwave photonic filter. Once the frequency response of the MPF is determined, the setup illustrated in Figure 13 is assembled for carrying out the fiber-radio transmission.

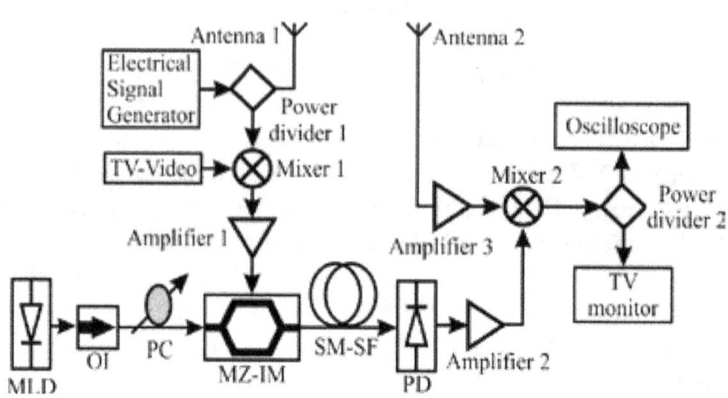

Figure 13. Experimental setup for optical and wireless transmission.

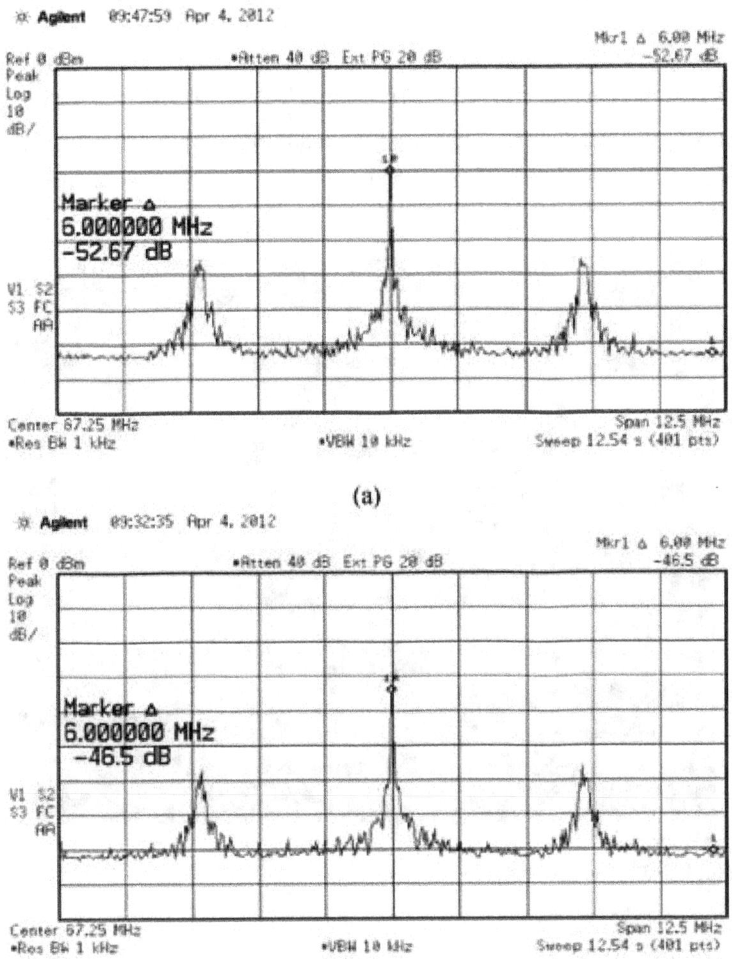

Figure 14. Electrical Spectrums for (a) Transmitted and (b) recovered TV signal.

Now, the electrical signal generator provides a signal of 2.8 GHz at 0 dBm that is used as the electrical carrier and demodulated signal. This signal is separated by using a power divider. Part of this signal is transmitted via radio frequency by the fabricated microstip antenna shown in Figure 5, and the rest is mixed with an analog NTSC TV signal of 67.25 MHz. The resulting mixed electrical signal is then applied to the electrodes of the MZ-IM for modulating the light emitted by the MLD. The modulated light is coupled into the 20.70 km SM-SF coil. At the end of the optical link, the signal is injected to a fast photo-detector (PD), and its electrical output is then amplified and launched to an electrical mixer. Another microstrip patch antenna placed at a distance of 10 meters is connected to a port of the mixer in order to recuperate the microwave signal that plays the role of the demodulated signal. Finally, by using another power divider, recovered analog TV signal can be launched to a digital oscilloscope or to the electrical spectrum analyzer in order to evaluate the quality of the recovered signal and at the same time display the TV signal on a TV monitor. Figure 14 (a) shows the measured electrical spectrum (Agilent, E4407B) corresponding to the transmitted TV signal where the SNR is 52.67 dB, whereas Figure (b) corresponds to the recovered TV signal with a SNR of 46.5 dB.

Finally, Figure 15 corresponds to a photograph of the screen of the oscilloscope where upper and lower traces are the waveforms of the transmitted and recuperated signal, respectively.

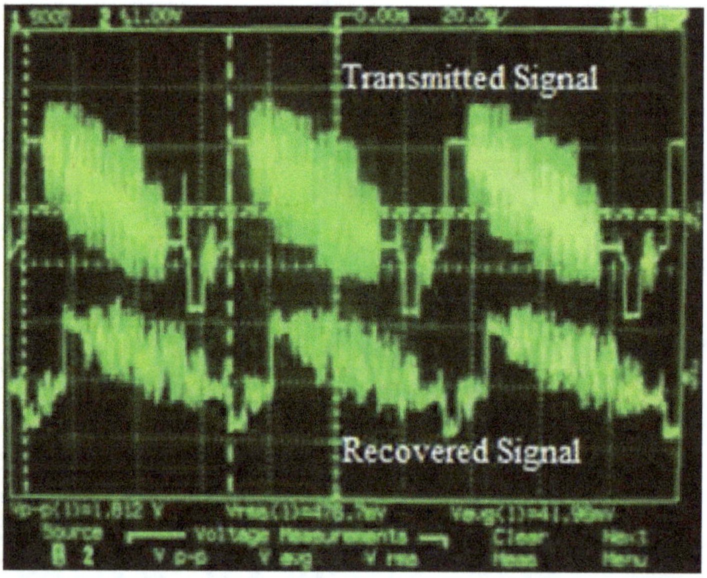

Figure 15. Transmitted and recovered TV signal.

9. Conclusions

Wireless communication systems require compact sources for the generation of mm-wave signals, that must have high spectral purity (linewidth < 100 kHz, phase noise < 100 dBc @100 kHz offset), tuneability, low power consumption and low cost, and although optical heterodyne of two DFB lasers has phase noise of –75 dBc/Hz even at an offset frequency of 100 MHz and it does not very compact, we have demonstrated in this chapter that by using optical heterodyne technique, a TV signal was transmitted and received satisfactory as a result of our proposed communication system generates a microwave carrier and a local oscillator simultaneously ensuring synchronization in frequency as well as in phase between microwave carrier and a local oscillator and avoiding in this case the use of an analog phase locked loop in the receiver to recover the TV information. The authors consider that the first proposed scheme in this chapter is not a truly wireless communication system, since an optical fiber is required to deliver the local oscillator in the receiver, however in order to obtain a wireless communication systems by using optical heterodyne technique, it is necessary to have collimated beams from optical fiber to photodetectors. On the other hand, due to the fact that the distribution of TV over microwave signals in the electrical domain presents loss associated with electrical distribution lines, the authors consider that the optical fiber is an ideal solution to fulfill this task because of its extremely broad bandwidth and low loss. In that case the distribution of TV over microwave can be directly by using optical fiber. In this way the second proposed experiment in this chapter represents a novel fiber-radio scheme to transmit an analog NTSC TV signal coded on a microwave band-pass located at 2.8 GHz. Filtering of microwave signal was achieved through the appropriate use of the chromatic fiber dispersion parameter, the physical length of the optical fiber, and the free spectral value of the multimode laser. Transmission of a TV signal was achieved over an optical link of 20.70 km, whereas a demodulated signal was transmitted via radiofrequency using the fabricated microstrip patch antennas. Although the distance between antennas was short, this distance can be lengthened if an array of antennas is used. Besides, a mathematical analysis corresponding to the microwave photonic filter was described demonstrating that the frequency response of the microwave photonic filter is proportional to the Fourier transform of the spectrum of the optical source used. The proposed microwave photonic filter represents an interesting technological alternative for transmitting information by using optoelectronic techniques. The results obtained in this work ensure that as an interesting alternative, several modulation schemes can be used for transmitting not only analog information but also digital information. Besides as optical heterodyne technique described here can generate microwaves continually tuned, we can use this feature to transmit several TV signals using frequency division multiplexing schemes FDM [19] and wavelength division multiplexing WDM techniques, not only point to point but also with bidirectional schemes by using simultaneous wired and wireless systems.

Acknowledgements

This work was supported by CONACyT (grants No 102046 and 154691).

Author details

Alejandro García Juárez[1], Ignacio Enrique Zaldívar Huerta[2], Antonio Baylón Fuentes[3], María del Rocío Gómez Colín[4], Luis Arturo García Delgado[1], Ana Lilia Leal Cruz[1] and Alicia Vera Marquina[1]

1 University of Sonora, Department of Physics Research, México

2 National Institute of Astrophysics, Optics and Electronics. Department of Electronics, México

3 Inst. FEMTO-ST, Université de Franche-Comté, Besançon, France

4 University of Sonora, Department of Physics, México

References

[1] A García-Juárez, I E Zaldívar-Huerta, G Aguayo-Rodríguez, J Rodríguez-Asomoza, M R Gómez-Colín, A G Rojas-Hernández. Coherent demodulation of microwave signals by using optical heterodyne technique with applications to point to point indoor wireless communications systems. Journal of Physics:Conference Series 274 (2011) 012008.

[2] Alejandro García Juárez, Ignacio E. Zaldívar Huerta, Antonio Baylón Fuentes, María del Rocío Gómez Colín, Jorge Rodríguez Asomoza, Armando G. Rojas Hernández. Reflected Power Measurement of Antennas between 0 and 4 GHz using Optical Mixing of Distributed Feedback Lasers. Radioengineering, vol. 22, no. 4, December 2013. pp. 1276-1274.

[3] Yang Jing Wen, Hai Feng Liu, Dalma Novak, and Yoh Ogawa. Millimeter-Wave Signal Generation from a Monolithic Semiconductor Laser via Subharmonic Optical Injection. IEEE Photonics Technology letters, vol. 12, no. 8, august 2000, pp. 1058-1060.

[4] Langley L. N., Elkin M. D., Edge C., Wale M. J., Gliese X., Huang X., and Seeds A. J. Packaged Semiconductor Laser Optical Phase-Locked Loop (OPLL) for Photonic Generation. Processing and Transmission of Microwave Signals", IEEE Transactions On Microwave Theory and Techniques, Vol. 47, No. 7, pp. 1257–1264, July. (1999).

[5] J. J. O'Reilly, P. M. Lane, R. Heidemann, and R. Hofstetter. Optical generation of very narrow linewidth millimeter wave signals. Electron. Letters, Vol. 28, No. 25, pp. 2309–2311, (1992).

[6] M. García Larrodé, A. M. J. Koonen, J. J. Vegas Olmos, and E. J. M. Verdurmen. Microwave Signal Generation and Transmission Based on Optical Frequency Multiplication With a Polarization Interferometer. Journal of Lightwave Technology, vol. 25, no. 6, June 2007. pp. 1372-1378.

[7] J. Capmany and D. Novack, "Microwave Photonics combine two worlds," Nature Photonics vol. 1, pp. 319-329, June 2007.

[8] J. Capmany, B. Ortega and D. Pastor, "A Tutorial on Microwave Photonic Filters," Journal of Lightwave Technology vol. 24, no. 1, pp. 201-229, January 2006.

[9] Ignacio E. Zaldivar-Huerta, Alejandro García-Juárez, Pablo Hernández-Nava, and Jorge Rodríguez-Asomoza. Experimental transmission in a fiber-radio system using a microwave photonic filter at 2.8 GHz. IEICE Electronics Express Vol. 10, No. 3, pp. 1-10, February 14 2013, ISSN: 1349-2543.

[10] Thang Tien Pham, Hyun-Seung Kim, Yong-Yuk Won, and Sang-Kook Han. Bidirectional 1.25-Gbps Wired/Wireless Optical Transmission Based on Single Sideband Carriers in Fabry–Pérot Laser Diode by Multimode Injection Locking. Journal of Lightwave Technology, vol. 27, no. 13, July 2009, pp. 2457-2464.

[11] Stavros Iezekiel Microwave Photonics Devices and Applications. John Wiley; 2009.

[12] Kai Chang RF and Microwave Wireless Systems. John Wiley; 2000.

[13] Ramesh Garg, R. B. Garg, P. Bhartia, Prakash Bhartia, Apisak Ittipiboon, and I. J. Bahl, Microstrip Antenna Design Handbook, New York: Artech House Inc; 2000.

[14] J. Liang, C. C. Chiau, X. Chen, and C. G. Parini, "Printed Circular Disc Monopole Antenna for Ultra-Wideband Applications", Electronic. Letter. Vol. 40, No.20, pp. 1246–1247, 2004.

[15] K. L. Wong, T. C. Tseng, and P. L. Teng, "Low-profile Ultra Wideband Antenna for Mobile Phone Applications", Microwave Optical Technolgy Letter, Vol. 43, pp.7-9, 2004.

[16] K. L. Wong, L. C. Chou, and H. T. Chen, "Ultra-wideband Metal-Plate Monopole Antenna For Laptop Application", Microwave Optical Technology Letter, Vol. 43, pp. 384-386, 2004.

[17] Constantine A. Balanis, Antenna Theory-Analysis and Design, Third Edition. John Wiley & Sons, Inc; 2005.

[18] G. Aguayo-Rodríguez, I. E. Zaldívar-Huerta, J. Rodríguez-Asomoza, A. García-Juárez and P. Alonso-Rubio "Modeling and performance analysis of an all-optical photonic microwave filter in the frequency range of 0.01-15 GHz" International Society for Optics and Photonics, 2010. p. 76200B-76200B-12.

[19] Alejandro García-Juárez, Ignacio E. Zaldívar-Huerta, Jorge Rodríguez-Asomoza, María del Rocío Gómez-Colín "Method to transmit analog information by using a long distance photonic link with distributed feedback lasers biased in the low laser threshold current region" Opt. Eng. 51(6), 065006 (2012).

4

Multi-User Visible Light Communications

Oswaldo González, Marcos F. Guerra Medina and
Inocencio R. Martín

1. Introduction

Nowadays, the demand for high capacity wireless connectivity is endless. Radiofrequency networks try to meet this demand but strict regulations and the increasing number of users mean service providers have to look for new alternatives to radio communications. Wireless optical communications could be a practical solution.

Among optical communication systems, visible light communications (VLC), first proposed by researchers at Keio University in Tokyo [1][2], have prompted great interest in the scientific community in the last few years [3][4][5]. There have also been regulatory efforts made in this technology that have led to the appearance of a standard [6]. These new VLC systems, based on the use of sustainable, energy-efficient, visible LED (light-emitting diode) lamps [7] to simultaneously transmit information together with their normal use as illumination devices, share the same advantages as their infrared counterparts [8]. They are also eye-safe (visible light is not harmful to the human eye), which enables the use of higher transmission powers. However, the main drawback is the limited transmission bandwidth of current LED devices, typically several MHz, and whose enhancement has been one of the main issues addressed by researchers [9][10][11][12].

Zeng *et al.* [13] have proposed the use of multiple-input multiple-output (MIMO) schemes based on imaging receivers to obtain high capacity VLC networks. Additionally, the orthogonal frequency division multiplexing (OFDM) technique has proved to be a feasible candidate to obtain these high-speed networks [14], demonstrating impressive experimental data rates for short-range communications [15][16][17][18]. In [19], it was forecast that combining OFDM technique and imaging reception could be a potential research field for the future, which was rapidly demonstrated experimentally by Azhar *et al.* [20]. Alternatively, Burton *et al.* have proposed the use of non-imaging receivers for MIMO-OFDM-based VLC, also demonstrating very interesting results [21].

This chapter describes the characteristics of MIMO-OFDM schemes applied to multi-user visible light communications, comparing the capacity of both non-imaging and imaging reception to separate the information corresponding to each individual user.

2. The indoor wireless visible-light optical channel

The indoor wireless visible-light optical channel is basically composed of three elements: the emitting sources (the visible-light LED lamps), the room where lamp emissions are enclosed and the optical receiver. In this section, the different components of the communication channel are described thoroughly, highlighting their impact on the effective channel bandwidth.

2.1. Emitting sources

Visible-light LED lamps are commonly made up of a significant number of single chips, each presenting a generalized Lambertian radiation pattern, typically of *mode number* n=1 or *radiation semiangle* (at half power) $\phi_{1/2}$=arccos$\sqrt[n]{1/2}$=60° (see Fig. 9(a)). The Lambertian radiation pattern, having uniaxial symmetry, is given by [22]:

$$R_E(\phi) = P_E \frac{n+1}{2\pi} \cos^n \phi, -\frac{\pi}{2} \leq \phi \leq \frac{\pi}{2} \tag{1}$$

where P_E is the total power emitted by the single LED chip. There are two major types of visible-light (white) LEDs: those consisting of a blue LED chip with a phosphor wavelength conversion, which enables re-emission at longer wavelengths thus leading to a combined broadband optical spectrum in the visible region, and those devices composed of three independent LED chips emitting at blue (B), green (G) and red (R) wavelengths, whose joint radiation is perceived as white light [23][24]. The first group of devices, after suppressing the slow phosphorescent component by means of a blue filter (see Fig. 1), experience a modulation bandwidth enhancement from ~2 MHz to ~20 MHz [4], whereas tri-chip RGB LEDs present a modulation bandwidth of ~25 MHz [24]. In this chapter, phosphor-based white LED will be mostly considered, although bearing in mind the results are directly transferable to the case when RGB LEDs are used.

Fig. 1 shows the *power spectral density* (PSD), which describes the radiant power per unit wavelength, of a typical phosphor-based white LED (WLED) [25] and the *eye sensitivity function* under daylight (well-lit) conditions [23], denoted as $p(\lambda)$ and $V(\lambda)$, respectively. The WLED optical power [W] (*radiant flux*) is given by

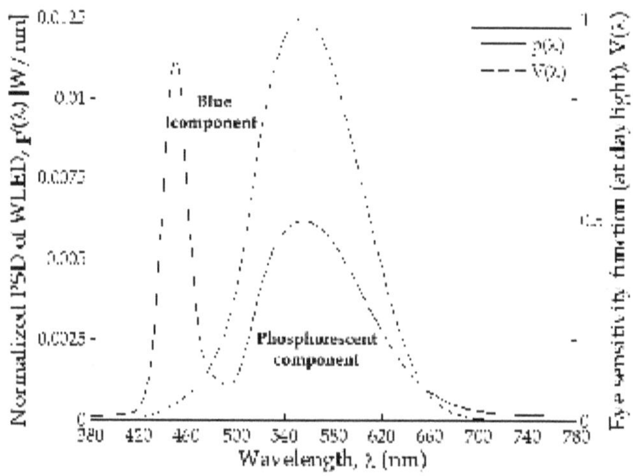

Figure 1. Normalized power spectral density (PSD) of a typical phosphor-based white LED (WLED) and eye sensitivity function under daylight (well-lit) conditions versus wavelength.

$$P_E = \int_{\lambda_{\min}(380\mathrm{nm})}^{\lambda_{\max}(780\mathrm{nm})} p(\lambda)\,d\lambda \tag{2}$$

In the PSD of a WLED, the blue component comprises approximately 50% of the power emitted by the device [4]. In most communication applications, this component will solely be detected by filtering the phosphorescent one to attain high data rates.

In visible-light communications, it is important to remember the relationship between photometric and radiometric quantities [23], as LED lamps are used as illumination devices too. The *luminous flux* [lm] can be obtained from the radiometric light power by [23]

$$\Phi_{lum} = 683\frac{\mathrm{lm}}{\mathrm{W}}\int_{\lambda_{\min}}^{\lambda_{\max}} V(\lambda)p(\lambda)\,d\lambda \tag{3}$$

The normalization factor 683 lm/W denotes that a monochromatic 1-W optical source emitting at 555 nm, where photopic eye sensitivity is maximal (see $V(\lambda)$ in Fig. 1), has a luminous flux of 683 lumens. The *luminous efficacy* is then given by

$$\eta_{lum} = \frac{\Phi_{lum}}{P_E} = \left[683\frac{\mathrm{lm}}{\mathrm{W}}\int_{\lambda_{\min}}^{\lambda_{\max}} V(\lambda)p(\lambda)\,d\lambda\right] \Big/ \int_{\lambda_{\min}}^{\lambda_{\max}} p(\lambda)\,d\lambda \tag{4}$$

This describes the conversion efficiency from the LED power to luminosity. The *illuminance* [lx=lm/m^2] is the incident luminous flux per unit area. The horizontal illuminance at a point (x,y) can be obtained as [2]

$$E_{hor} = I(\phi)\cos\varphi / d^2 = I_0 \cos^n \phi \cos\varphi / d^2 \tag{5}$$

where d is the distance between the optical source (LED) and the illuminated surface, ϕ is the angle of irradiance, φ is the angle of incidence on the surface (see Fig. 9(a)), and I_0 is the center *luminous intensity* of the LED [4]

$$I_0 = I(\phi = 0) = \frac{n+1}{2\pi}\Phi_{lum} \tag{6}$$

and n being the Lambert index of the optical source, which typically equals unity in WLEDs.

Fig. 2 depicts the direct illuminance at the working plane (0.75-m height) of a room with two large windows (see parameters for characterization study in Table 1), due to the line-of-sight (LOS) component, i.e. the one generated by the LED lamps in their direct paths to the illuminated point. When considering the reflection of light on walls and windows (see Fig. 3), the values of illuminance are significantly enhanced and this becomes more uniform at the working plane. In this last case, we can verify that standard levels for workplaces (> 400 lx) are clearly achieved [26].

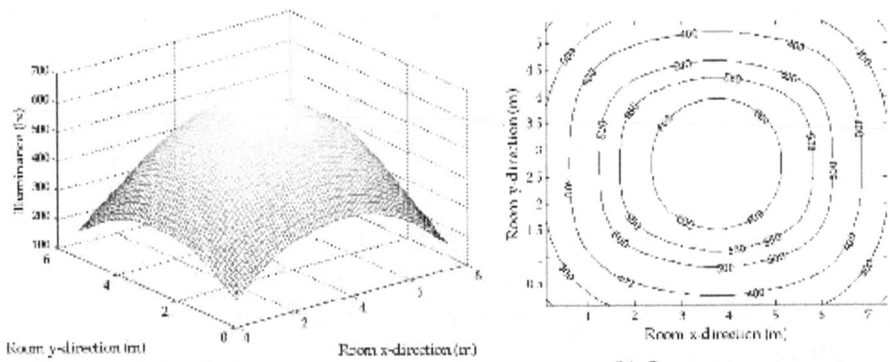

(a) 3D illustration (b) Contours map

Figure 2. Direct (line-of-sight) illuminance at the receiver plane (E_{min}=185 lx; E_{max}=640 lx; $\bar{E}$=460 lx).

2.2. Reflective surfaces

Rough reflective surfaces commonly present a purely diffuse reflection pattern, as shown in Fig. 4(a), i.e. they do not favor any particular direction after reflection regardless of the incidence angle. Moreover, this reflection pattern follows Lambert's model with mode number n=1:

$$R_S(\theta,\lambda) = \rho(\lambda) P_I \frac{1}{\pi}\cos\theta \tag{7}$$

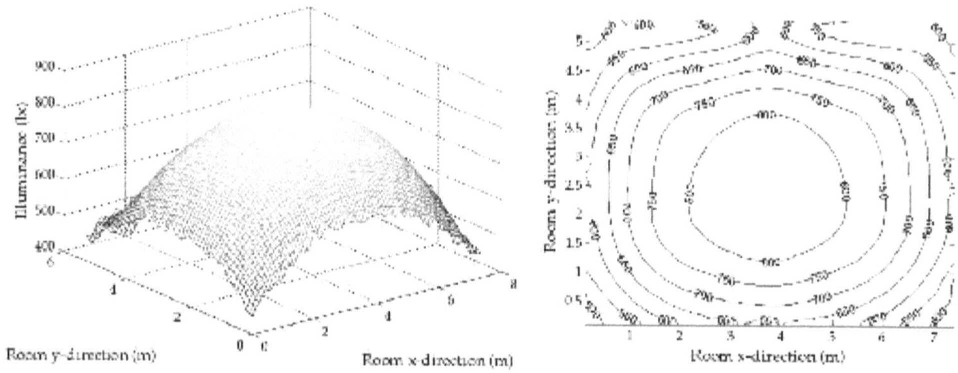

(a) 3D illustration (b) Contours map

Figure 3. Illuminance at the receiver plane, considering contributions of reflections on walls and windows (E_{min}=445 lx; E_{max}=845 lx; $\bar{E}$=690 lx).

where P_I is the incident optical power, $\rho(\lambda)$ the surface reflection coefficient, which depends on the wavelength λ, and θ is the observation angle. However, some smooth surfaces, e.g. glass, additionally exhibit a strong specular component, which is dependent upon the incidence angle of the incoming ray, as illustrated in Fig. 4(b), therefore they are more properly modeled by Phong's model, where the reflection pattern behaves as the sum of both diffuse and specular components [27]

$$R_S(\theta,\theta',\lambda) = \rho(\lambda)P_I\left[r_d\frac{1}{\pi}\cos\theta + (1-r_d)\frac{m+1}{2\pi}\cos^m(\theta-\theta')\right] \tag{8}$$

Here, r_d is the fraction ($0 \le r_d \le 1$) of the incident signal that is reflected diffusely, m denotes the directivity of the specular reflection component (mode number), θ' is the incidence angle with respect to the normal of the reflective surface, and the remaining parameters maintain the same meaning as those of eqn. (7).

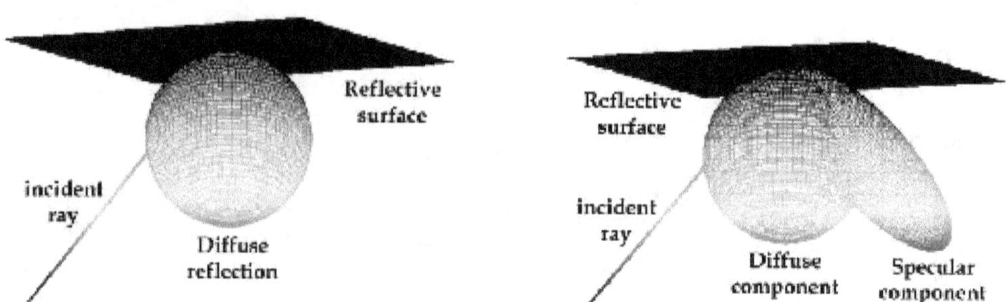

(a) Lambert's model (b) Phong's model

Figure 4. Purely diffuse (Lambert's model) and combined diffuse-specular (Phong's model) reflections.

Room size (length × width × height):	7.5 m × 5.5 m × 3.5 m		
Number of LED arrays (lamps):	6 (3 × 2)		
Number of LEDs per array:	900 (30 × 30)		
Dimensions of each LED array:	0.6 m × 0.6 m		
Positions of LED arrays (central point)(x, y, z):	array 1: (1.50, 1.50, 3.50) array 2: (3.75, 1.50, 3.50) array 3: (6.00, 1.50, 3.50) array 4: (1.50, 4.00, 3.50) array 5: (3.75, 4.00, 3.50) array 6: (6.00, 4.00, 3.50)		
Power of a single LED (P_E):	20 mW		
LED Lambertian mode number (n):	1		
LED transmission bandwidth:	~ 15 MHz		
Receiver plane height:	0.75 m		
Surface materials parameters:	$\overline{\rho(\lambda)}$	r_d	m
Ceiling	0.35	1	-
Floor	0.55	1	-
Walls (plaster)	0.69	1	-
Windows (glass)	0.04	0	280
Windows dimensions (width × height):	2.5 m × 1.5 m		

Table 1. Parameters for environment characterization.

2.3. Optical receivers

In this section, two kinds of angle-diversity receivers will be considered [28]: *imaging diversity receivers*, which use a lens for projecting an image of the optical sources (lamps) onto a pixelated detecting surface, and *non-imaging diversity receivers*, which consist of several detecting branches, each with its own optical detector, oriented in different directions which provides angle-diversity reception.

2.3.1. Imaging receiver

In a VLC scenario based on an imaging receiver, the lamps' images are projected on an array of photodetectors (PD) by means of a lens, as depicted in Fig. 5. This projection can illuminate several detecting pixels. However, when the photodetecting surface is divided into a significant number of pixels, the lamps impressions will be spread out over different detecting areas, therefore being able to separate the information relative to each individual lamp. In this study, we will consider a paraxial optics approach, as in [13].

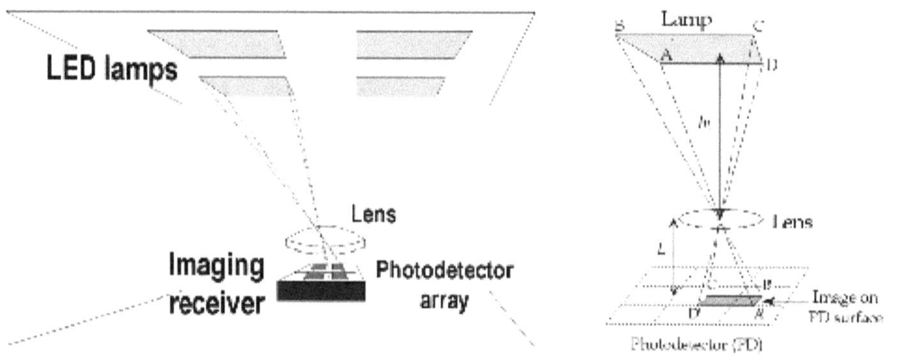

(a) Pictorial representation (b) Image projection on PD surface

Figure 5. Imaging-receiver-based visible-light communication scenario.

In a lens of *diameter* D and *f-number* $f_{\#}$, its *focal length* L is given by $L=D\ f_{\#}$. Then, the *magnification* of the system, when the lens is located at a vertical distance h_L from the ceiling (see Fig. 5(b)), is $M=h_L/L$, i.e. any dimension d_o of the original object in the ceiling creates a replica (image) on the PD surface of size $d_i=d_o/M$. For example, for the width of the lamp, we have AD=M A′D′. Geometrically, we can determine the image that is drawn on the PD surface by projecting every environment point along its path towards the imaging receiver through the lens center, similar to that shown in Fig. 5(b) for points A, B, C and D of the LED lamp. Therefore, a square $N\times N$ photodetector array of width w_r (area $A=w_r^2$) will only be able to observe any point inside of its square-shape equivalent *field of view* (FOV), as shown in Fig. 6 for a 4×4-pixel imaging receiver. The total equivalent FOV, *observation semi-angle,* varies between $FOV_{min}=\mathrm{arctg}(w_r/2L)$, in the center of a PD side, and $FOV_{max}=\mathrm{arctg}(w_r/\sqrt{2}L)$, in a PD corner. For example, for the 4×4-pixel imaging receiver whose characteristics are specified in Table 2, we have $56.3° \leq FOV \leq 64.8°$. However, this total equivalent FOV is shared by the $N\times N$ detector elements. For example, for an inner single detector element, this FOV reduces to $FOV_{min}=\mathrm{arctg}(w_r/NL)/2=18.4°$ and $FOV_{max}=\mathrm{arctg}(\sqrt{2}w_r/NL)/2=23.3°$. Therefore, this kind of receiver is very directed, where every pixel only observes a small portion of its tridimensional environment. Owing to the dimensions of the receiver structure with respect to the room size, a single LED is seen as a point by the receiver too, i.e. it will only illuminate a specific pixel of the PD array. The line-of-sight contribution of a single LED to the received optical power is given by

$$P_R = R_E(\phi)\int_\lambda p'(\lambda)T(\varphi,\lambda)A_{\text{eff}}(\varphi)d\lambda / d^2 = P_E\frac{n+1}{2\pi d^2}\cos^n\phi A_r\cos\varphi\int_\lambda p'(\lambda)T(\varphi,\lambda)d\lambda \tag{9}$$

where $R_E(\phi)$ is the LED radiation pattern in the outgoing angle ϕ defined in (1) and illustrated in Fig. 9(a), A_{eff}=$A_r$cos φ is the *effective area* of the photodetecting element (pixel), whose physical area is A_r=A/N^2, for an observation angle φ and d is the distance from the LED to the PD. Moreover, $T(\varphi,\lambda)$ is the signal transmission of the filter, if this is incorporated (e.g. a blue filter to enhance the modulation bandwidth as described in section 2.1), and $p'(\lambda)$=$p(\lambda)/P_E$ is the normalized PSD of the WLED. When no filter is attached, $T(\varphi,\lambda)$=1 and $\int_\lambda p'(\lambda)T(\varphi,\lambda)d\lambda=1$. As we will describe in section 2.4, the light reflected on walls can also contribute to the total received optical power.

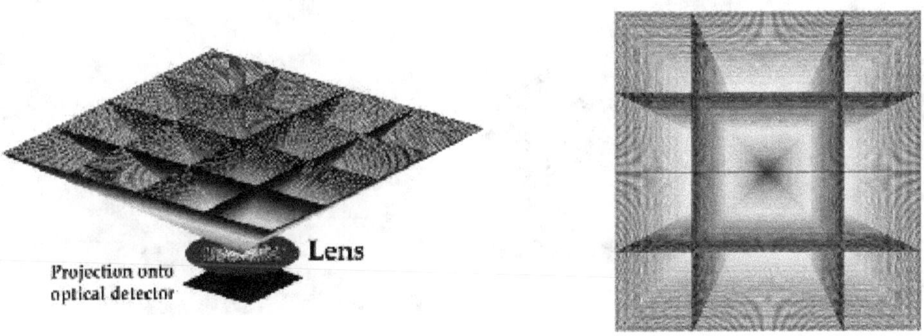

(a) Lateral view of FOV (b) Top view of FOV

Figure 6. Equivalent field of view (FOV) of the 4×4-pixel imaging receiver.

Fig. 7 depicts the received light intensity images projected, through the lens, onto the photodetector surface for three different positions of the imaging receiver in the room, as specified in Fig. 7(a). Here, not only LOS image is considered but also the reflections of light on the room's surfaces. In Fig. 7(d), we can clearly recognize the reflections of lamps 5 and 6 on one of the windows, whereas for positions 1 and 2, the windows are out of the field of view of the receiver.

Actually, each photodetecting element (pixel) integrates the total light arriving to it; therefore, there will eventually be a 4×4 image. However, this figure is interesting to illustrate which lamps directly illuminate a certain pixel and to see how two lamps never illuminate a single pixel without at least one of these lamps also illuminating another pixel. As we will see, this characteristic of imaging receivers is what makes them very appropriate for visible light communications.

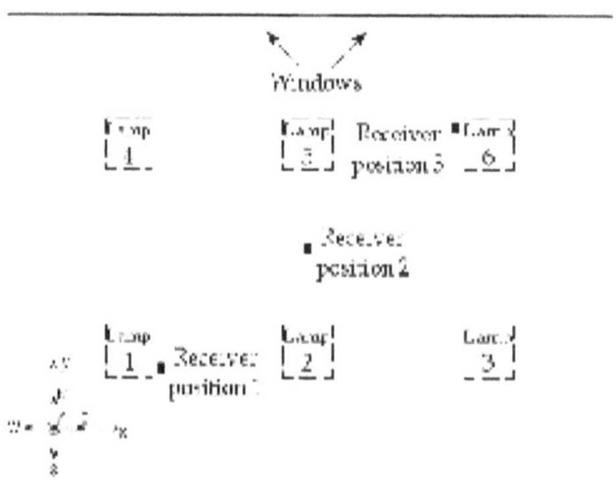

(a) Top view of receiver locations

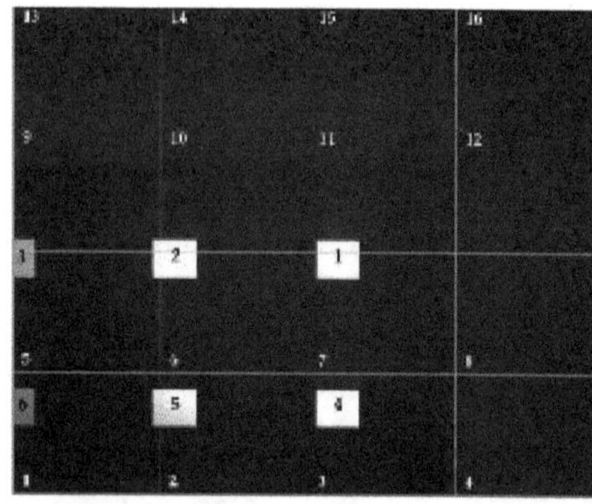

(b) Image at receiver for location 1

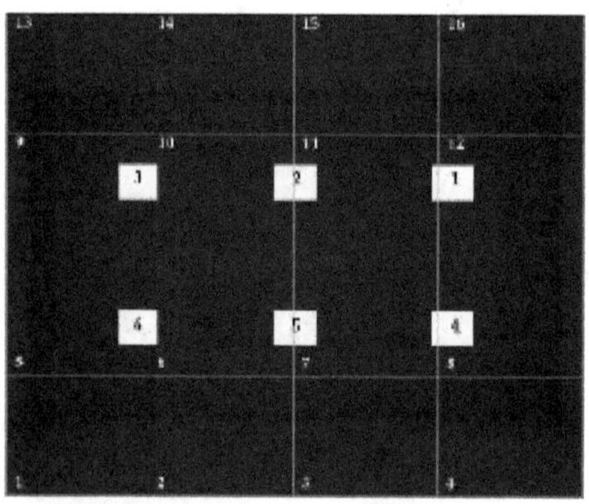

(c) Image at receiver for location 2

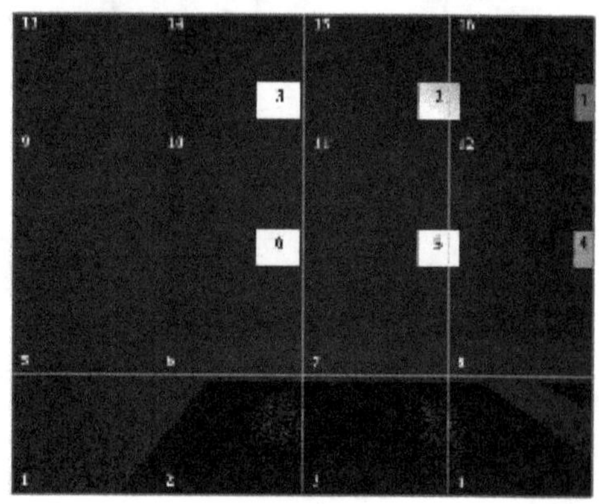

(d) Image at receiver for location 3

Figure 7. Images impressed on photodetector surface for three different receiver locations.

2.3.2. *Angle-diversity non-imaging receiver*

An angle-diversity receiver makes use of multiple receiving elements (branches), oriented in different directions, to collect the power emitted from optical sources. Since each element has a different view of its environment, this diversified received signal can be used to separate the information relative to several users by using an appropriate multi-user detection scheme. Moreover, each receiving branch can be equipped with a *concentrator*, increasing the gathering of received optical power. In this case, for a specific branch, the *effective signal-collection area* becomes [8]

$$A_{\text{eff}}(\varphi) = \begin{cases} A_r g(\varphi)\cos\varphi, & 0 \le \varphi \le \varphi_c \\ 0 & \varphi > \varphi_c \end{cases} \tag{10}$$

where A_r is the physical area of the photodetector placed in that branch, φ the observation angle, φ_c the concentrator FOV, and $g(\varphi)$ its gain, which, for an idealized non-imaging concentrator with internal refractive index n, is given by [8]

$$g(\varphi)=\begin{cases} n^2/\sin^2\varphi_c & 0\le\varphi\le\varphi_c \\ 0 & \varphi>\varphi_c \end{cases} \tag{11}$$

Fig. 8 depicts the equivalent FOV of a 7-branch non-imaging receiver with each branch having an individual FOV of 25° (the remaining parameters of the receiver are summarized in Table 2). Compared to the imaging receiver of section 2.3.1, its total equivalent FOV is 75°, and considering the individual single-element FOVs, the non-imaging receiver becomes very directed too. However, this structure, which requires a separate optical concentrator for each photodetector, is bulky and more costly than the imaging receiver, and, additionally, it does not provide so much diversity as the latter.

Imaging receiver		**Non-imaging receiver**	
Detector physical area (A):	36 cm^2	Physical area of each PD (A_r):	2.25 cm^2
Number of pixels (P):	16 (4×4)	Number of branches (P):	7
Pixel physical area (A_r):	2.25 cm^2	Concentrator FOV (φ_c):	25°
Lens f-number ($f_\#$):	1	Concentrator refractive index (n):	1.5
Lens diameter (D):	2 cm	Orientation of detector branches (elevation, azimuth):	(0°,0°), (50°,0°), (50°,60°), (50°,120°), (50°,180°), (50°, 240°), (50°,300°)

Table 2. Parameters of imaging and non-imaging receivers.

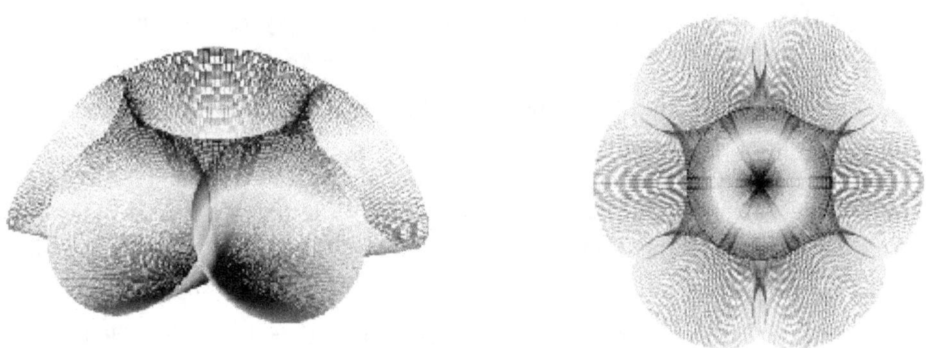

(a) Lateral view of FOV (b) Top view of FOV

Figure 8. Equivalent field of view (FOV) of the 7-branch non-imaging receiver.

2.4. Simulation algorithm

In wireless optical communications, the optical link is typically established by means of *intensity modulation* (IM), in which the desired waveform is modulated onto the instantaneous power of the carrier, in conjunction with *direct detection* (DD) as a down-conversion technique at the receiver end. Therefore, the transmitted waveform $x(t)$ is the instantaneous optical power of the emitter, and the received waveform $y(t)$ is the instantaneous current in the receiving photodetector. In this way, the optical channel with IM/DD can be modeled as a baseband linear system with impulse response $h(t)$ or, alternatively, it can be described in terms of the frequency response

$$H(f) = \int_{-\infty}^{\infty} h(t) e^{-j2\pi ft} dt \tag{12}$$

which is the Fourier transform of $h(t)$. This channel model $h(t)$ is practically stationary because it only varies when emitter, receiver or objects in the room are moved by tens of centimeters. In many applications, optical links are operated in the presence of intense infrared and visible background light. The received background light adds *shot noise*, which is usually the limiting noise source in a well-designed receiver. Due to its high intensity, this shot noise can be modeled as white, Gaussian, and independent of $x(t)$. When little or no ambient light is present, the dominant noise source is receiver preamplifier noise, which is also signal-independent and Gaussian (though often nonwhite) [8]. Thus, the noise $n(t)$ is usually modeled as Gaussian and signal-independent, and the instantaneous output current at the receiver can be represented as

$$y(t) = Rx(t) \otimes h(t) + n(t) \tag{13}$$

where the "⊗" symbol denotes convolution and R is the detector *responsivity* (A/W). According to (13), the optical link can be completely characterized by means of the impulse response $h(t)$ and noise sources $n(t)$. Knowing $h(t)$ allows us to determine the multipath penalty, which limits the maximum baud rate. The second term is related to the *signal-to-noise ratio* (SNR), which determines the performance of the digital link.

In order to evaluate the impulse response on indoor wireless optical channels, several deterministic methods were first proposed [22]. However, these methods can only be implemented to determine the impulse response up to the third reflection due to their computational complexity. Later on, modified Monte Carlo-based ray-tracing algorithms were introduced, which present a lower computational cost and no limit to the number of reflections that can be considered [29][30].

In these algorithms, ray directions are randomly generated according to the radiation pattern from the emitter. The contribution of each ray from the source or after a bounce to the receiver is computed deterministically. Consequently, the discretization error is due to the number of random rays. The line-of-sight (LOS) and multiple-bounce impulse responses are considered when calculating the total impulse response.

The LOS contribution to the received optical power at a certain detector, illustrated in Fig. 9(a), can be directly determined by using (9). In the case of non-imaging receivers, when they are equipped with a concentrator, their effective area $A_{eff}(\varphi)$ is also affected by its gain as described in (10) an (11).

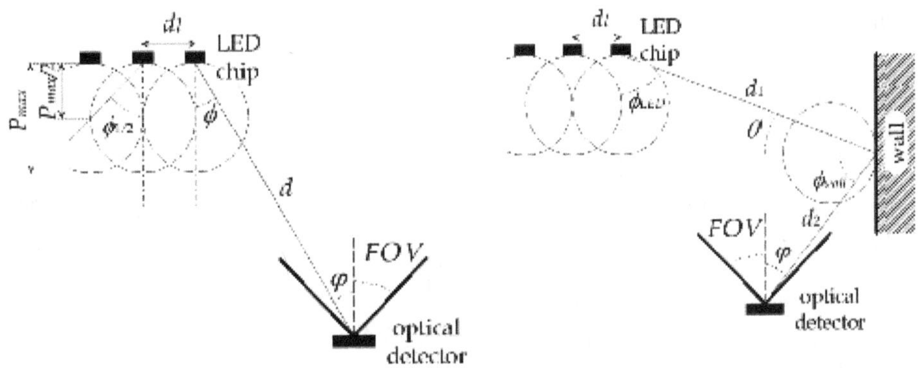

(a) LOS contribution (b) Power contribution after a reflection

Figure 9. Line-of-sight (LOS) contribution of a single LED chip to received power and its power contribution after reflection off a wall.

Additionally, when the communication is established in a room with reflectors, the radiation from the emitter can reach the receiver after any number of reflections, as depicted in Fig. 9(b) for the first-order reflection. In a ray-tracing algorithm, many rays are generated at the emitter position with a probability distribution equal to its normalized radiation pattern $R_E(\phi,n)/P_E$ (see eqn. (1)). The power of each generated ray is initially P_E/N, where N is the number of rays used to discretize the source. When a ray impinges on a surface, the reflection point is converted into a new optical source, thus, a new ray is generated with a probability distribution provided by the normalized reflection pattern of that surface, $R_S'(\theta,\theta')=R_S(\theta,\theta',\lambda)/P_I\rho(\lambda)$, described by Phong's model according to (8). The process continues throughout the simulation time t_{max}. After each reflection, the reflected power reaching the receiver is computed by

$$P_R^{(k)} = R_S'(\theta,\theta')A_{eff}(\varphi)\frac{P_E}{N}\int_\lambda \rho_1(\lambda)\rho_2(\lambda)\ldots\rho_k(\lambda)p'(\lambda)T(\varphi,\lambda)d\lambda / d^2 \tag{14}$$

where (k) denotes k^{th} reflection of the ray since it originated from an emitter, $\rho_1(\lambda)\rho_2(\lambda)\ldots\rho_k(\lambda)$ are the reflection coefficients of the surfaces against which the ray collided, d is the distance between the last reflection point and receiver and $A_{eff}(\varphi)$ the effective area for the observation angle φ. The remaining parameters were previously described for eqn. (8)-(11).

In the VLC multi-user application, each lamp is composed of a significant number M of single emitters (LEDs) and we require an angle-diversity, imaging or non-imaging, receiver, composed of multiple receiving elements. By using (14), the contribution of the ith ray emitted by the mth LED (of the lth lamp) to the received power reaching each jth branch of the receiver during a certain time interval ($p_{i,j,k}^{(l,m)}$, k^{th} time interval) can be computed. The total received power due to the lth LED lamp at the jth branch of the optical detector in the kth time interval (width Δt) is computed as the sum of the power of the $N_{j,k}^{(l)} = \sum_m N_{j,k}^{(l,m)}$ rays that contribute in that interval

$$p_{j,k}^{(l)} = \sum_{m=1}^{M} \sum_{i=1}^{N_{j,k}^{(l,m)}} p_{i,j,k}^{(l,m)} \tag{15}$$

At the receiver end, the impulse responses $h_j^{(l)}(t)$ at all branches (j=1,..., P) due to the lth lamp (l=1,..., L) are given by

$$h_j^{(l)}(t) = \sum_{k=0}^{K-1} p_{j,k}^{(l)} \delta(t - k\Delta t) \tag{16}$$

where $K = t_{\max}/\Delta t$, and where we have assumed as the time origin the instant when the rays are generated from the emitter (lamp). This process must be repeated in order to obtain the different impulse responses between each emitter and each receiving branch in the multi-user scenario.

Fig. 10 shows the power balance at the receiver plane for the imaging receiver (see parameters for the study by referring to Table 1 and Table 2) when a blue filter is incorporated to enhance modulation bandwidth. Fig. 11 presents the same study results when, in its place, a non-imaging receiver is used (see Table 2 for parameters of this receiver). In both results, five reflections of light on walls have been considered, which ensures that at worst we are only neglecting less than 1% of received power if a greater number of reflections were considered. We can observe that the changes in power level when moving around the room are smoother for an imaging-receiver, but, more importantly, a non-imaging receiver offers a power gain of about 12 dB due to the use of the concentrator.

Fig. 12 and Fig. 13 show the impulse responses for imaging and non-imaging receivers, respectively, when these are located at position 1, as indicated in Fig. 7(a). In the case of the imaging receiver, we can observe a clear connection between image of Fig. 7(b) and the obtained impulse responses. We can also verify that they are responses with a substantial LOS component (more than 80% of the received power, in the blue region, is due to this factor), in contrast with those at the non-imaging receiver, where multiple-bounce components, after the first large impulse, are more noteworthy (in this second case, the LOS components represent approximately 66% of the total received power). In spite of these components corresponding to reflections on walls, we can observe in Fig. 13 that users 3 and 6 are overshadowed by users 2 and 5, respectively, i.e. the first ones are practically delayed replicas of smaller intensity of the latter ones, which, as we will see, will limit maximum achievable joint data rates. This is something that does not occur in the imaging receiver case.

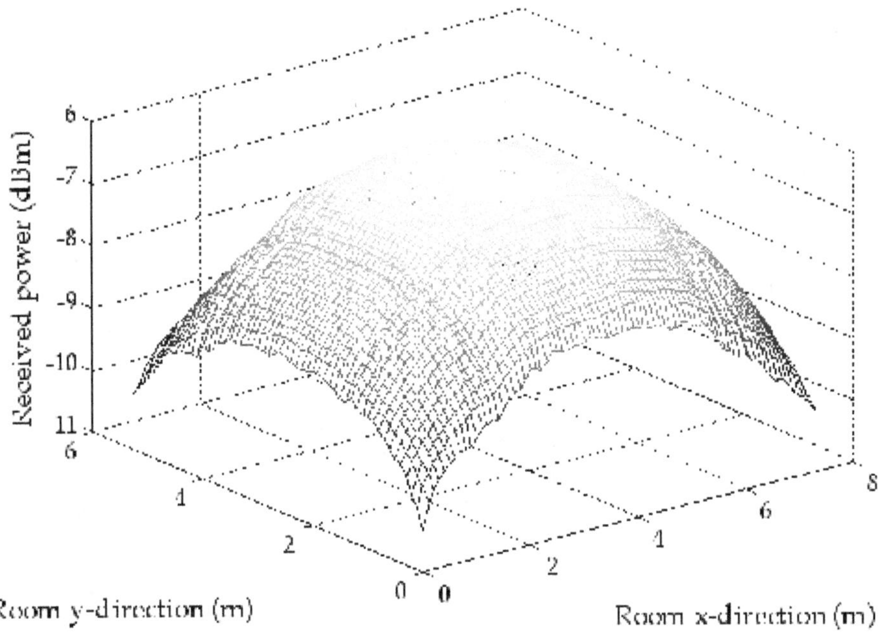

Figure 10. Power balance at the receiver plane for the imaging receiver (P_{min}=-10.4 dBm; P_{max}=-6.2 dBm; $\bar{P}$=-7.4 dBm).

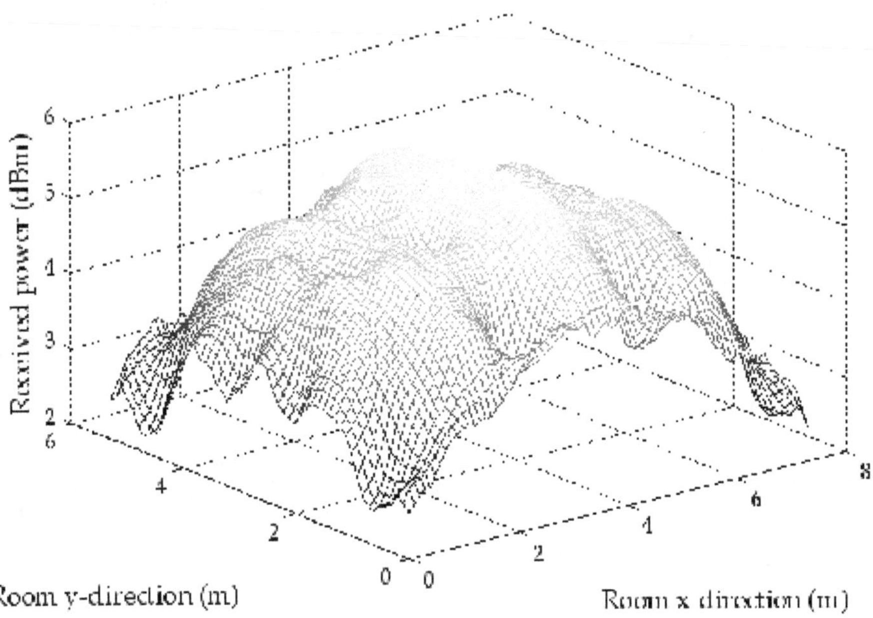

Figure 11. Power balance at the receiver plane for the non-imaging receiver (P_{min}=2.2 dBm; P_{max}=5.8 dBm; $\bar{P}$=4.5 dBm).

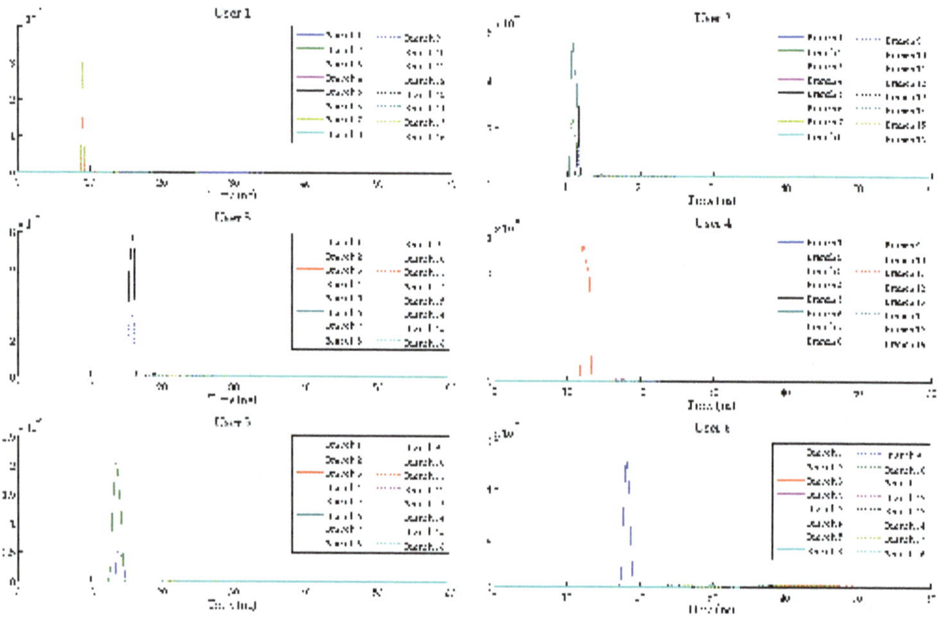

Figure 12. Impulse responses between transmitting users (lamps) and each receiving branch (pixel) for imaging receiver at position 1 (see Fig. 7(a)).

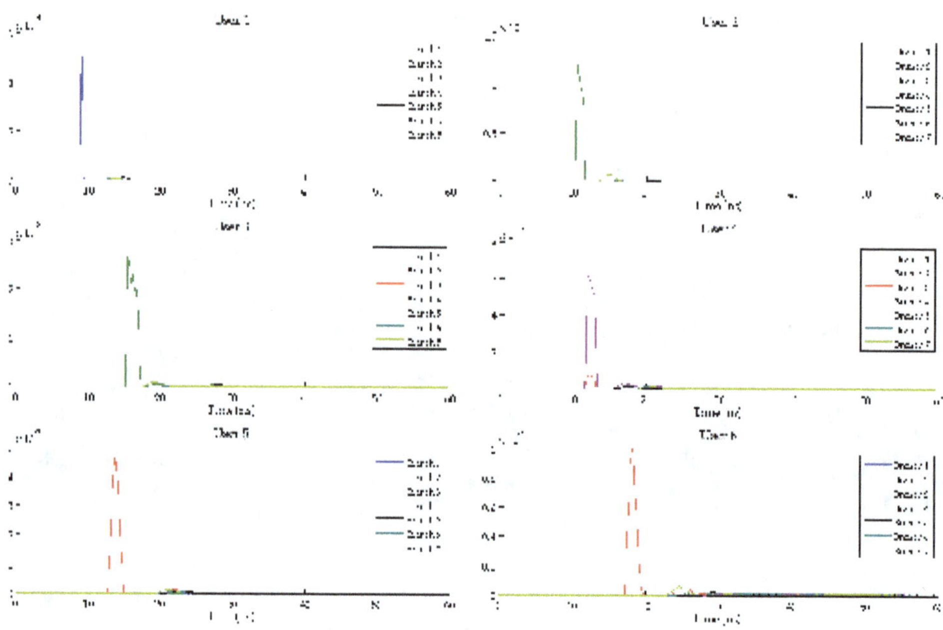

Figure 13. Impulse responses between transmitting users (lamps) and each receiving branch for non-imaging receiver at position 1 (see Fig. 7(a)).

Finally, we have determined the available signal-to-noise ratio (SNR) throughout the room at the receiver plane. For that, we have considered the windows as planar Lambertian ambient light (noise) sources with spectral radiant emittance S_w=0.2 W/(nm m[2]), as in [31]. The SNR is expressed as [8]

$$\mathrm{SNR} = \frac{(RP_s)^2}{\sigma_{\mathrm{total}}^2} \tag{17}$$

where R is the detector responsivity, P_s is the average received signal power and σ_{total} is the total noise variance, which is the sum of the contributions of shot and thermal noises

$$\sigma_{\mathrm{total}}^2 = \sigma_{\mathrm{shot}}^2 + \sigma_{\mathrm{thermal}}^2 \tag{18}$$

In a well-designed receiver, and in the presence of intense background light, the shot noise is the dominant term in (18), and is given by [8]

$$\sigma_{\mathrm{total}}^2 \approx \sigma_{\mathrm{shot}}^2 = 2qR\left(P_s + P_n\right)B \tag{19}$$

where q is the electron charge, R the detector responsivity, P_s and P_n the average received optical power corresponding to the desired signal and the ambient light, respectively, and B is the channel bandwidth (20 MHz).

Fig. 14 shows the available SNR throughout the room at the receiver plane for imaging and non-imaging receivers. The SNR values descend when approaching the windows, as expected, but they are considerably above 50 dB throughout the room for both types of receivers, with a not insignificant SNR gain in the case of non-imaging receivers.

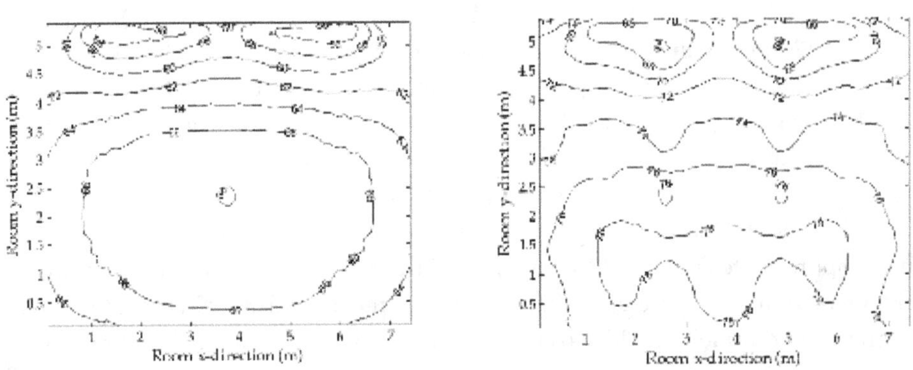

(a) SNR for imaging receiver (b) SNR for non-imaging receiver

Figure 14. Signal-to-noise ratio (SNR) throughout the room at the receiver plane.

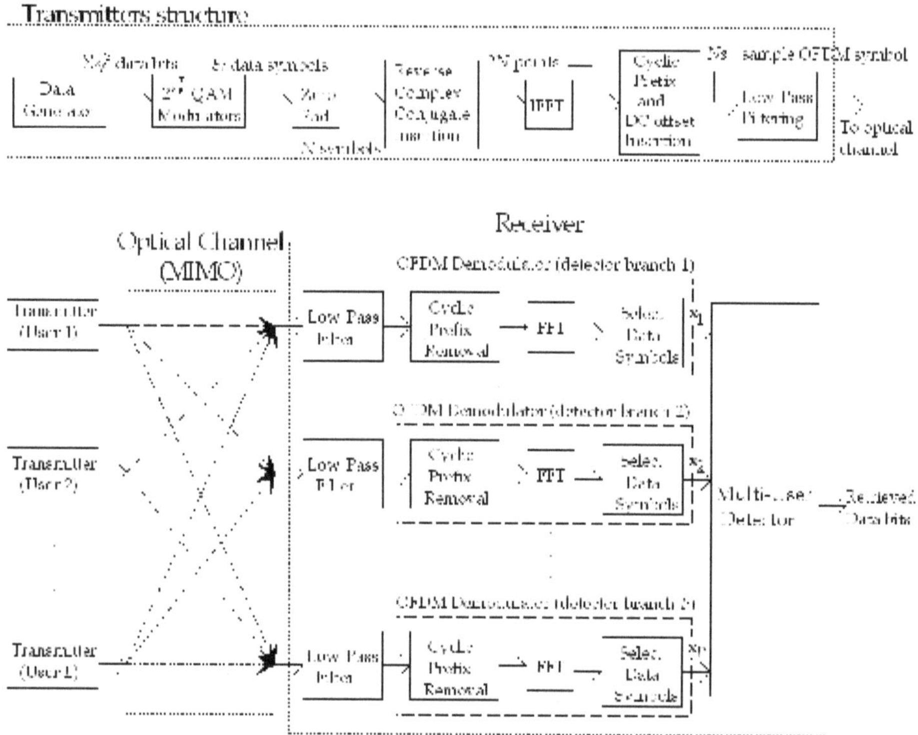

Figure 15. Multi-user MIMO-OFDM system for optical wireless communications.

3. The MIMO-OFDM system

Fig. 15 shows the block diagram of the MIMO-OFDM system for multi-user communications over an indoor wireless optical channel. As we can see in the transmitter structure, $\Sigma_p b_p^{(l)}$ data bits are generated by l^{th} user and modulated by using an appropriate $2^{b_p^{(l)}}$-QAM (*quadrature amplitude modulation*) modulator, $b_p^{(l)}$ being the number of bits conveyed by the pth subcarrier ($1 \leq p \leq S_I$), according to a rate adaptive algorithm [32][33]. Then, $N - S_I$ zeros are padded to this set of S_I symbols generated by the modulators, the first zero being added before, corresponding to the direct-current (DC) value, and the remaining ones after. Finally, the complex conjugate of the mirror of the word of N symbols is added before computing the IFFT. The outgoing IFFT symbol is a real sequence of $2N$ points, which can be made non-negative by adding an appropriate DC level. Moreover, a cyclic prefix (N_e-sample extension) is inserted to combat inter-symbol interference (ISI), leading to a transmitted signal with a total of N_S=2N +N_e samples and duration T_S. This positive and real signal can modulate the intensity of light emitted by the optical source. Note that the cyclic extension prefix will be ISI-corrupted due to the low-pass channel response and will be discarded at receiver. Next, the FFT of the remaining 2N points will be computed at each receiving branch, and only the first S_I points (after the DC value) will be considered by the multi-user (MU) detector.

As we have L transmitters and the receiver possesses P receiving branches, the optical channel constitutes a MIMO system, whose individual impulse responses are given by (16). We can consider that, after the FFT processing carried out by each branch of a new received OFDM symbol, the P-branch detector array provides the MU detector with a vector of complex symbols$\boldsymbol{x}_p$ at p^{th} subcarrier. This is the superposition of the independently distorted signals associated with the L users sharing the same space-frequency resource and also corrupted by noise at the detector array elements

$$\boldsymbol{x}_p = \boldsymbol{H}_p \boldsymbol{s}_p + \boldsymbol{n}_p \tag{20}$$

where the vector $\boldsymbol{x}_p \in \mathbb{C}^{P\times 1}$ of received signals at the pth subcarrier, the vector $\boldsymbol{s}_p \in \mathbb{C}^{L\times 1}$ of transmitted signals and the array noise $\boldsymbol{n}_p \in \mathbb{C}^{P\times 1}$, respectively, are given by

$$\begin{aligned} \boldsymbol{x}_p &= \left(x_{1p}, x_{2p}, \ldots, x_{Pp}\right)^T \\ \boldsymbol{s}_p &= \left(s_p^{(1)}, s_p^{(2)}, \ldots, s_p^{(L)}\right)^T \\ \boldsymbol{n}_p &= \left(n_{1p}, n_{2p}, \ldots, n_{Pp}\right)^T \end{aligned} \tag{21}$$

where $()^T$ means the transpose of (), thus $\boldsymbol{x}_p$, $\boldsymbol{s}_p$ and $\boldsymbol{n}_p$ are all column vectors. The frequency domain channel transfer function factor matrix $\boldsymbol{H}_p \in \mathbb{C}^{P\times L}$ is composed of the set of channel transfer vectors $\boldsymbol{H}_p^{(l)} \in \mathbb{C}^{P\times 1}$, l=1,..., L of the L users

$$\boldsymbol{H}_p = \left(\boldsymbol{H}_p^{(1)}, \boldsymbol{H}_p^{(2)}, \ldots, \boldsymbol{H}_p^{(L)}\right) \tag{22}$$

each of which hosts the frequency domain channel transfer factor between the single emitter source associated with a particular user l and the receiving branches j=1,..., P at the pth subcarrier band

$$\boldsymbol{H}_p^{(l)} = \left(H_{1p}^{(l)}, H_{2p}^{(l)}, \ldots, H_{Pp}^{(l)}\right)^T \tag{23}$$

with l=1,..., L. Note that the frequency domain channel transfer factors $H_{jp}^{(l)}$, $\forall p$ between the lth emitter and the jth receiving branch are obtained by computing the Fourier transform (12) at the subcarrier frequency $f=pf_0$ over the corresponding impulse response (16), where $f_0=1/T$. Here, T is the time duration of the part of the OFDM symbol which actually conveys information, i.e. the OFDM symbol duration excluding the cyclic prefix extension, $T=2NT_S/N_S=2NT_S/(2N+N_e)$.

3.1. Least squares error detector

Using a linear detector, an estimate $\hat{s}_p \in \mathbb{C}^{L\times 1}$ at the pth subcarrier band of the vector of transmitted signals s_p of L simultaneous users is generated by linearly combining the signals received by P different receiving branches with the aid of a weight matrix $H_p \in \mathbb{C}^{P\times L}$

$$\hat{s}_p = W_p^H x_p \tag{24}$$

where W_p^H denotes the complex conjugate matrix of matrix W_p. When least-squares (LS) error detector is considered, also called zero-forcing (ZF) combiner, the weight matrix $W_{p,\text{LS}} \in \mathbb{C}^{P\times L}$ is given by [34]

$$W_{p,\text{LS}} = H_p (H_p^H H_p)^{-1} \tag{25}$$

By substituting the received signal's model of (20) and the LS estimation based weight matrix (25) into (24), we obtain

$$\hat{s}_{p,\text{LS}} = s_p + W_{p,\text{LS}}^H n_p \tag{26}$$

which indicates that the LS-estimate $\hat{s}_{p,\text{LS}}$ of the transmitted signal vector s_p of L simultaneous users is an unbiased noise-contaminated version of s_p, so that $E\{\hat{s}_{p,\text{LS}}\}=s_p$. When using the LS combiner, the lth user's associated vector component of estimate (24) of the vector of transmitted signals at the pth subcarrier can be expressed as

$$\hat{s}_{p,\text{LS}}^{(l)} = W_{p,\text{LS}}^{(l)H} x_p \tag{27}$$

where the lth user's associated weight vector $W_{p,\text{LS}}^{(l)H} \in \mathbb{C}^{P\times 1}$ coincides with the lth column vector of the matrix $W_{p,\text{LS}}$. The complex symbol that is most likely to have been transmitted by the lth user can be determined by minimizing the Euclidean distance between estimate (27) of the transmitted signal obtained at the lth user's combiner output and all the constellation points associated with the specific modulation scheme employed. This only requires $M=2^{b_p^{(l)}}$ evaluations compared to the M^L ones of the optimum *maximum likelihood* (ML) detector [34], and the performance degradation is not significant [19].

3.2. Frequency-domain channel transfer factor matrix estimation

From (25) and (27), we can observe that the LS detector requires knowledge of the transfer factor matrix H_p. The method described in [35], where TS known training sequences are used to estimate the channel transfer function between a single user and the receiver, can be used to estimate matrix $\boldsymbol{H}_p$, assuming that only one user is transmitting in a specific instant during the training periods. If a certain known training sequence X_p is transmitted by the lth user over the slowly time-varying wireless optical channel, the channel response of each sub-band $H_{jp}^{(l)}$ for this user at the jth receiving branch can be estimated from the S_l received symbols ($Y_{jp}^{(l)}$) as follows

$$\hat{H}_{jp}^{(l)} = Y_{jp}^{(l)} / X_p \tag{28}$$

Note that $\hat{H}_{jp}^{(l)}$ values are distorted by noise and therefore several training sequences should be transmitted, and then the mean values of the estimates can be used to obtain a better channel characterization. In fact, by using ten training sequences, the channel response estimation given by (28) is very close to the actual channel response, and we will obtain a system performance very similar to that of the ideal case when a perfect knowledge of the channel response is available [35].

3.3. Rate adaptive algorithm

In this chapter, we will apply the rate adaptive algorithm described in [32]. Here, we will only describe it briefly. Effective demodulation SNR can be computed at receiver as follows (Fig. 16): after each OFDM symbol demodulation, the retrieved data bits are modulated again and the average SNR of received QAM symbols is computed, using outgoing QAM modulators symbols as reference (we are assuming error-free transmission). The calculation of the lth user's effective demodulation SNR is given by

$$\overline{\mathrm{SNR}}_{\mathrm{eff}}^{(l)} = \overline{\left|s_p^{(l)}\right|^2} / \overline{\left|r_p^{(l)} - s_p^{(l)}\right|^2} \tag{29}$$

where $s_p^{(l)}$ and $r_p^{(l)}$ are the lth user's transmitted and received (before demodulation) data symbols, respectively. From (29), it can be deduced that the effective SNR is an average of all constituent data symbols (p=1,..., S_l) of each received OFDM frame. The channel response estimation and the mean effective demodulator SNR of each received OFDM symbol can be used to determine the SNR at the pth subcarrier band as follows [32]

$$\mathrm{SNR}_p^{(l)} = \frac{\overline{\left[\left(\boldsymbol{H}_p^H \boldsymbol{H}_p\right)^{-1}\right]_{[l,l]}}}{\left[\left(\boldsymbol{H}_p^H \boldsymbol{H}_p\right)^{-1}\right]_{[l,l]}} \overline{\mathrm{SNR}}_{\mathrm{eff}}^{(l)} \tag{30}$$

The sub-band SNR value can be compared with switching levels for picking out the modulation mode (including 'no transmission', i.e. $b_p^{(l)}$=0) that ensures the instantaneous bit error rate (BER) always remains below a certain threshold [35]. In addition, disabled subcarriers, owing to their low SNR values, must carry on sending dummy data so as to compute their current sub-band

SNR by means of (30). These dummy symbols should be known *a priori* by receiver in order to avoid an erroneous calculation of the effective demodulation SNR.

A further improvement can be made if subcarriers with higher SNR values between two switching levels are prompted to use the next modulation mode, whenever the average error probability does not exceed the imposed threshold [32]. Let $b_p^{(l)}$ be the number of bits conveyed by the pth subcarrier, and $P_p^{(l)}$ the bit error probability of the subcarrier when the modulation mode $2^{b_p^{(l)}}$-QAM is used by the lth user, thus the average error probability is given by

$$P_{\text{avg}}^{(l)} = \frac{1}{B^{(l)}} \sum_{p=1}^{S_l} b_p^{(l)} P_p^{(l)}\left(b_p^{(l)}, \text{SNR}_p^{(l)}\right) \tag{31}$$

where $B^{(l)}$ is the total throughput of the lth user's adaptive system: $B^{(l)}=\Sigma_p b_p^{(l)}$. The bit error probability $P_p^{(l)}(b_p^{(l)},\text{SNR}_p^{(l)})$ can be determined from the estimated signal-to-noise ratio of each sub-band ($\text{SNR}_p^{(l)}$) given by (30) and the BER curves against SNR obtained over AWGN channel [34][35]. Initially, the modulation modes of subcarriers are set to those that exceed the switching levels for a certain target BER, and then the modulation modes are successively increased for the 'best subcarriers' trying to enhance $B^{(l)}$, while ensuring that $P_{\text{avg}}^{(l)}$ does not exceed the imposed BER threshold value [32].

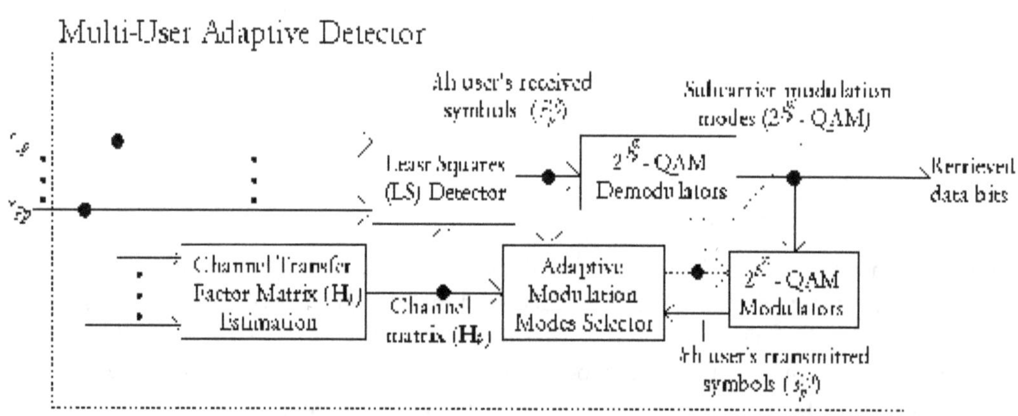

Figure 16. Multi-user adaptive detector.

4. Results and discussion

In this section, we will consider two communication scenarios. The first is concerned with jointly demodulating the information of all the users, which we will denote as *joint detection*, and a second scenario in which, despite having a multi-user environment (several lamps transmitting information simultaneously), the receiver is only concerned, at a specific receiver point, with the data corresponding to a particular user lamp, which we will refer to as *single-user detection.*

4.1. Joint detection

In this section, we show the results obtained with an adaptive MIMO-OFDM system based on LS detection, which can select the most appropriate modulation mode for each subcarrier from the 5-ary group ('no transmission', BPSK, QPSK, 16-QAM, 64-QAM). The main parameters of this adaptive MIMO-OFDM system are summarized in Table 3. In all the results presented below, the number of subcarriers is N=64 (128-point FFT), but only S_I=48 convey data (QAM modulated). A cyclic prefix extension of N_e=8 samples was used and the transmission symbol rate was 250 ksymbol/s, which leads to a maximum total system throughput of ($B_{\max} \times S_I \times L$)/ T_S bit/s (e.g. 432 Mbit/s for L=6 users –lamps– when every subcarrier is 64-QAM modulated). This OFDM symbol extension ($T_S\, N_e/(2N+N_e) \approx$ 235 ns) is large enough to compensate for ISI in most of the non-LOS diffuse-link scenarios. The separation between subcarriers is Δf=$f_0 \approx$ 266 kHz, leading to a transmission bandwidth of approximately B_T=$S_I\Delta f$=12.75 MHz plus a convenient out-of-band guard [35], which is extremely practical for typical optical WLED devices. Finally, we used twenty training sequences to estimate (using the method described in section 3.2) the frequency-domain channel transfer function factors of matrix H_p required by LS detector to obtain the weight matrix (25).

Total number of subcarriers (N):	64
Number of information subcarriers (S_I):	48
Available modulation modes:	('no transmission', BPSK,QPSK, 16-QAM, 64-QAM)
Maximum number of bits per subcarrier ($B_{\max}$):	6 (64-QAM)
OFDM symbol period (T_S):	4 μs
Cyclic prefix extension (N_e):	8
Maximum aggregate throughput:	432 Mbit/s
Number of training sequences (TS):	20

Table 3. Parameters of adaptive MIMO-OFDM system.

Fig. 17 and Fig. 18 show the system performance for imaging and non-imaging receivers, respectively; when the information from each user (lamp) is jointly demodulated at a particular position of the receiver. Fig. 17(a) presents the achievable average throughput (BPS, *bits per symbol*) at every information subcarrier as a function of the average SNR at the receiving branches (pixels), when three different target bit error rates (P_e) are sought and for the receiver located at position 1 inside the room (see Fig. 7(a)). We can see how the BER is maintained nearly constant around the specified target BER in all the cases until maximum achievable throughput (6 BPS in all the subcarriers) is reached. From that point on, BER decreases rapidly

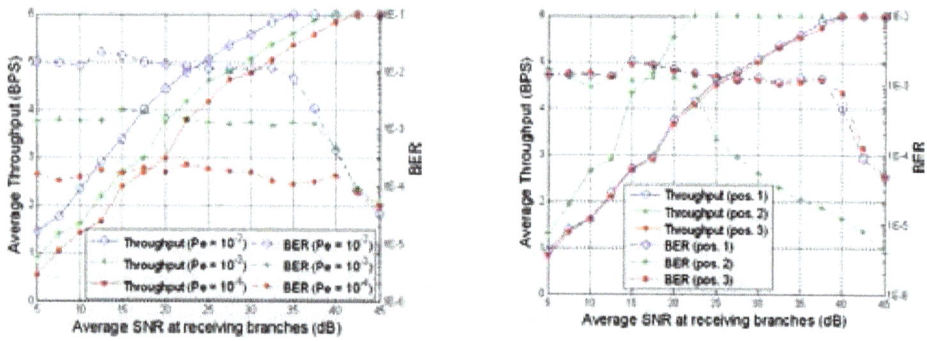

(a) Performance for several target P_e (pos. 1) (b) Performance at different positions (P_e=10^{-3})

Figure 17. Performance (aggregate throughput, bits per symbol (BPS)) of MIMO-OFDM system with imaging receiver at several positions versus average SNR at receiving branches (pixels).

as SNR grows. Equally, demanding a lower target BER requires greater SNR values to attain a specific average throughput, as expected.

Fig. 17(b) compares the system performance at the three receiver locations specified in Fig. 7(a). Once again, we observe a near-constant BER around the imposed threshold (P_e=10^{-3} for these simulations), while the maximum achievable throughput is not reached. The BER drops dramatically when maximum throughput is attained and SNR continues increasing. Additionally, we can see an improved system performance at position 2 (center of the room) compared to positions 1 and 3. This is due to the better signal quality received at that location (from Fig. 7, for positions 1 and 3, there seems to be a clear degradation of the signals coming from farther emitting lamps; these lamps appear less illuminated in Fig. 7(b) and Fig. 7(d), which eventually deteriorates the achievable aggregate throughput).

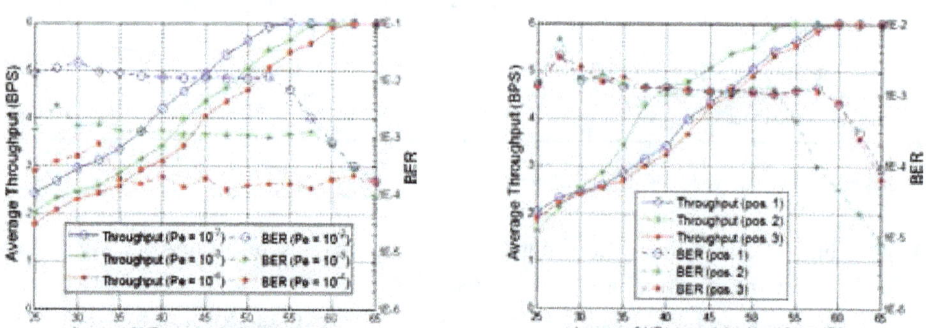

(a) Performance for several target P_e (pos. 1) (b) Performance at different positions (P_e=10^{-3})

Figure 18. Performance (aggregate throughput, bits per symbol (BPS)) of MIMO-OFDM system with non-imaging receiver at several positions versus average SNR at receiving branches.

Fig. 18 presents the same results as Fig. 17, but for a non-imaging receiver. In general, we can see similar system performance behavior, although requiring much larger SNR values (about 20 dB in performance degradation) to attain an identical objective. As before, we observe how receiver position affects system performance (see Fig. 18(b)). However, the larger field of view of a non-imaging receiver means position 2 (center of the room) is not so advantageous compared to locating the receiver closer to the room corners.

4.2. Single-user detection

In many applications, we are not concerned with jointly demodulating the information coming from all the lamps, but only with decoding the data corresponding to one lamp which has been assigned to a specific receiver. Fig. 19 presents the performance of the adaptive MIMO-OFDM system in this second situation (single-user detection) as compared with the joint detection case described in the previous section. In the simulations, the receiver is located in positions 1, 2 or 3, as defined in Fig. 7(a), and only demodulates the data coming from its nearest lamp (for position 2, in the center of the room, lamps 2 and 5 are symmetrically equivalent, thus either of them can indistinctly be considered, both delivering identical results). Moreover, detection complexity reduction described in section 4.3 of [19] was applied to optimize and accelerate single-user detection.

For all cases, except for the non-imaging receiver at position 2, single-user detection requires much lower SNR values to obtain a certain average throughput per subcarrier concerning the data corresponding to that lamp. This is logical because the receiver is positioned very close to the emitting lamp, thus receiving very directive and intense optical signals from its associated emitter. Therefore, each individual receiving user, if located appropriately close to its assigned lamp, would benefit from a high-quality signal level while sharing the room with other simultaneous users.

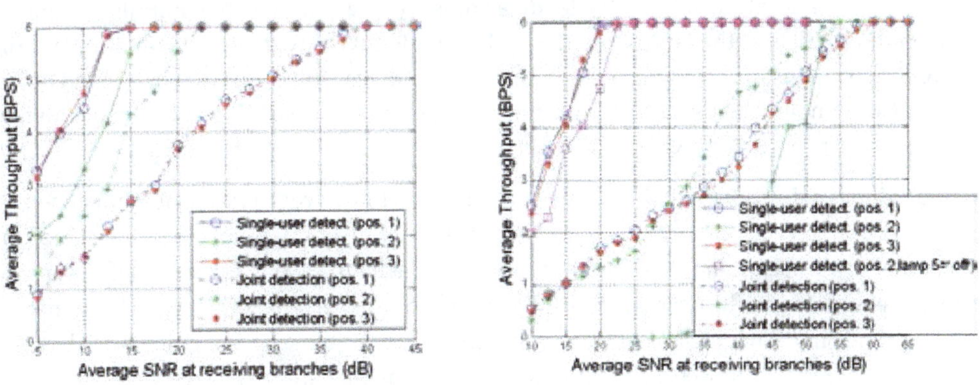

(a) Performance for imaging receiver

(b) Performance for non-imaging receiver

Figure 19. Performance comparison of MIMO-OFDM system for single-user detection and joint detection at several positions versus average SNR at receiving branches when P_e=10^{-3}.

Additionally, the system performance degradation of a non-imaging receiver compared to an imaging one is not so significant, except when the former is positioned in the center of the room. This is because a non-imaging receiver does not provide as much diversity as an imaging-based type. This is more noticeable at the center of the room where emissions from lamps 2 and 5 can scarcely be distinguished, leading to the observed important SNR losses experienced by a non-imaging detector at position 2. Fig. 19(b) also shows the single-user performance in the center of the room when lamp 5 is 'disabled' (the result shown by a pink line), i.e. this lamp, although maintaining its functionality as an illumination device, stops sending data information. In this situation, the single-user performance is enhanced greatly, which demonstrates the problems of the non-imaging receiver to separate the information coming simultaneously from lamps 2 and 5. However, the cost of this performance improvement is the new maximum achievable aggregate throughput, since, with only five active lamps, it falls to 360 Mbit/s (5 users × 6 bits/subcarrier symbol × 48 subcarriers/OFDM symbol $\times 250 \cdot 10^3$ OFDM symbols/second) as compared with the maximum throughput of 430 Mbit/s when all six lamps were active. Therefore, it seems evident that positioning the non-imaging receiver in the center of the room must be avoided to prevent an excessive system performance penalty.

5. Conclusions

In this chapter, the use of multi-user LS detection in conjunction with angle-diversity imaging and non-imaging receivers and adaptive OFDM modulation technique for visible light communications has been evaluated. The indoor wireless visible-light channel model, which can be determined by a Monte Carlo-based ray-tracing algorithm, has been described thoroughly, detailing the main features associated with the different elements constituting this kind of optical channel: white light-emitting diodes, reflective room surfaces and optical receivers. This algorithm allows us to determine very accurately the impulse responses between the WLEDs of the lamps and the optical angle-diversity detector, while considering LOS and also not insignificant multiple-bounce reflection contributions to the received optical power at the photodetector. This algorithm accuracy is essential to enable a more reliable analysis of the proposed MIMO-OFDM performance.

A rate adaptive MIMO-OFDM scheme, based on a linear combination of the incoming signals at its receiving branches and dynamic throughput adaptation to channel occupation (number of users) and signal quality (SNR at each subcarrier for a specific user), has been proposed for multi-user visible-light communications. The system performance for imaging and non-imaging reception, and considering joint and single-user detection scenarios, has been assessed. The results have shown that imaging receivers provide an improved performance with SNR gains of about 20 dB with respect to non-imaging ones when evaluating joint detection scenario, i.e. the former offer greater diversity in a VLC environment. Additionally, for single-user detection only demodulating data coming from a specific lamp, we find that although not so evident at positions closer to the room walls, non-imaging receivers present considerable performance degradation when moving into the room's central area. Therefore, it can be concluded from the previous results that, in general terms, imaging receivers provide the best solution as an angle-diversity detection method for multi-user visible light communications.

Acknowledgements

This work has been funded by Spanish Government (MINECO) and the European Regional Development Fund (ERDF) programme under projects TEC2013-47682-C2-2-P, MAT2013-46649-C4-4-P, MAT2010-21270-C04-02 and Malta Consolider Ingenio 2010 (CSD2007-0045).

Author details

Oswaldo González[1*], Marcos F. Guerra Medina[1] and Inocencio R. Martín[2]

*Address all correspondence to: oghdez@ull.es

1 Department of Industrial Engineering, University of La Laguna, Spain

2 Department of Physics, University of La Laguna, Spain

References

[1] Komine T., Nakagawa, M. Integrated system of white LED visible-light communication and power-line communication. IEEE Transactions on Consumer Electronics 2003;49(1): 71–79.

[2] Komine T., Nakagawa M. Fundamental analysis for visible-light communication system using LED lights. IEEE Transactions on Consumer Electronics 2004;50(1): 100–107.

[3] O'Brien D.C., Zeng L., Minh H.L., Faulkner G.E., Walewski J.W., Randel S. Visible light communications: challenges and possibilities. In: PIMRC 2008: proceedings of the IEEE 19th International Symposium on Personal, Indoor and Mobile Radio Communications, 15-18 September 2008, Cannes, France.

[4] Grubor J., Randel S., Langer K.-D., Walewski J.W. Broadband information broadcasting using LED-based interior lighting. Journal of Lightwave Technology 2008;26(24): 3883–3892.

[5] Elgala H., Mesleh R., Haas H. Indoor optical wireless communication: potential and state-of-the-art. IEEE Communications Magazine 2011;49(9) 56–62.

[6] IEEE Std 802.15.7-2011. IEEE standard for local and metropolitan area networks–part 15.7: Short-range wireless optical communication using visible light. IEEE Standards Association 2011; 1–286.

[7] Kavehrad M. Sustainable energy-efficient wireless applications using light. IEEE Communications Magazine 2010;48(12): 66–73.

[8] Kahn J.M., Barry J.R. Wireless infrared communications. Proceedings of IEEE 1997;85(2): 265–298.

[9] O'Brien D.C., Minh H.L., Faulkner G.E., Zeng L., Lee K., Jung D., Oh J. High-speed visible light communications using multiple-resonant equalization. IEEE Photonics Technology Letters 2008;20(15): 1243–1245.

[10] Komine T., Lee J.H., Haruyama S., Nakagawa M. Adaptive equalization system for visible light wireless communication utilizing multiple white LED lighting equipment. IEEE Transactions on Wireless Communications 2009;8(6): 2892–2900.

[11] Minh H.L., O'Brien D.C., Faulkner G.E., Zeng L., Lee K., Jung D., Oh J., Won E.T. 100-Mb/s NRZ visible light communications using a postequalized white LED. IEEE Photonics Technology Letters 2009;21(15): 1063–1065.

[12] Haigh P.A., Ghassemlooy Z., Minh H.L., Rajbhandari S., Arca F., Tedde S.F., Hayden O., Papakonstantinou I. Exploiting equalization techniques for improving data rates in organic optoelectronic devices for visible light communications. Journal of Lightwave Technology 2012;30(19): 3081–3088.

[13] Zeng L., O'Brien D.C., Minh H.L., Faulkner G.E., Lee K., Jung D., Oh J., Won E.T. High data rate multiple input multiple output (MIMO) optical wireless communications using white LED lighting. IEEE Journal on Selected Areas in Communications 2009;27(9): 1654–1662.

[14] Li X., Vučić J., Jungnickel V., Armstrong J. On the capacity of intensity-modulated direct-detection systems and the information rate of ACO-OFDM for indoor optical wireless applications. IEEE Transactions on Communications 2012;60(3): 799–809.

[15] Vučić J., Kottke C., Nerreter S., Langer K.-D., Walewski J.W. 513 Mbit/s visible light communications link based on DMT-modulation of a white LED. Journal of Lightwave Technology 2010;28(4): 3512–3518.

[16] Vučić J., Kottke C., Habel K., Langer K.-D. 803 Mbit/s visible light WDM link based on DMT modulation of a single RGB LED luminary. In: OFC/NFOEC 2011: proceedings of the Optical Fiber Communication Conference and Exposition, and the National Fiber Optic Engineers Conference, 6-10 March 2011, Los Angeles, USA.

[17] Cossu G., Khalid A.M., Choudhury P., Corsini R., Ciaramella E. 3.4 Gbit/s visible optical wireless transmission based on RGB LED. Optics Express 2012;20(26) B501–B506.

[18] Tsonev D., Chun H., Rajbhandari S., McKendry J.J.D., Videv S., Gu E., Haji M., Watson S., Kelly A.E., Faulkner G., Dawson M.D., Haas H., O'Brien D. A 3-Gb/s single-LED OFDM-based wireless VLC link using a Gallium Nitride μLED. IEEE Photonics Technology Letters 2014;26(7) 637–640.

[19] González O. Multiple-input multiple-output (MIMO) optical wireless communications. In: Das N. (ed.) Optical Communication. Rijeka: InTech; 2012. p393–414. Available from http://www.intechopen.com/books/optical-communication/multiple-input-multiple-output-mimo-optical-wireless-communications (accessed 3 October 2012).

[20] Azhar A.H., Tran T.-A., O'Brien D. A gigabit/s indoor wireless transmission using MIMO-OFDM visible-light communications. IEEE Photonics Technology Letters 2013;25(2) 171–174.

[21] Burton A., Minh H.L., Ghassemlooy Z., Bentley E., Botella C. Experimental demonstration of 50-Mb/s visible light communications using 4×4 MIMO. IEEE Photonics Technology Letters 2014;26(9) 945–948.

[22] Barry J.R., Kahn J.M., Krause W.J., Lee E.A., Messerschmitt D.G. Simulation of multipath impulse response for wireless optical channels. IEEE Journal on Selected Areas in Communications 1993;11(3) 367-379.

[23] Schubert E.F. Light-emitting diodes. Cambridge: Cambridge University Press; 2006.

[24] Cui K., Chen G., He Q., Xu Z. Indoor optical wireless communication by ultraviolet and visible light. In: Majumdar A.K., Davis C.C. (eds.) Proc. SPIE 7464: Free-Space Laser Communications IX: proceedings of the International Society for Optics and Photonics (SPIE), 2-3 August 2009, San Diego, USA.

[25] Lee K., Park H., Barry J.R. Indoor channel characteristics for visible light communications. IEEE Communications Letters 2011;15(2) 217-219.

[26] European Standard EN 12464-1. Lighting of indoor work places. European Committee for Standardization (CEN) 2002.

[27] Lomba C.R., Valadas R.T., de Oliveira Duarte A.M. Experimental characterisation and modelling of the reflection of infrared signals on indoor surfaces. IEE Proceedings-Optoelectronics 1998;145(3) 191–197.

[28] Kahn J.M., You R., Djahani P., Weisbin A.G., Teik B.K., Tang A. Imaging diversity receivers for high-speed infrared wireless communication. IEEE Communications Magazine 1998;36(12) 88-94.

[29] López-Hernández F.J., Pérez-Jiménez R., Santamaría A. Modified Monte Carlo scheme for high-efficiency simulation of the impulse response on diffuse IR wireless indoor channels. Electronics Letters 1998;34(19) 1819-1820.

[30] González O., Rodríguez S., Pérez-Jiménez R., Mendoza B.R., Ayala A. Error analysis of the simulated impulse response on indoor wireless optical channels using a Monte Carlo based ray-tracing algorithm. IEEE Transactions on Communications 2005;53(1) 124-130.

[31] Carruthers J.B., Kahn J.M. Angle diversity for nondirected wireless infrared communication. IEEE Transactions on Communications 2000;48(6) 960-969.

[32] González O., Rodríguez S., Pérez-Jiménez R., Delgado F., Ayala A. Muti-user adaptive orthogonal frequency-division multiplexing system for indoor wireless optical communications. IET Optoelectronics 2007;1(2) 68–76.

[33] Park K.-H., Ko Y.-C., Alouini M.-S. On the power and offset allocation for rate adaptation of spatial multiplexing in optical wireless MIMO channels. IEEE Transactions on Communications 2013;61(4) 1535-1543.

[34] Hanzo L., Münster M., Choi B.-J., Keller, T. OFDM and MC-CDMA for broadband multi-user communications, WLANs and broadcasting. West Sussex (England): John Wiley & Sons; 2003.

[35] González O., Pérez-Jiménez R., Rodríguez S., Rabadán J., Ayala A. OFDM over indoor wireless optical channel. IEE Proceedings-Optoelectronics 2005;152(4) 199–204.

5

Application of Metal-Semiconductor-Metal Photodetector in High-Speed Optical Communication Systems

Farzaneh Fadakar Masouleh and Narottam Das

1. Introduction

Recently, there is a very strong interest towards the miniaturization of optical and electrical components with faster and more efficient performance which has incorporated new capabilities in various aspects including high-speed telecommunication systems [1]. Optical communication technology has greatly developed during recent years that affect all areas of the modern telecommunication systems. Photodetectors along with optical sources and fibers are regarded as integral part of all optical fiber communication systems [2]. The high bandwidth and/or the gain of photodetectors with the wavelength in the near-infrared region (0.8 to 1.6 μm) are quite important because of their ideal commercial and industrial usage in optical fiber communication systems. The photodetectors are known as optoelectronic devices that can convert the absorbed optical energy into electrical energy which usually appears as a photocurrent that can be used by telephone systems, computers, or other terminals at transmitting and receiving part of the communication systems [3]. There are many types of photodetectors depending on their application in optical communication systems, imaging systems, and so on. The sensitivity of detectors varies through different optical spectra, such as the infrared and ultraviolet. Semiconductor detectors are commonly used in optical fiber communication systems because they rely on internal photoelectric effect but there is no photoemission effect. They either work in photovoltaic mode as solar cells or in photoconductive mode as revers biased photodetectors [4]. Metal-semiconductor-metal photodetectors (MSM-PDs), positive-intrinsic-negative (pin) photodetectors, avalanche photodiodes and heterojunction PDs as solid state semiconductor devices are most widely used in high-speed optical communication systems. The MSM-PDs are attractive devices compared with other photodetector structures, for their remarkable high sensitivity-bandwidth product, ease of fabrication and ease of integration with other components into monolithic receiver circuits. The MSM-PD consists of two identical Schottky contacts with interdigitated electrode configuration on top of an undoped semiconductor substrate, one of the contacts being forward and the other reverse biased. With fabrication of interdigitated electrodes the closely spaced fingers provide smaller transit time for the carriers as well as allowing a larger photo-absorption area for the device [5, 6]. The two Schottky barriers associated with the presence of contacts block current flow from the metal to the semiconductor which is the cause for the extremely low dark current observed in MSM-PDs. One important feature of the MSM photodetector is its low capacitance compared with a pin photodetector, (with an intrinsic region (i.e., undoped semiconductor) in between the n-and p-type semiconductors). The capacitance of a MSM photodetector with interdigitated electrodes is always smaller than that of a pin photodetector of equal light sensitive area and leads to very high-performance operation.

High speed photodetection manifests an exciting new paradigm for modern telecommunication systems. Advanced or modern optical systems, i.e., the ultrahigh-speed optical telecommunication systems, such as any typical fiber optic communication system consist of a transmitter, a data transmission media or channel (usually optical fiber, waveguide, and free space air-gap), and a receiver (pin photodiodes and photodetectors). The major part of the optical transmitter is a light source (laser or light-emitting diode (LED)), whose function is to convert an information signal from its electrical form into light. The photodetectors as an important part of receiver are used to convert an optical information signal back into an electrical signal. For higher speed and broader bandwidth applications in optical communications and interconnects, high-performance optical receivers are required and the MSM photodetector as the heart of optical receiver has many advantages such as wide bandwidth and low capacitance. MSM-PDs are promising candidates in optoelectronic integrated circuits. Also they have fast time response and very low dark current, as compared with other types of photodetectors.

The new field of plasmonics has received particular attention due to unique optical features of nano scale architectures in noble metals particularly silver and gold. The coherent oscillations of electrons as surface plasmons are strongly localized in the nanoscale at the metal-dielectric interface, and metal nanoparticles. When the losses are small enough, the surface plasmon resonances can occur. The metals with good quality for plasmonic applications should satisfy two properties such as $\varepsilon'_m<-1$, and $\varepsilon''_m<<|\varepsilon'_m|$, where ε'_m, and ε''_m are real and imaginary parts of metal dielectric permittivity, respectively [7].

Dielectric and magnetic properties of the noble metals with nano-textured structure can easily be determined by implementation of Lorentz-Drude model [8]. Below plasma frequency, the good conductors like silver (Ag) and gold (Au) have negative values for the real part of complex dielectric constant. Therefore, it is necessary to define an appropriate model to specify the dielectric properties of the materials. A complex dielectric function for some metals and surface plasmas which have good agreement with the experimentally measured results can be expressed in the following form [9]:

$$\varepsilon_r(\omega) = \varepsilon_r^f(\omega) + \varepsilon_r^b(\omega) \tag{1}$$

where, the term $\varepsilon_r(\omega)$ is the complex dielectric function for metals, $\varepsilon_r^f(\omega)$ is referred to as free electron effects, and $\varepsilon_r^b(\omega)$ is associated with bound electron effects. This model takes both the intraband, $\varepsilon_r^f(\omega)$, and interband, $\varepsilon_r^b(\omega)$, effects into the account for simulations. The former, Drude model, can describe the transport properties of electrons in good conductors and the later, Lorentz model, is a semi-quantum model describing bound electron effects. The Drude and Lorentz model in frequency domain can be written in the following form of (2) and (3), respectively.

$$\varepsilon_r^f(\omega) = 1 + \frac{\Omega_p^2}{i\omega\Gamma_0 - \omega^2} \tag{2}$$

$$\varepsilon_r^b(\omega) = \sum_{m=1}^{M} \frac{G_m \omega_p^2}{\omega_m^2 - \omega^2 + i\omega\Gamma_m} \tag{3}$$

where, $\Omega_p = G_0^{1/2}\omega_p$ is the plasma frequency associated with intraband transitions with oscillator strength G_0 and damping constant Γ_0, while m is the number of oscillators with frequency ω_m and $1/\Gamma_m$ is the oscillator lifetime [10]. The following equation accounts for the complex index of refraction and dielectric constant of materials and can be represented as a combined form [11]:

$$\varepsilon_r(\omega) = \varepsilon_{r,\infty} + \sum_{m=0}^{M} \frac{G_m \Omega_m^2}{\omega_m^2 - \omega^2 + i\omega\Gamma_m} \tag{4}$$

where, $\varepsilon_{r,\infty}$ is the relative permittivity at infinite frequency, G_m is the strength of each resonance term, Ω_m is the plasma frequency, ω_m is the resonant frequency, and Γ_m is the damping factor or the collision frequency.

The frequency dependent Gold dielectric permittivity is complex and is obtained from Lorentz-Drude model, $\varepsilon_m = \varepsilon'_m + i\varepsilon''_m$. It consists of a large negative real part ε'_m and a small positive imaginary part ε''_m responsible for light absorption. The letter i is an imaginary unity. Gold dielectric constant varies for different frequency ranges.

For our simulation of plasmonic-base MSM-PD, the Lorentz-Drude model for gold is solved with 6 resonant frequencies (multi-pole dispersion). The simulation is performed over a constant plasma frequency which depends on the density of charge carriers with the amount of 0.137188E+17. The resonant frequency changes according to the resonance strength. The Lorentz-Drude material parameters are listed in the following Table-I.

Term	Strength	Resonant Frequency	Damping Frequency
0	0.7600	0.000000E+00	0.805202E+14
1	0.0240	0.630488E+15	0.366139E+15
2	0.0100	0.126098E+16	0.524141E+15
3	0.0710	0.451065E+16	0.132175E+16
4	0.6010	0.653885E+16	0.378901E+16
5	4.3840	0.202364E+17	0.336362E+16

Table 1. Lorentz-Drude parameters for gold (measured in radians per second).

Investigation on the optical properties of the nano-structures is crucial for the design of optical devices. Next generation information and communication technologies are under direct influence of nano-plasmonic devices because of their unique properties that hold promise for potential applications in various fields of technology, properties such as overcoming the diffraction limit, efficiency in confinement of light at subwavelength scale, and ultrahigh speed signal transport with the same order as the speed of light make them vigorous devices.

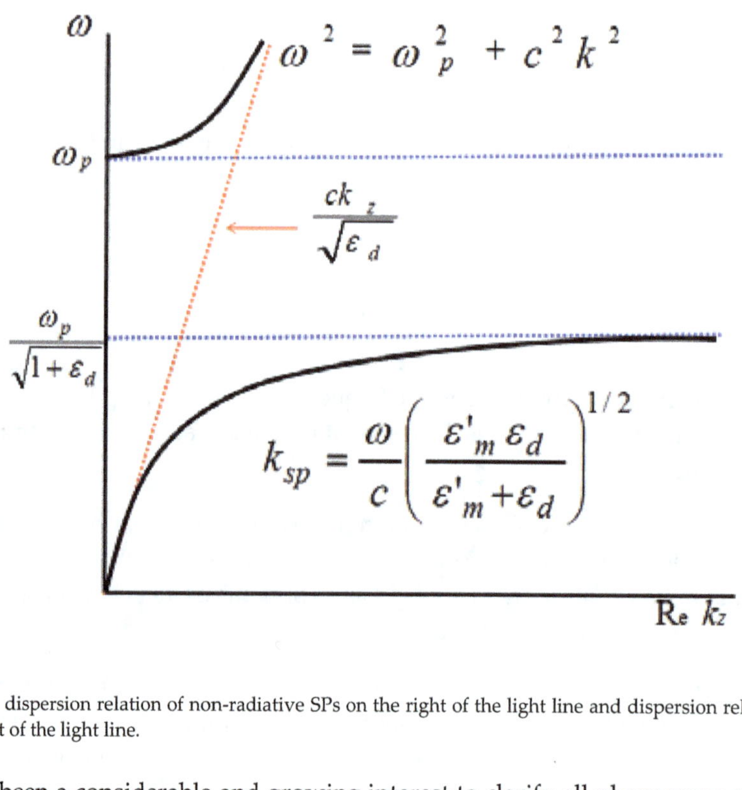

Figure 1. The dispersion relation of non-radiative SPs on the right of the light line and dispersion relation of radiative SPs on the left of the light line.

There has been a considerable and growing interest to clarify all phenomena corresponding to the improvement in light absorption via nano-scale structuring recently [12-15]. Over the decade, an interesting effect of light interacting with metallic structures has been revealed. For sufficiently small nano-grating period, higher order diffractions are suppressed and only the zero order diffraction is present. The main motivation is to come up with a practical method to avoid the undesirable light reflection from the surface of the electrodes and enhance the light transmission efficiency through the subwavelength aperture.

It is useful to describe the SP dispersion relation as shown in Fig. 1. To have propagating bound SPs, the wave vector component must be real along the interface. Any light cannot be used for SP generation, because the real part of the wave vector in the z-direction exceeds that of free light ($\omega=ck$) as shown in Fig. 1 and a coupling mechanism is required. In the presence of a transverse magnetic (TM) polarized light, the SPs exist along a metal-dielectric interface. There are several techniques to excite the SP waves, such as prism coupling and grating coupling. For the case of semiconductors, prism coupling is not very much advantageous. The attention should be paid to prism refractive index which is hardly higher than popular semiconductors used for corresponding researches [16, 17].

2. Nano-gratings structures and operation principles

The metallic nano-gratings can exhibit absorption anomalies. One of these particularly remarkable anomalies is the surface plasmon polariton (SPP) excitations and is observed for p-polarized light only. The incident light illuminated on top of the one dimensional metallic nano-gratings and subwavelength slits is converted into the propagating SPPs that can absorb the light efficiently in extremely thin (10-nm to more than 100-nm's thick) layers. The coupling of light to the structure in the form of the SPPs is obtained with the utility of periodicity. There are two mechanisms to produce the transmission of light in one-dimensional (1-D) metal nano-gratings with narrow slits which are the excitation of horizontal and vertical surface resonances however there is not always a clear distinction between these two ways of transferring light from the upper surface to the lower one and that their existence strongly depends on the gratings geometry. The horizontal surface resonances are cavity modes excited by the periodic structure of the nano-gratings at the grating's interfaces. Two vertical resonances existing on the slit walls can compose a fundamental waveguide mode which satisfies the Fabry-Perot condition and reflects repeatedly from both ends of the slit. Therefore, the vertical surface resonances have relations with Fabry–Perot resonances of the fundamental TM guided wave in the slits [18], and the guided modes can be excited along vertical direction, evanescent coupling mechanism accounts for this activation [19]. The responsibility of these two effects on extraordinary optical transmission (EOT) is still under debate. However, it is indicated that both mechanisms are important in the EOT [20].

Nowadays, it is clear that the real metals play a very important role in EOT, because their effect is essential for the SP excitation. The condition for the existence of the SP on a flat air-metal interface, $\varepsilon'_m<-1$, $\varepsilon''_m \ll |\varepsilon'_m|$, is fulfilled for many metals, including the silver and gold. Concentrating the light in small areas with the assistance of extremely thin layers (plasmonic lenses), the EOT is a unique phenomenon which is introduced as the SP assisted multiple diffractions, coupling the SPs into the aperture, and conversion back into near-field at the exit side of the aperture. The role of SP modes is essential to clarify the EOT of light through the slits. In the EOT, the aperture transmits more light than the standard aperture theory. The EOT can occur whenever some specific conditions are satisfied, i.e., the slit width must be much smaller than the incident light wavelength, the periodicity has to be in the range of wavelength and it can lead to outstanding results if the light is normally incident to the structure's surface. Subwavelength apertures have also been used to concentrate light efficiently into the deep subwavelength regions.

A considerable development has been reported for the EOT through metallic gratings with very narrow slits. The SPPs and subwavelength slits in metallic thin films are always involved with this phenomenon but the details are still under investigation [21, 23]. The SPs can be efficiently excited in the nano-structured noble metals since they almost have free-electron behavior. The noble metal nano-textured structures have special properties to produce localized regions of high energy concentration and show larger enhancement for EOT. This effect and its underlying mechanism have important applications in photolithography and near-field microscopy.

Specifically, in this chapter we present the effects of nano-grating structures on the MSM photodetector performance. Optimization of subwavelength nano-gratings shape and pitch is very much important to generate the zero-order diffraction waves. Thus, the subwavelength nano-gratings can be represented as a homogeneous medium with optical properties determined by the nano-grating geometry. When the nano-grating period is within the order of the light wavelength, the light wave may be resonant and reflects into the structure, hence the higher diffraction orders will be suppressed and resonant reflection occurs for the zero-order diffraction waves.

3. Modeling and characterization of plasmonic-based MSM-PDs

For proper design of MSM photodetector structure, it is necessary to define an appropriate model to specify the dielectric properties of the materials. In our modeling, semiconductor substrate is made up of gallium arsenide (GaAs) which is a direct band gap compound semiconductor. Usually, the GaAs substrate is preferred for the design of electronic and photonic devices, because it has unique electrical properties. Being a direct bandgap semiconductor, it can collect and emit light more efficiently than indirect band gap semiconductors such as Si and Ge. The GaAs substrate as an isotropic material has a constant refractive index of $3.666+i6.12\times10^{-2}$ in our simulation. GaAs is suitable for infrared range applications and optical fiber communication because of its wide bandgap and fast electron conduction. It also has short absorption length that enables the detector to combine wide range bandwidth with good responsivity characteristics. The GaAs has a conduction-band structure that leads to fast electron conduction.

Periodic nano-structures can produce an efficient light transmission and absorption by excitation of surface plasmons (SPs). The continuing progress in plasmonic interaction with nano-structures and their outstanding effects in development of MSM-PDs design have developed a unique context for future-generation optoelectronic systems, such as, optical fiber communication systems.

The conventional MSM-PD is a symmetrical device equivalent to two back-to-back connected Schottky diodes on a semiconductor substrate. To create a Schottky junction, along with the shape and size of the interface, some essential properties, such as type and quality of metal and semiconductor must be satisfied.

A conventional MSM-PD structure is shown in Fig. 2. By impinging the light from the top on the conventional device surface, a considerable amount of illumination on top is reflected, so the light absorption inside the MSM photodetector (or substrate) is significantly reduced in device's active region. Fabrication of nano-gratings on the metal fingers of the conventional MSM-PD structure avoids this unwanted phenomenon (or light reflection), compare Fig. 2 and 3. Such nanostructure then can capture most of the reflecting light inside the active region of the device (or substrate). Hence, the individual subwavelength apertures can exhibit notable light transmission when surrounded by periodic nano-structures that harvest the externally incident light into the slit. This proposed mechanism is shown in Fig. 3.

By tailoring the electrodes structure surface with metallic nano-gratings, MSM-PDs can be modified for the light absorption and the modification process strongly depends on the corrugation parameters. Recently, different shapes of 1-D nano-structured surfaces have been developed in which noble metals, such as gold (Au) are used for nano-grating structuring. When an electric field (or a voltage) is applied between the electrodes and the device's active region is under illumination then the electric carriers (i.e., electrons and holes) are generated and drifted towards the opposite electrodes due to the electric field and can form a photocurrent. Improvement of recombination between electrons and holes can lead to enhance the light absorption in the subwavelength aperture region.

The TM polarized wave is required to excite the propagating SP waves, when the *z*-component of the light k-vector matches with the SP k-vector, because only the TM polarization has electric field component in the same direction as surface normal. The polarization is described with regard to the electric field configuration of the incident light with respect to the surface normal (angle β) as shown in Fig. 3. The wave is purely TE polarized, when $\beta=\pi/2$, and purely TM polarized, when $\beta=0$. The angle ϕ represents the direction of the plane of incidence, such that the nano-grating is in the classical mount in this model, where all the diffracted orders remain in the *x-z* plane and the plane of incidence is normal to the nano-grating grooves ($\phi=0$). Besides, when the grooves are parallel to the plane of incidence, the nano-grating is in the conical mount (ϕ=90). In our simulation, we used the nano-grating period of Λ=810-nm. The grating period has to be optimized to effectively couple surface plasmon resonances to the light wave and trigger the quality absorption process.

It is very important to build a new type of plasmonic-based MSM-PD with the characterization of high speed and high responsivity application which is useful in optical interconnect and communication systems. An MSM-PD speed in operation is intrinsically limited by the carrier transit time between the fingers or by the carrier recombination time. Although reducing the

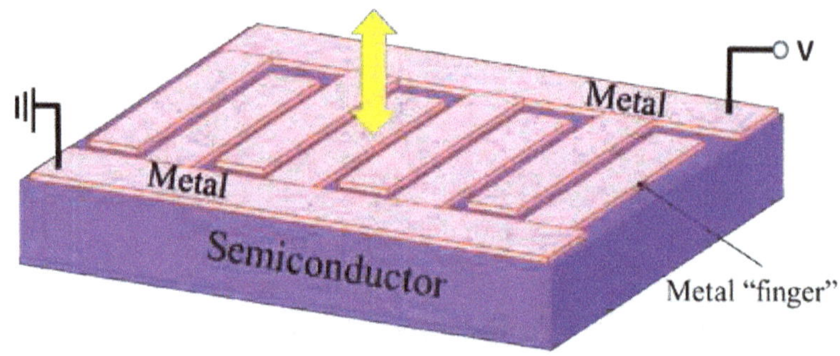

Figure 2. Schematic diagram of conventional MSM-PD structure with interdigitated electrodes (metal fingers) and semiconductor substrate (GaAs). For this symmetric device under illumination, there are undesirable light reflections.

interdigitated electrodes spacing as well as scaling down the entire MSM-PD dimensions are common ways to increase the photodetector speed, as transit time is quite small, undesirable light reflection from the surface of the electrodes and the metal fingers shadowing the detection area which leads to decreased active area, lowers the MSM-PD responsivity. Employing metal nano-gratings on interdigitated electrodes demonstrates substantial transmission enhancement through the excitation and guidance of surface plasmon polaritons into the photodetector's semiconductor region [24].

4. Plasmonic-based MSM-PD simulation model

Metallic nano-grating structures produce efficient and ultrafast photodetection properties. In this section, a plasmonic-based MSM-PD is introduced which utilizes electromagnetic and optical properties of nanostructures to enhance the light harvesting inside the device active region. The plasmonic-based MSM-PD consists of a semiconductor absorbing layer on which the interdigitated electrodes have been deposited to form two back-to-back connected Schottky diodes. The interdigitated electrodes are designed to resolve the conventional MSM-PD's degraded efficiency problem because of the metallic electrodes opacity. Also, the MSM-PDs are patterned by nano-gratings to improve the light capturing capacity into the device active region. Electromagnetic fields coupled to a charge density wave propagating at the metal-dielectric interface produce transverse-magnetic optical surface waves namely surface plasmon polaritons. The coupling condition for the SPPs is provided by metallic nano-gratings with optimized dimensions and geometry. This structure has the feasibility to create a plasmonic lens which utilizes plasmonic effects to produce surface plasmon resonances and funnel the energy toward the central focal point.

The design of high performance plasmonic-based MSM-PD is shown in Fig. 3. A plasmonic-based MSM-PD structure has three separate regions or parts, namely, the incident region (the metal nano-gratings), the under layer and the subwavelength slit region, and the transmission region (the semiconductor substrate). The nano-scale gaps between the interdigitated electro-

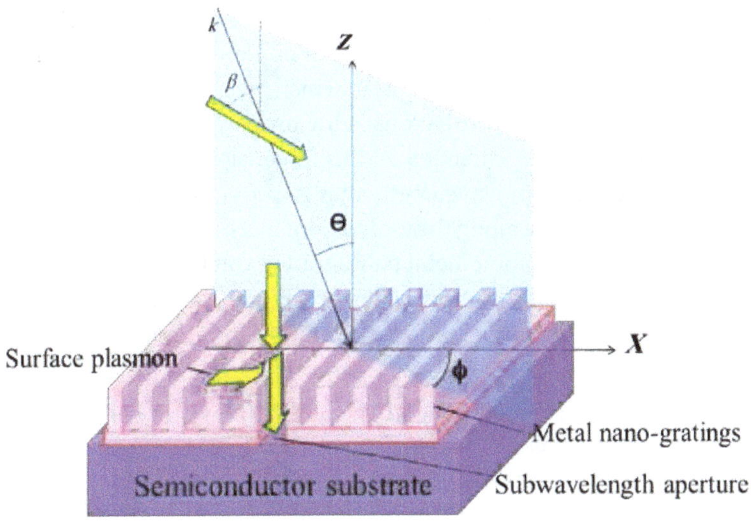

Figure 3. Schematic diagram of MSM-PD structure with rectangular shaped nano-gratings on top of the subwavelength slit. The subwavelength slit layer is just on top of the semiconductor (GaAs) substrates. The excited SPPs travel along the interface to reach the subwavelength slit/aperture.

des in the MSM-PDs result in a huge increase in bandwidth and reduction in dark current, whereas the conventional pin photodetectors with similar sized active areas are unable to achieve that amount of light absorption.

In addition, the near-field characteristics and associated field enhancements can be achieved for periodic nano-structures and subwavelength apertures with aid of the SPPs interactions. This suggests that, properly designed metallic nano-grating grooves trigger surface plasmon polaritons under illumination and carry them toward the central slit. By using a subwavelength central slit, a well-directed source of light could be generated, an exciting development that is being pursued as a source for a variety of optical technologies. The light continuously re-emits from a very small area surrounding the central aperture which is associated with properties of the Fabry-Perot cavity resonances for symmetric SPP modes of the slit. Particularly, The SPPs which are supported by the active region of the device show a great potential of subwavelength photonic phenomenon. The excitation of the SPP waves causes a resonance absorption, which can be observed as partial or total absorption of the incident light. The absorption enhancement caused by the excitation of SPPs is associated with the incident photons and their interaction with the nano-gratings, while Fabry-Perot like resonances are included in transmission absorption process through the subwavelength slits.

In periodic subwavelength structures, the nano-gratings are deposited on top of the under layer from the same metal (such as gold). The absorption enhancement can be achieved by the SPP resonant excitations in the subwavelength region. Here, the FDTD method is used to specify the TM polarized plane wave via Poynting vector. The SPPs are evanescent waves, which are generated by the interaction between the surface electron densities and the electromagnetic fields whilst trapping the light power inside the surface. Since the SP modes have longer wave vectors than the light waves with the same energy, the SP waves are non-radiative on smooth metallic surfaces and cannot propagate in non-metallic media. One way to excite the SPPs is the nano-grating coupling technique in which the incident radiation is coupled into the SPPs using periodic surface corrugations with proper dimensions. The nano-grating grooves are perpendicular to the x-direction and its dimensions and geometry are optimized to couple the light near the design wavelength, that is, providing the missing momentum in order to make the SPPs propagate along the z-direction. In a metal-dielectric interface, the SPP wave vector matching condition for a metal nano-grating can be defined with some changes to the well-known prism resonance condition. Hence, in a metal-dielectric interface, the SPP propagating constant or wave vector matching condition for a metal nano-grating with the period of Λ is given by [22, 24]:

$$k_{spp} = \frac{\omega}{c}\sin(\theta) \pm \frac{2\pi l}{\Lambda} = \frac{\omega}{c}\sqrt{\frac{\varepsilon'_m \varepsilon_d}{\varepsilon'_m + \varepsilon_d}} \quad (5)$$

where, ω is the angular frequency of the incident light wave, c is the speed of light in vacuum. l is an integer number i.e., l=1, 2, 3, ..., N and θ is the light angle of incidence to the device normal. This relation illustrates that the wave vector of a given frequency is smaller than the SPP wave vector, therefore the light wave vector should increase with the support of a coupling mechanism to provide SPPs which in this case is satisfied by nano-grating structures.

The wave vector has to be complex as the metal permittivity is complex, $\varepsilon_m=\varepsilon'_m+i\varepsilon''_m$. To trigger SPPs, the dielectric permittivity has to change sign in the metal-dielectric interface. The values of dielectric constant vary for different frequency ranges. To represent the influence of electric field in organizing electric charges and dipoles in the medium, we introduce electric displacement field, $D=\varepsilon_0\varepsilon_r E$, between two isotropic media where, ε_0 is the vacuum permittivity and ε_r is the relative (dielectric) permittivity. Here, the real part of complex dielectric permittivity of gold is used. The dielectric permittivity of air as the incidence medium is denoted as ε_d. The electric displacement field derived from the Maxwell's equation is continuous across the interface. With the continuous normal component of D across the interface and the permittivity sign difference for metal and dielectric, the electric field changes direction passing through two different media. This characteristic will only be satisfied if there is a normal component for electric field across two regions that is TM polarization.

We are interested in metals with the large negative real part and a very small imaginary contribution to the dielectric constant for the design wavelength, such as gold. The SPP damping while propagating along the interface will be determined by the imaginary part of the wave vector parallel component. As the researchers demonstrated their results for noble metals, such as gold and silver in metal-dielectric interface, there will be a high field confinement at the interface while the losses remain minimum. When the plasmonic excitations occur, the left side of equation (5) matches the wave vector of the excited SPP (k_{spp}), that is the equivalence of the interaction of incident radiation and l_{th} diffracted order with the wave vector of the SP at the interface.

The FDTD as a powerful engineering tool allows for the effective and powerful simulation and analysis of sub-micron devices with very fine structural details. The FDTD algorithm was originally proposed by K. S. Yee in 1966 [25]. In order to investigate the optical response of plasmonic-based MSM-PD, finite difference time-domain (FDTD) numerical method is used as a premier solution for the simulation of propagating electromagnetic field by solving Maxwell's curl equations in time domain. The computational mesh points (grids) are made up of unit cells, and the electric (E) and magnetic (H) fields are arranged at special places of the computational domain denoted by (i, j, k) with respect to Ampere and Faraday's laws. The FDTD method is able to model light propagation, scattering, diffraction, reflection, and polarization effects.

The Opti-FDTD software package developed by Optiwave Inc. was used to perform a 2D simulation, and it is the first software to employ the Lorentz–Drude model into the FDTD algorithm to calculate the transmission and reflection spectra. The FDTD simulation results have demonstrated significant light-capture performance through periodic 1-D slit arrays, which is useful for the design of ultrafast MSM-PDs.

In this model, the light wave hits on the nano-grating structures perpendicularly and passes through the subwavelength aperture and finally reaches to the semiconductor substrate.

This chapter's focus is on nano-gratings architectures that can increase the MSM-PD responsivity by exciting the SPPs and manipulation of light in the subwavelength slit. It has been reported that the rectangular nano-gratings produce the best absorption process, however, we will present some new more qualified nano-gratings with more efficient performance. The most influential parameters controlling the device light absorption can be categorized in two general groups. One can be the basic structural differences for nano-gratings clearly distinguishable with respect to their cross sections such as nano-gratings shape and dimensions, metal nano-grating heights, duty cycle, and number of nano-gratings on each side of the central slit. The other one is optimization process for subwavelength aperture region as subwavelength slit width, and electrodes (under layer) thickness. Furthermore the incident light polarization and angle of incidence play an important role to produce quality light absorption in optimized plasmonic-based MSM-PD. To verify the importance of these features, the power flow through the central subwavelength slit for plasmonic-based MSM-PD is compared with the amount of power reaching the active region for conventional MSM-PD. Improved light interaction process with different nano-grating shapes and geometries results in verification of the simulated results for the design and development of high responsivity MSM-PDs which have applications in high-speed optical fiber communication, high-speed sampling, and chip to chip interconnectors. The light absorption enhancement factor (LAEF) is introduced as a dimensionless quantity to measure the optimal absorbed radiation ratio to the whole incident power [26]. Therefore the impact of nano-gratings implementation on the quality of light flux transmitted into the active region of the MSM-PD is well approved compared with the conventional MSM-PD without the nano-gratings.

5. Simulation results and discussion

5.1. Electric field distribution inside the GaAs substrate

In this subsection, local electric field intensity distribution for conventional and plasmonic-based MSM-PD structures will be clarified through simulations to evaluate the adequacy of plasmonic-based structure. We use a custom designed Matlab code to show the density plot and the transmitted power into the substrate for two different structures. While the radiation reaches to the structure normally, Fig. 4(a) represents the electric field distribution in a conventional device without nano-gratings but with the Au contacts just on the GaAs substrate. The slit width is 100-nm and the under layer height is 60-nm. Therefore, in this situation, the light transmission inside the substrate is not influenced by the surface plasmon excitations and the incident light normally passes through the subwavelength aperture. Different colors represent each point's electric field strength in the density plots. Also, rectangular nano-grating structures are designed and deposited on the metal contacts to take the plasmonic effects into account. Fig. 4(b) demonstrates the electric field confinement in the GaAs substrate at cross section of plasmonic-based MSM photodetector device and the field concentration is due to SPP coupling with nano-structures. The maximum intensity appears for the part of the substrate located just under the slit [27].

Due to the plasmonic interactions and the confinement of light into the central subwavelength slit, it is obvious from Fig. 4 that the grating assisted MSM-PD tendency is to concentrate the power (or energy) into the photoactive region which is just below the central slit. Therefore, implementation of the nano-gratings enables the photodetector to lead the light into the central aperture quite effectively. Hence, it is important to obtain optimized geometrical parameters for efficient light confinement, and it is what we describe in the following subsections.

5.2. Effect of duty cycles on the LAEF

For 2 similar nano-gratings with the discrepancy in profile shapes, the light harvesting ability can be different. The Au nano-grating profile and geometry on the GaAs substrate can make changes in the absorption enhancement spectrum. Therefore it is interesting to discuss the effects of duty cycle on the LAEF of the nano-structured MSM-PDs. Duty cycle (DC) of corrugations is the percentage of ridges width to the nano-grating period, i. e. 60% DC refers to the ratio of the nano-gratings grooves width to the ridges width of six to four in one period. We specify the optimum duty cycle for trapezoidal and rectangular-shaped nano-grating profiles which are designed with optimum heights [28].

Under normal incidence, the LAEF spectra in rectangular-shaped nano-grating structures is calculated for different duty cycles, such as from 30% to 100%, as shown in Fig. 5, while the subwavelength aperture width has been kept constant at 50-nm, the under layer thickness and the nano-grating height were 20-nm and 120-nm, respectively. It can be inferred that the peak wavelength is different for each specific duty cycle and the maximum LAEF, with the amount of ~32.7, occurs for 40% duty cycle. It is clear that the duty cycle can affect the peak wavelength as well as the amount of light flux transmitted into the active area of the MSM-PDs.

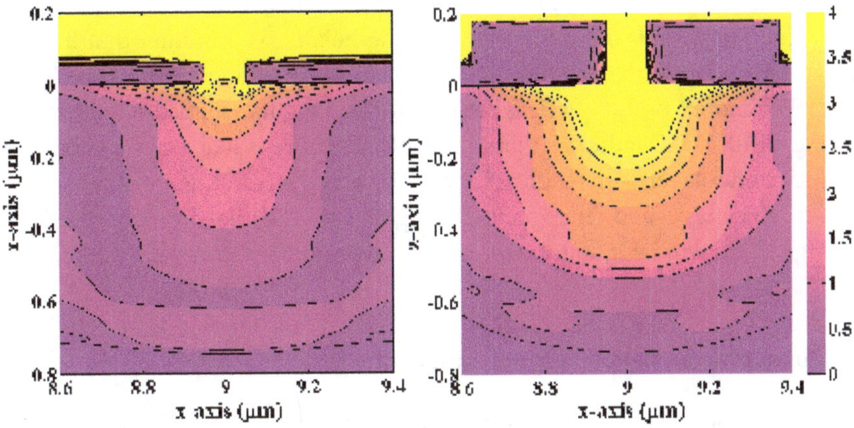

Figure 4. Field distribution at the cross section (a) conventional MSM-PD, (b) rectangular plasmonic-based MSM photodetector. The calculated total electric field intensity distribution inside the GaAs substrate is shown using the following parameters: the subwavelength slit width of 100-nm, gold under layer thickness of 60-nm, and the nano-gratings height of 100-nm for (b).

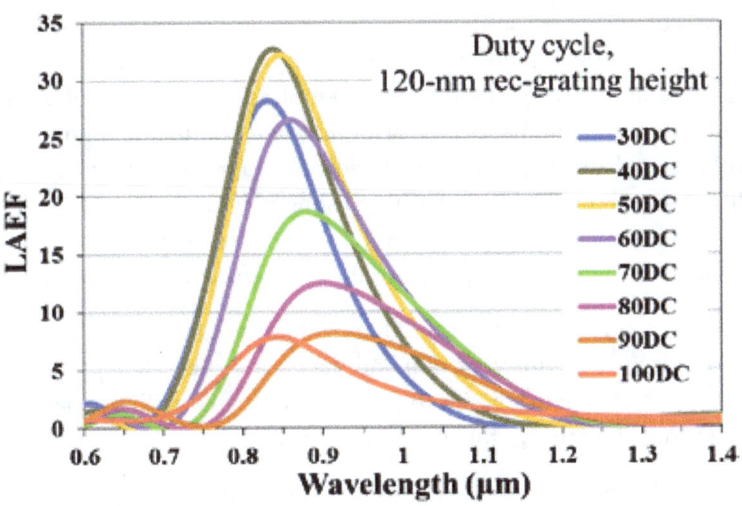

Figure 5. Light absorption enhancement factor spectra of MSM-PDs with rectangular-shaped nano-gratings for various duty cycles ranging from 30% to 100%.

The optimum duty cycle has also been specified for trapezoidal-shaped nano-grating profile with the subwavelength aperture width and under layer thickness of 50-nm and 20-nm, respectively. In this case, the maximum LAEF is obtained~31.5 for 40% DC which is the same duty cycle as the optimum DC for rectangular-shaped nano-gratings.

The results shown in Figs. 5 and 6 illustrate that the amount of LAEF is a function of duty cycle which grows gradually towards the 40% DC and drops down dramatically towards higher DCs and the peak wavelengths are also red-shifted for the LAEF curves of the higher duty cycles for both the structures. In the case of the rectangular and trapezoidal nano-grating profiles designed with their optimum nano-grating height, we can infer that the amount of light transmitted into the active region not only depends on the nano-gratings height and shape but also on the amount of duty cycles. Besides, for rectangular-shaped nano-grating profile shown in Fig. 5, 100% DC indicates that the whole structure is like a conventional MSM-PD having a thick under layer with height of about 140-nm, the resulting thickness is the sum of under layer thickness and the nano-grating height, while for trapezoidal shaped nano-gratings, there will be triangular grooves between the trapezoidal ridges for 100% DC, Fig. 7.

5.3. Impact of subwavelength aperture width on the LAEF

We discuss the impact of subwavelength aperture width on the light absorption and reflection for MSM-PDs and the results are discussed for different duty cycles. Compared with the incident light wavelength (λ_0), the aperture width is very small, hence only symmetric and fundamental SP modes will propagate into the slit. When the subwavelength slit width is much smaller than the incident light wavelength (λ_0), in addition to the light transmission and absorption caused by manipulation of metal nano-gratings, the light harvesting and confinement in the semiconductor substrate can be obtained via optimization of subwavelength aperture. Fig. 8 shows the simulation results of the absorption spectrum for several subwavelength aperture widths when the number of nano-gratings on each side of the slit (subwavelength aperture) is 9, and the nano-grating period and the nano-grating height are kept constant at 810-nm, and 100-nm, respectively. However, a portion of the lights is reflected in the central slit area, as shown in Fig. 9. The interesting point is that the range of wavelengths in the spectrum corresponding to the minimum reflection for the LRF curves in Fig. 9 is equivalent to the range of maximum LAEF for the same structure.

The results for less than 50-nm and more than 500-nm width slits are not presented here because the LAEF is reduced for very thin and very wide slits. The optimized slit width is selected as 50-nm which is much easier to fabricate compared with thinner slits and shows very promising results to improve the device performance.

The simulated results show that the LAEF decreases rapidly with the increase of the subwavelength aperture width from 500-nm to 50-nm. Figure 8 shows clearly that the LAEF is more than 12-times with 50% DC and about 13.5-times with 60% DC for a 50-nm subwavelength aperture width, the narrowest slit in this simulation, also with presentation of LAEF curves for 40% and 70% DCs of the aforementioned slit width, we show that 60% DC is optimized for this special structure. However, the LAEF is ~4-times with 50% DC but a little lower with 60 DC for a slit width of 250-nm and the LAEF wavelength is red shifted for all bigger DC.

The effective refractive index is a function of slit width for symmetrical SP modes when the slit experiences the TM incident wave. Therefore, with reducing the subwavelength aperture width, the effective refractive index increases and leads to an enhancement of the light absorption inside the active region.

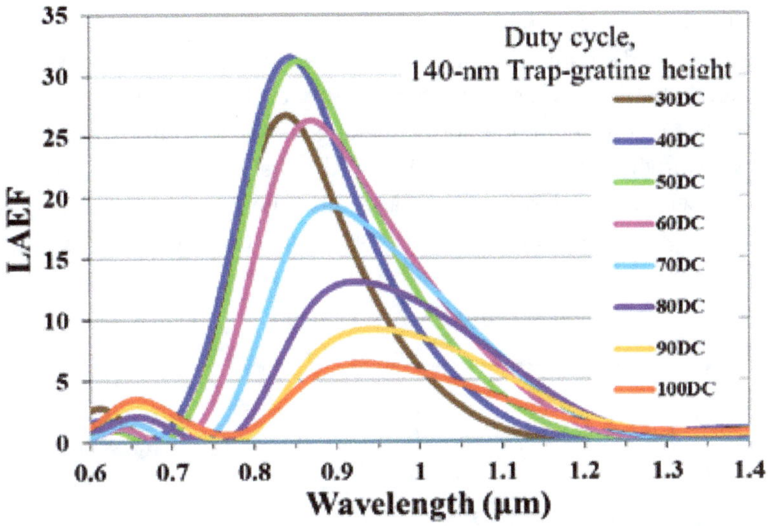

Figure 6. Light absorption enhancement factor spectra of MSM-PDs with trapezoidal-shaped nano-gratings for various duty cycles ranging from 30% to 100%.

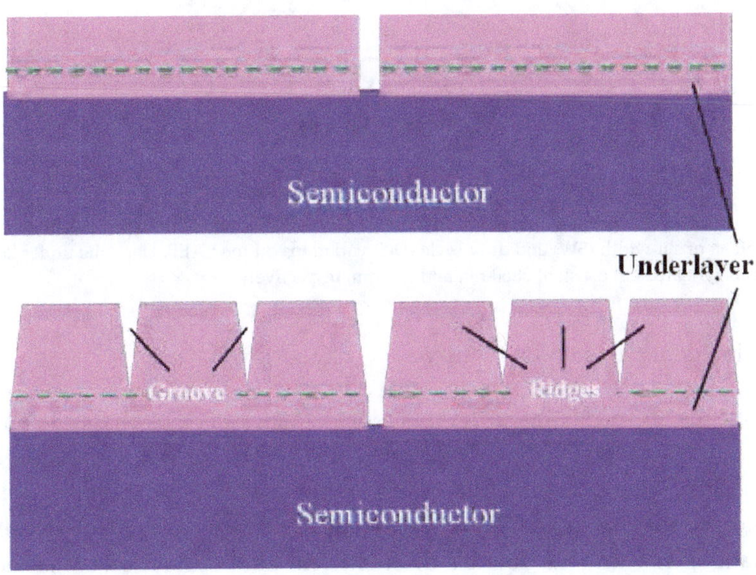

Figure 7. Cross-section of rectangular and trapezoidal-shaped nano-gratings while the duty cycle is 100%.

However, the idea for optimization of the MSM-PDs can be made more precisely by modifying the amount of transmitted light power through the subwavelength slit in the GaAs substrate [29]. Therefore, at normal incidence, we made this amendment by the subtractions of localized power near the top of the substrate and the power propagating to the bottom of the substrate, that is the LAEF values at the slit opening and beyond the slit opening at the outer edge of the substrate. The subtracted amount for LAEF is presented in Fig. 10 for the optimized geometry presented in Fig. 8, that is 50-nm slit width, 60% duty cycle. The maximum LAEF is almost unchanged and the fact that a minor amount of energy is lost from the substrate is justified as long as the device dimensions are optimized. Also the absolute values for Poynting vector in x direction are presented for 3 depths of the GaAs substrate, top, 0.1 μm from top and bottom. The $|S_x|$ value is negligible at the bottom of the substrate compared with the slit opening.

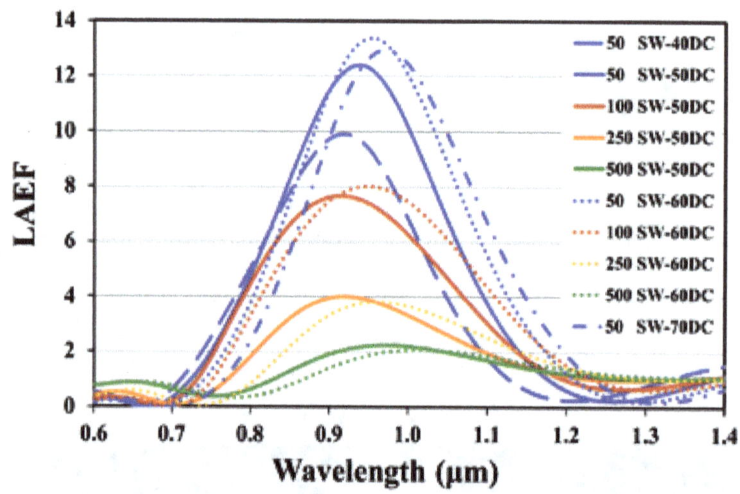

Figure 8. Light absorption enhancement factor spectra of MSM-PDs with rectangular-shaped nano-gratings. Different curves show the effect of slit width (SW) and duty cycle (DC) variations on the LAEF. Here, the under layer thickness and nano-gratings height are kept constant at 60-nm and 100-nm, respectively.

5.4. Impact of incident angle on LAEF curves

We present some results to investigate the effects of incidence angle upon the maximum LAEF for plasmonic-based MSM-PD device. The incident angle varies through a straight angle with negative and positive values, ranging from-90° to 90°, representing inclination from the normal incidence to left and right, respectively. While changing the angle of incidence for illuminated light in simulation, we show the device's most efficient light absorption enhancement for a specific angle. Fig. 11 shows maximum LAEF curve for various incident angles for the nano-grating structures with the subwavelength slit width of 50-nm and nano-grating height of 100-nm while the DC is kept constant at 60%, they are the parameters for the optimized curve producing the maximum LAEF in Fig. 8. The resonant wavelength is constant for most of the angles at 947-nm. The presented results indicate that the optimized incident angle is-46° for this geometry.

5.5. Nano-grating height optimization in plasmonic-based MSM-PD

With the variation of nano-gratings height, different sets of results show significant changes in the amount of light transmitted into the active area of the MSM-PD. Hence the height of the ridge is an effective parameter in optimization of the detector performance [30]. We present the simulation results for rectangular and trapezoidal MSM-PDs with different nano-grating heights. The incident light with TM polarization was perpendicularly illuminated on the groove profiles and we have calculated the amount of light flux transmitted into the slit for four different heights in rectangular and trapezoidal nano-grating assisted MSM-PDs.

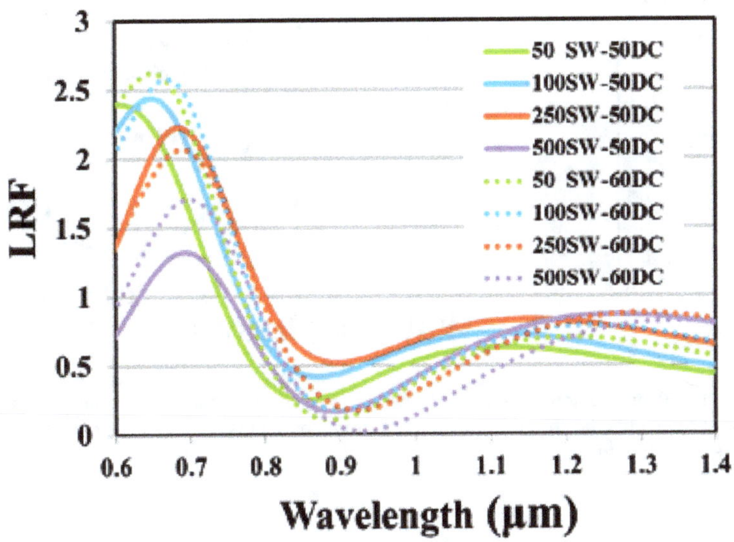

Figure 9. Light reflection factor spectra of MSM-PDs with rectangular-shaped nano-gratings. Different curves show the effect of slit width (SW) and duty cycle (DC) variations on the Light reflection factor.

Figure 12 shows the LAEF spectrum for different heights of the rectangular shaped nano-gratings in plasmonic-based MSM-PD, such as 80, 100, 120, 140-nm. Simulation results confirm that 120-nm is the optimized height for this design. The peak wavelength behaves like a sinusoidal manner and wavelength (λ) is red shifted as the ridge's height increases. The duty cycle is 60% while the subwavelength aperture width, and subwavelength aperture thickness are kept constant at 50-nm, and 20-nm, respectively. There are some interpretations to analyze the curves. The SPPs coupling process and the following expected absorption can easily occur for higher gratings and reduces after certain heights. This light absorption has a maximum for a specific wavelength in the spectra which varies in different heights, and for the wavelengths higher than the peak, the amount of LAEF decreases because the incident light might be coupled into radiative SPs rather than bound SP modes.

Also, several sets of numerical analysis are carried out to illustrate the effect of trapezoidal-shaped nano-gratings on the optimized height at which the maximum resonance transmission occurs. The simulations are performed for 4 different nano-grating heights of 100, 120, 140, and 160-nm with the subwavelength aperture width of 50-nm and the under layer thickness of 20-nm. The results are shown in Fig. 13. The LAEF is 26.3 times in its maximum for the curve representing the 140-nm nano-grating height for the duty cycle of 60%.

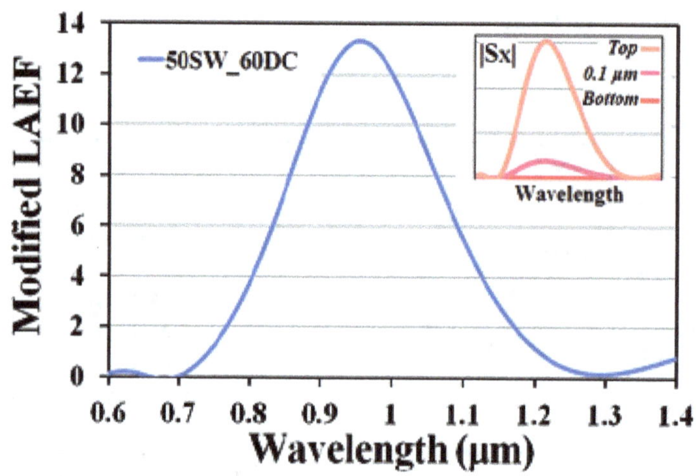

Figure 10. Modified LAEF for plasmonic-based MSM-PDs with 50-nm slit width, 100-nm nano-grating height, and 60% duty cycle. The inset represents the absolute value of S_x in different depths of GaAs substrate, the top, 0.1 μm depth from the substrate surface, and the bottom.

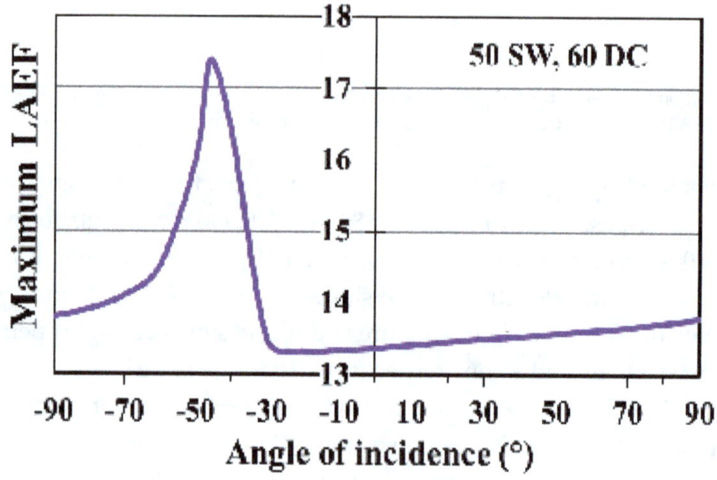

Figure 11. Maximum LAEF versus incidence angle characteristics for MSM-PDs with 50-nm slit width and 60% DC. Here, the subwavelength aperture height and nano-gratings height are kept constant at 60-nm and 100-nm, respectively.

From Figs. 12 and 13, it can be recognized that the optimum wavelength is red shifted for higher nano-gratings in both figures. However, trapezoidal nano-gratings height at maximum light absorption, 140-nm, is higher than the optimum height of its rectangular counterpart, 120-nm, while its resonant wavelength is blue shifted.

5.6. Nano-grating geometries effect in plasmonic-based MSM-PD

There is no doubt about the effects of nano-grating textured structures and geometries on light trapping inside the device active region as they are responsible for the creation of the SPPs which can assist for the light confinement in the subwavelength regions. Hence we analyze the MSM-PD device performance and its enhanced responsivity for different nano-grating shapes.

By involving plasmonics (i.e., metallic nano-gratings), the device performance has been improved due to the advances in nano-technology fabrication methods. Focused ion beam (FIB) lithography, electron beam (E-beam) lithography, and nano-lithography are the new approaches to fabricate nano-scale devices. These techniques can be used to obtain the state-of-the-art for very small nano-structures in order of tens of nano-meters which are the accepted dimension for visible and near infrared regions [31].

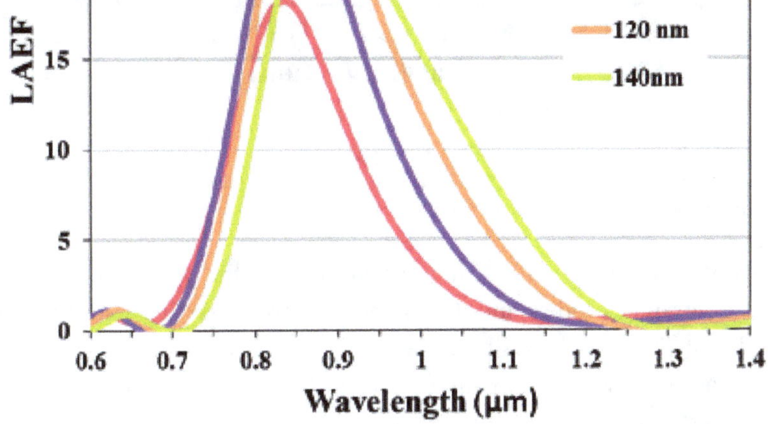

Figure 12. Light absorption enhancement factor spectra of the rectangular nano-gratings with different heights. Here, DC is 60% and subwavelength aperture width is 50-nm.

In general, a good light absorption performance is achieved for rectangular-shaped nano-gratings. It has been proved that the rectangular nano-gratings are the best option for the design of nano-gratings to improve the MSM-PDs performance. The normal wall nano-gratings are easily fabricated with lithography and etching, while in focused-ion beam (FIB) milling, the nano-gratings are rather taper than the rectangular one. The observation of scanning electron microscopy images represents the nano-gratings with taper walls rather than the rectangular nano-gratings. The metallic nano-gratings fabrication is not an easy process but the FIB lithography is an approved technique to develop elaborate structures with the nano-gratings. This nano-scale fabrication technology is very promising to access near field excitations and detection of the plasmon polaritons at air/Au interface.

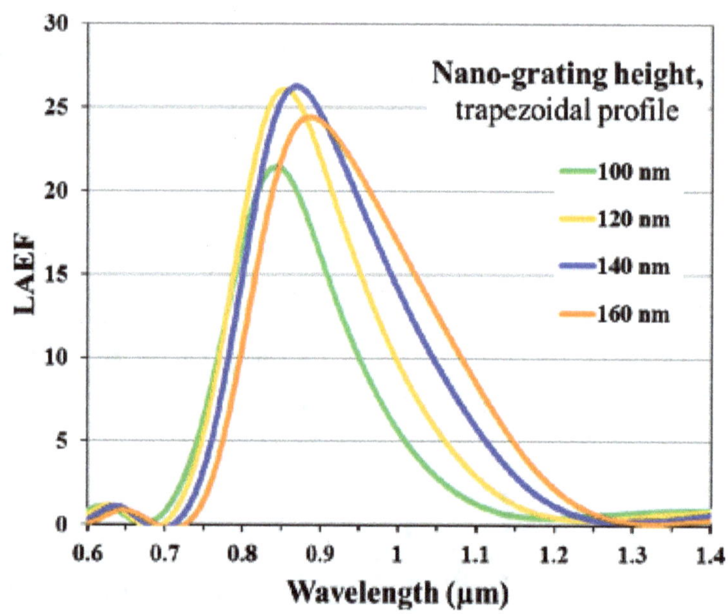

Figure 13. Light absorption enhancement factor spectra of the trapezoidal profile nano-gratings with different heights. Here, DC is 60% and subwavelength aperture width is 50-nm.

Here the plasmonic-based MSM-PD structures are set with the rectangular, trapezoidal, and ellipse-wall nano-grating profiles and their behaviors are compared with each other. It has been reported that the straight wall nano-gratings produce an optimum light absorption for plasmonic-based MSM-PDs, but under practical device manufacturing situations, the rectangular shaped profile is closer in appearance to semi-trapezoidal nano-gratings. Then we introduce and characterize grating-assisted MSM-PD which utilizes ellipse-wall nano-gratings and prove that it achieves better efficiency than rectangular and trapezoidal counterpart.

Metallic nano-grating geometries affect the LAEF in MSM-PDs. Figure 14 shows the light absorption enhancement spectra evolution for MSM-PDs with triangular, rectangular, trapezoidal with 0.4, 0.5, 0.8, and 0.9 aspect ratios, and the ellipse-wall nano-gratings with 0.5, and 0.9 aspect ratios. Aspect ratio (AR) is introduced to define a relationship between the width of upper and lower bases for taper and ellipse wall structures as a dimensionless coefficient smaller than unity. For ordinary structures, the lower base and for inverted ones the upper base is always bigger. These groove shapes along with the differences between their aspect ratios are shown in Fig. 15. The simulation results illustrate a strong confinement for the rectangular shaped structures, although the realistic subwavelength nano-gratings do not have the normal walls. The maximum LAEF for taper and rectangular profiles are quite close.

However, we introduce a novel grating structure for our design with ellipse walls which has recently been stated to have more light capturing efficiency than the rectangular and taper nano-grating profiles [32]. As stated the experimental results in [31], the analysis of atomic force microscope (AFM) systems and scanning electron microscope (SEM) images demonstrated the trapezoidal structures rather with curved walls than the linear walls.

Therefore, we have designed the ellipse-wall nano-gratings with legs satisfying the exponential equation having an exponential coefficient of 0.5. The exponential coefficient is supposed to satisfy the exponential function of $z = C\, e^{x}$ for nano-gratings lateral walls, where C is the exponential coefficient. Performing the simulations on these ellipse-wall nano-structures gives a better view of more realistic condition for nano-scale devices. The simulation is performed for the under layer thickness of 20-nm, the nano-grating height of 120-nm, and the subwavelength aperture width of 50-nm which are optimized values for a MSM-PD with rectangular symmetric nano-gratings [33].

Besides, the results of light absorption for trapezoidal structures with different aspect ratios are presented. In the case of slanted walls, the increase of slit opening width is a drawback for reflection of gap plasmon from the upper termination resulting in a weak LAEF, because the cavity nature which is responsible for resonant absorption faded away with variation of the slit opening width. However, the decrease of taper aspect ratio leads to a blue shift resonance position. Figure 14 shows this slight blue shift while this parameter changes from 0.9 to 0.4 for tapered nano-grating structures.

As shown in Fig. 14, the plasmonic interactions are more efficient for ellipse wall nano-gratings. We know that the localization of optical energy around sharp corners is remarkable. The nonlinear design of ellipse-wall nano-gratings enables the possibility to improve energy concentration in the active region of the MSM-PD device because the energy flow is facilitated through the interface in comparison with the rectangular nano-gratings. Depending on the aspect ratio parameter, that is the normalized value of bottom side to the top side width as shown in Fig. 15, nano-gratings can increase their LAEF. Ellipse-wall nano-gratings improve their performance from LAEF of 20 (for 0.5 aspect ratio) to 28 (for 0.9 aspect ratio). Even the maximum LAEF of rectangular nano-grating, 26.5, is less than 0.9 aspect ratio ellipse-wall structures. In addition, the maximum peak wavelength is red shifted for rectangular structure. While the 0.9

aspect ratio ellipse-wall nano-grating is the most suitable structure in light absorption, it is worth noting that the 0.5 aspect ratio ellipse-wall nano-gratings offer a better transmission in comparison to their trapezoidal counterparts. They almost doubled their efficiency compared with the 0.5 aspect ratio taper profiles from 10.5 to 20.3.

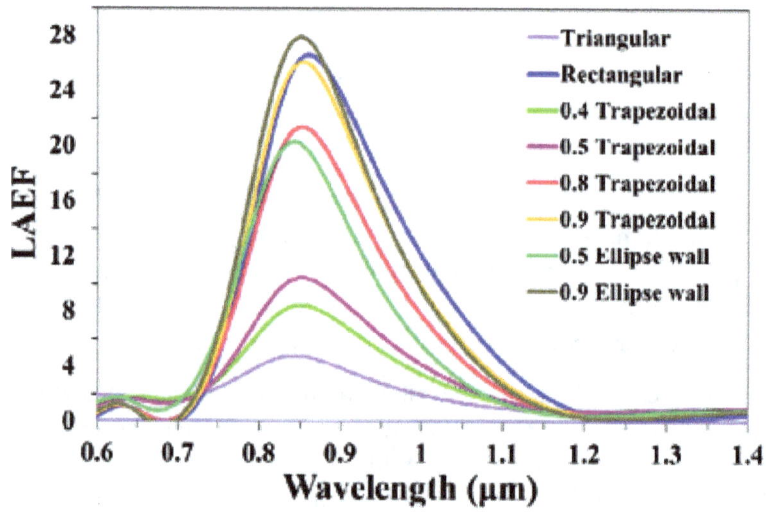

Figure 14. Light absorption enhancement factor spectra of MSM-PD with triangular, rectangular, taper nano-grating structures with 0.4, 0.5, 0.8, and 0.9 aspect ratios, and ellipse wall nano-gratings with 0.5 and 0.9 aspect ratio and exponential coefficient of+5. The duty cycle of corrugations, subwavelength aperture thickness, and the nano-gratings thickness are 60%, 20-nm, and 120-nm, respectively.

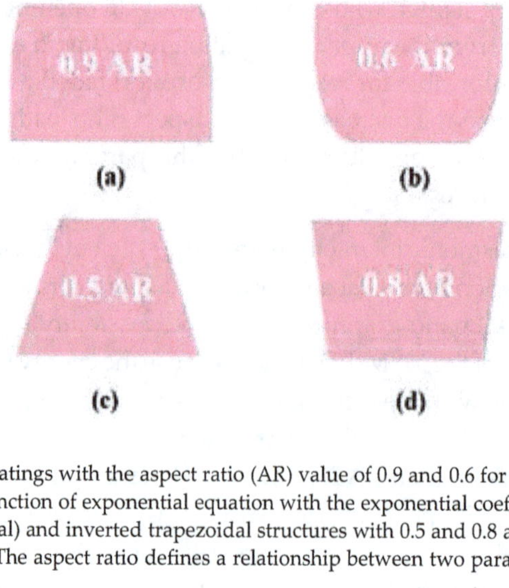

Figure 15. Ellipse-wall nano-gratings with the aspect ratio (AR) value of 0.9 and 0.6 for (a) and (b), respectively. For (a) and (b), the lateral walls are function of exponential equation with the exponential coefficient of+5 for (a) and-5 for (b). (c) and (d) are taper (trapezoidal) and inverted trapezoidal structures with 0.5 and 0.8 aspect ratios. For inverted structure, the lower base is bigger. The aspect ratio defines a relationship between two parallel bases widths of trapezoidal and ellipse-wall nano-gratings.

5.7. Impact of under layer thickness

Here, we discuss the impact of under layer or subwavelength aperture thickness (Uth) on the amount of transmitted light through the active region of the MSM-PD. Under layers are the photodetector's metallic electrodes which metallic nano-gratings can be designed on them. Optimization of this parameter also affects the light absorption enhancement in the MSM-PDs. The thick under layers absorb a proper amount of the incident lights and the carrier recombination become more prominent which reduces the internal quantum efficiency. The thinner under layer leads to better electrical properties moreover they have advantages of material saving and higher carrier collection efficiency; however they lose light harvesting proficiency which can be resolved by implementation of nano-grating structures of the same metal. The reduction of the under layer thickness enhances the LAEF, because in this case the EOT can also occur in the flanked slits other than the central aperture. Here, the under layer thickness attenuation effect is shown, which assists the light absorption in the device in addition to the light absorption enhancement caused by optimization of other parameters, i.e. nano-gratings shape. As expected, an ideal rectangular nano-grating shape with the thinnest under layer contributes more effectively in the light absorption in comparison to the thicker under layer heights. However, the amount of LAEF for ellipse wall structure is higher than the rectangular grooves, because in this design, the central slit opening has no corner which avoids useless light confinement at sharp edges, also the slit opening is broader, so there is less reflection for the illuminated light at the top due to the grooves' non-linear walls and a greater part of the energy involves in the SPP coupling at metal-air interface. Design of the nano-gratings with the thickness of 120-nm, which is the optimized height for rectangular nano-grating as shown in Fig. 12, results in about 32.8 fold enhancement for ellipse-wall nano-gratings with 0.9 aspect ratio and exponential coefficient of 0.5 when the duty cycle is fixed at 40% and the subwavelength slit width is 50-nm, but ellipse-wall nano-gratings quite close absorption peak to rectangular nano-grating with the height of 120-nm is the result of doing simulations for optimized rectangular nano-grating.

While absorption is enhanced for thinner under layers, useful results will be revealed through comparison of absorption curves for optimized nano-grating thickness of ellipse-wall nano-gratings and rectangular nano-gratings, 140-nm and 120-nm, respectively. Our simulated results indicate better enhancement of LAEF for ellipse-wall nano-gratings with optimized thickness of 140-nm that is 33.6 compared with maximum LAEF of 120-nm rectangular nano-grating which is 32.7. These results are shown in Fig. 16.

In addition to the influence of under layer thickness on the amount of transmitted light flux into the device active region, we can also demonstrate under layer's direct influence on other structural parameters like duty cycle. Looking back at sections 5.2 and 5.3, we notice the LAEF curves calculated in different duty cycles. In Fig. 8, for the slit widths of 50-nm, the 60% is the optimized duty cycle while in Fig. 5 and 6, the optimized duty cycle is 40%. We mentioned this to prove the fact that the optimized duty cycle varies for the structure with different under layer thickness, the under layer thickness of 20-nm and 60-nm in Fig. 5 and Fig. 8, respectively.

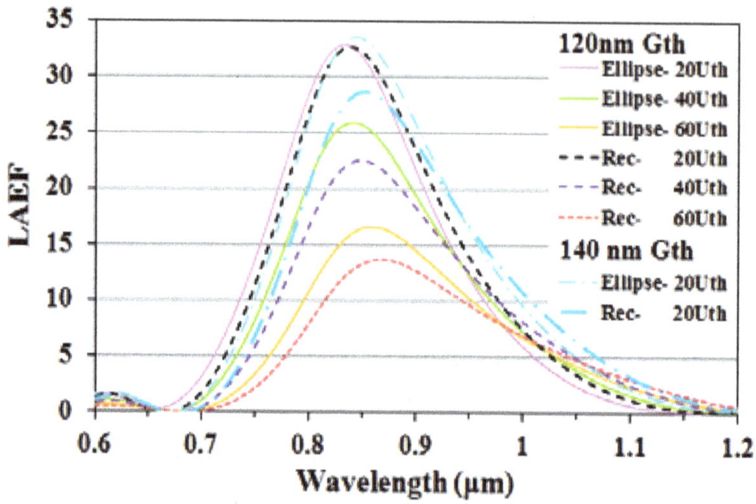

Figure 16. Light absorption enhancement factor (LAEF) calculation of different under layer thicknesses (Uth) for the rectangular and 0.9 aspect ratio ellipse-wall nano-grating structures with exponential coefficient of+5 for two different nano-grating thicknesses (Gth).

6. Conclusions

We discussed the interaction of illuminated light as electromagnetic waves through the central sub-wavelength slit on a metallic thin film surrounded by periodic nano-gratings. The concept of SPPs has been introduced and the excited SPPs are generated at metal-dielectric interface which are used for plasmonic-based applications. Plasmonics offers the ability to concentrate light into subwavelength volumes in ultra-small optoelectronic devices utilizing high-speed and broad bandwidth. Plasmonics has also made impression in nano-scale photodetector development. Photodetectors play key role in development of modern optical communication technology. Surface plasmon resonances have found practical applications in sensitive photodetectors recently. We have analyzed the performance and advantages of plasmonic-based MSM photodetectors and modeled their light absorption enhancement. The FDTD simulation tool has been used to analyze and optimize the impact of the physical and geometrical parameters on the amount of transmitted light into the MSM-PD structures. In order to maximize the flow of energy into the device active region, the performance of different nano-grating profiles and their effects on the efficiency of the photodetector have be evaluated using Drude-Lorentz dielectric function via FDTD algorithm. The corrugations on the surface lead to an effective impedance for the surface modes which favors to the resonant coupling of the SPs with incident electromagnetic wave and hence facilitates the enhancement of light transmission. The main motivation to design the nano-gratings on the MSM-PD electrodes is to assist and improve the light transmission into the slit. With the grating assisted MSM-PDs, the device could benefit the small spacing between the two top contacts, namely the central slit, for a fast response of optical pulses. The simulated results reveal that the amount of LAEF is much better than the conventional MSM-PDs. A substantial light absorption enhancement has been found for the symmetrical MSM-PD devices. Two distinct mechanisms targeting the absorption enhancement in MSM-PDs are namely, the metal nano-gratings assisted light absorption and the subwavelength slit Fabry-Perot resonances. Feasibility of developing MSM photodetectors with high-responsivity and high-speed characteristics has made them reliable choices for high-speed optical communication systems. We have studied the transmission of TM-polarized light through subwavelength apertures in metallic films flanked by metallic nano-gratings in plasmonic-based MSM-PDs and optimized the interdigitated electrodes thicknesses and the nano-grating shapes. Simulation results confirm that the plasmonic–based grating–assisted MSM-PD is more efficient in light absorption compared with a conventional MSM-PD. The light energy confinement in the nano-scale and the produced focal point of the incident beam in plasmonic device compared with conventional MSM-PD confirm the characteristics of a nano-plasmonic lens. The demonstration of device optimization results assists to improve the concept of novel high responsivity, plasmonic-based MSM-PDs for high speed applications in optical communication. These results provide useful information for the design and fabrication of nano-scale optoelectronic devices.

Author details

Farzaneh Fadakar Masouleh[1] and Narottam Das[2,3]

1 Physics Department, University of Guilan, Rasht, Guilan, Iran

2 Department of Electrical and Computer Engineering, Curtin University, Perth, WA, Australia

3 Department of Electrical and Computer Engineering, Curtin University, Miri, Sarawak, Malaysia

References

[1] G. P. Wiederrecht. Handbook of nanoscale optics and electronics, Elsevier, Amsterdam, First edition 2010. ISBN: 978-0-12-375178-2.

[2] C. DeCusatis, and I. Kaminow. The Optical Communications Reference, Academic Press 2009. ISBN: 978-0-12-375163-8.

[3] H. Zimmermann. Silicon Optoelectronic Integrated Circuits, Springer Berlin Heidelberg 2004; 13, 1-23, doi: 10.1007/978-3-662-09904-9_1.

[4] B. P. Pal. Fundamentals of Fibre Optics in Telecommunication and Sensor Systems, bohem press, 1992.

[5] G. P. Agrawal. Fiber-Optic Communication Systems, John Wiley & Sons, Inc., Fourth Edition, 2002. doi: 10.1002/9780470918524.

[6] S. Y. Chou, Y. Liu, and P. B. Fischer. Tera-hertz GaAs metal-semiconductor-metal photodetectors with nanoscale finger spacing and width, Electron Devices Meeting, 1991. IEDM '91. Technical Digest., International, 745-748. doi: 10.1109/IEDM.1991.235316.

[7] W. Zhang, A. K. Azad, and J. Han. Resonant Excitation of Terahertz Surface Plasmons in Subwavelength Metal Holes, Hindawi Publishing Corporation, Active and Passive Electronic Components, 2007, Article ID 40249, 8 pages, doi: 10.1155/2007/40249.

[8] R. Umeda, C. Totsuji, K. Tsuruta, and H. Totsuji. Dispersion Models and Electromagnetic FDTD Analyses of Nanostructured Metamaterials using Parallel Computer, *Memoirs of the Faculty of Engineering*, Okayama University January 2009; 43, p. 8.

[9] A. D. Rakic, A. B. Djurišic, J. M. Elazar, and M. L. Majewski. Optical Properties of Metallic Films for Vertical-Cavity Optoelectronic Devices, Applied Optics 1998; 37(22), 5271-5283. http://dx.doi.org/10.1364/AO.37.005271.

[10] M. Bordovsky et al. Waveguide design, modeling, and optimization: from photonic nanodevices to integrated photonic circuits, in Proceedings of the SPIE 5355, Integrated Optics: Devices, Materials, and Technologies VIII, 65, doi: 10.1117/12.526976, May 28, 2004.

[11] B. Ung. Study of the interaction of surface waves with a metallic nano-slit via the finite-difference time-domain method, *M.Sc. Thesis*, Laval University, Quebec, Canada, Ch. 3, 2007. http://theses.ulaval.ca/archimede/fichiers/24879/24879.html.

[12] F. F. Masouleh, N. Das, H. Mashayekhi. Impact of duty cycle and nano-grating height on the light absorption of plasmonics-based MSM photodetectors, in Proceedings of the 12th IEEE Int. Conf on Numerical Simulation of Optoelectronic Devices 2012; Shanghai, China, 13-14. doi:10.1109/NUSOD.2012.6316521.

[13] T. W. Ebbesen, H. J. Lezec, H. F. Ghaemi, T. Thio, P. A. Wolff. Extraordinary optical transmission through sub-wavelength hole arrays, Nature 1998; 391, 667-669. doi: 10.1038/35570.

[14] H. J. Lezec, A. Degiron, E. Devaux, R. A. Linke, L. Martin-Moreno, F. J. Garcia-Vidal, T. W. Ebbesen. Beaming light from a subwavelength aperture, Science 2002; 297, 820-822. doi: 10.1126/science.1071895.

[15] Y. Ding, J. Yoon, M. H. Javed, S. H. Song, R. Magnusson. Mapping surface-plasmon polaritons and cavity modes in extraordinary optical transmission, IEEE Photonics Journal 2011, 3, 365-374. doi:10.1109/JPHOT.2011.2138122.

[16] S. A. Maier. Plasmonics: Fundamentals and Applications, Springer, 2007; ISBN 978-0-387-37825-1.

[17] G. T. Reed and A. P. Knight. Silicon photonics: an introduction, John Wiley and Sons 2004; ISBN 0-470-87034-6.

[18] S. Collin, F. Pardo, R. Teissier, J. L. Pelouard. Horizontal and vertical surface resonances in transmission metallic gratings, Journal of Optics A: Pure and Applied Optics 2002; 4, 154-160. doi:10.1088/1464-4258/4/5/364.

[19] D. de Ceglia, M. A. Vincenti, M. Scalora, N. Akozbek and M. J. Bloemer. Enhancement and inhibition of transmission from metal gratings: Engineering the Spectral Response, at http://arxiv.org/abs/1006.3841, 2010.

[20] J. A. Porto, F. J. García-Vidal and J. B. Pendry. Transmission resonances on metallic gratings with very narrow slits, Physical Review Letters 1999; 83, 2845-2848. doi: http://dx.doi.org/10.1103/PhysRevLett.83.2845.

[21] A. Barbara, P. Quemerais, E. Bustarret and T. Lopez-Rios. Optical transmission through subwavelength metallic gratings, Physical Review *B* 2002; 66, Article ID 161403. doi: http://dx.doi.org/10.1103/PhysRevB.66.161403.

[22] H. Raether. Surface Plasmons on Smooth and Rough Surfaces and on Gratings, Springer-Verlag, Berlin, 1988. doi:10.1007/BFb0048323.

[23] L. Martín-Moreno, F. J. García-Vidal, H. J. Lezec, A. Degiron, and T. W. Ebbesen. Theory of highly directional emission from a single subwavelength aperture surrounded by surface corrugations, Physical Review Letters 2003; 90, Article ID 167401. doi: 10.1103/PhysRevLett.90.167401.

[24] J.A. Shackleford, R. Grote, M. Currie, J.E. Spanier, B. Nabet. Integrated plasmonic lens photodetector Appl. Phys. Lett. 2009; 94, 083501.http://dx.doi.org/10.1063/1.3086898.

[25] K. S. Yee, Numerical solution of initial boundary value problems involving maxwell's equations in isotropic media, Antennas and Propagation, IEEE Transactions on 1966; 14(3), 302-307. doi:10.1109/TAP.1966.1138693.

[26] E. Chen, and Y. S. Chou. Polarimetry of thin metal transmission gratings in the resonance region and its impact on the response of metal-semiconductor-metal photodetectors, Applied physics letters 1997; 70(20), 2673-2675. http://dx.doi.org/10.1063/1.118990.

[27] F. F. Masouleh, N. K. Das, and H. R. Mashayekhi. Assessment of amplifying effects of ridges spacing and height on nano-structured MSM photo-detectors, Journal of Optical and Quantum Electronics April 2014; 46(4). doi:10.1007/s11082-014-9900-8.

[28] N. Das, F. F. Masouleh, and H. R. Mashayekhi. A Comprehensive Analysis of Plasmonics-Based GaAs MSM-Photodetector for High Bandwidth-Product Responsivity, Advances in OptoElectronics, 2013, Article ID 793253, 10 pages, 2013. doi: 10.1155/2013/793253.

[29] N. K. Das, F. F. Masouleh, and H. R. Mashayekhi. Light Absorption and Reflection in Nano-Structured GaAs Metal-Semiconductor-Metal Photo-Detectors, IEEE Transactions on Nanotechnology Sep. 2014; 13(5), 1-8, doi: 10.1109/TNANO.2014.2336857.

[30] F. F. Masouleh, N. K. Das, and H. R. Mashayekhi. Optimization of light transmission efficiency for nano-grating assisted MSM-PDs by varying physical parameters, The Journal of Photonics and Nanostructures-Fundamentals and Applications February 2014; 12(1), 45-53. doi: 10.1016/j.photonics.2013.07.011.

[31] N. Das, A. Karar, M. Vasiliev, C.L. Tan, K. Alameh, Y.T. Lee. Analysis of nano-grating-assisted light absorption enhancement in metal–semiconductor–metal photodetectors patterned using focused ion-beam lithography, Optics Communications 2011; 284(6), 1694–1700. doi: 10.1016/j.optcom.2010.11.065.

[32] Y. Liang, W. Peng, R. Hu, and H. Zou. Extraordinary optical transmission based on subwavelength metallic grating with ellipse walls, Optics Express 2013; 21(5), 6139-6152. doi: 10.1364/OE.21.006139.

[33] F. F. Masouleh, N. K. Das, and H. R. Mashayekhi. Comparison of different plasmonic nano-grating profiles for quality light absorption in nano-structured MSM photo-detectors, Optical Engineering 2013; 52(12), 127101. doi:10.1117/1.OE.52.12.127101.

6

Design and Implementation of WDM-PON Solutions

Vjaceslavs Bobrovs, Jurgis Porins,
Sandis Spolitis and Girts Ivanovs

1. Introduction

Internet traffic is growing very rapidly. Based on Cisco Visual Networking Index (VNI) forecast and methodology, global IP traffic has increased more than 4 times in the past 5 years, and will increase 3 times over the next 5 years. Moreover, it will reach 1 zettabyte (ZB) per year or 83.8 exabytes (EB) per month in year 2015 and exceed 1.4 ZB per year or 120.6 EB per month threshold by the end of year 2017. Video-on-demand traffic will triple by year 2016. For these reasons there will be need for optical access networks which are capable to handle with large data amounts, consume less energy and are cost effective [3].

PONs has been considered to be one of the most promising solutions for access networks due to its broad bandwidth and low-cost infrastructure [4]. Especially, spectrum sliced wavelength division multiplexed passive optical network can be an attractive and cost effective solution to satisfy the growing worldwide demand for transmission capacity in the next generation fiber optical access networks [1, 2]. In principle there are two major factors that will influence the telecommunication networks of the future. The first one is the need to support high bandwidths and low latency and the second one is to use an architecture that is both cost and energy efficient [5]. Recent studies have uncovered that a large amount of electricity is consumed by telecom equipment's in broadband enable countries. Therefore it is important to put research effort for minimizing energy consumption in fiber optical access networks [6].

Traditional WDM systems have multiple transmitter lasers operating at different wavelengths, which need to be wavelength selected for each individual channel operated at a specific wavelength [1, 2]. It increases complexity of network architecture, cost and wavelength management [7]. The strength of spectrum sliced WDM-PON technology is use of one common broadband seed light source and its ability to place electronics and optical elements in one central office (CO), in that way simplifying the architecture of fiber optical network [1]. By using only one light source instead of one for every user we can make optical access system more energy efficient or in other words "green". It is reported that the total energy consumed by the infrastructures of communication networks including Internet take up more than 3% of the current worldwide electric energy consumption. In specific, the access networks contribute to a larger portion of the overall energy consumption when compared to the core and transport networks [8]. Spectrum sliced WDM PON systems benefit from the same advantages as traditional WDM, while employing low cost incoherent light sources like amplified spontaneous emission (ASE) source or light-emitting diode (LED) [7, 9].

The optical bandwidth per channel of SS-WDM PON system is large compared to the bit rate. Therefore, dispersion significantly degrades the performance of this system more than it is observed in conventional laser-based systems [10]. The influence of dispersion needs to be studied in order to understand the characteristics of a spectrum sliced WDM PON system employing standard single mode optical fiber (SMF).

It is very important to build a new type optical system based on widely used frequency grid, recommended by international standards because of such a system potentially is much more compatible with other already existing WDM-PON optical systems. The main benefit includes the reduction of network architecture complexity as well as cost per one user. It is possible by replacing the classic WDM-PON system (where one laser source is used for each user) with our proposed spectrum sliced dense WDM PON system with CD compensation (where one seed broadband ASE source is spectrally sliced and used for multiple users) [11, 12].

The more information we are transmitting the more we need to think about parameters like available bandwidth and latency. Bandwidth is usually understood by end-users as the important indicator and measure of network performance. It is surely a reliable figure of merit, but it mainly depends on the characteristics of the equipment. Unlike bandwidth, latency and jitter depend on the specific context of transmission network topology and traffic conditions. As network latency we understand delay from the time of packet transmission at the sender to the end of packet reception at the receiver [13]. If latency is too high it spreads data packets over the time and can create an impression that an optical network is not operating at data transmission speed which was expected. Data packets are still being transported at the same bit rate but due to latency they are delayed and affect the overall transmission system performance [14].

It should be pointed out, that there is need for low latency optical networks in almost all industries where any data transmission is realized. It is becoming a critical requirement for a wide set of applications like financial transactions, videoconferencing, gaming, telemedicine and cloud services which requires transmission line with almost no delay performance. These industries are summarized in references [15, 16] and shown below (see Table 1).

In fiber optical networks latency consists of three main components which adds extra time delay: the optical fiber itself, optical components, and opto-electrical components. Therefore, for the service provider it is extremely important to choose best network components and think on efficient low latency transport strategy [17, 18].

Industry	Applications and services
Education	• video conferencing • live-streaming • rich learning content • dynamic e-learning platforms • presentation applications • dynamic administration tools • cloud-based applications
Healthcare	• Picture Archiving Communications Systems (PACS) • telemedicine, telehealth applications • diagnostic imaging • Electronic Medical Records (EMR) • patient portals • mobile healthcare applications and equipment
Media and Entertainment	• live-streaming breaking news • television shows • videoconferencing • movies over Internet • transfer large files, images, and videos from the field to studios around the world • real-time gaming
Government	• interaction between communities and their governments • transportation management, • emergency response and general commerce • circulation of documents • self-service portals
Legal	• sharing large, bandwidth-intensive files quickly and securely • secure and high speed access to critical files "24 hours a day, 7 days a week"
Finance	• High-Frequency Trading (HFT) and high speed information exchange • financial transactions • connections to brokers, dealers, exchanges, hedge funds and information feeds

Table 1. Industries where low latency services are very important.

There are many researches about improvement of transmission speed in fiber optical networks while impact and sources of latency is not investigated sufficiently. As mentioned before, latency is a critical requirement for a wide set of applications. Even latency of 250 ns can make the difference between winning and losing a trade [19]. Surely "latency reduction" is among the hot key-words for vendors, end-users and telecommunication service providers. According to a recent market analysis, the request for ultra-low latency services is increasing dramatically and opening many opportunities especially in the field of end-to-end high-speed optical services [20]. Latency reduction is very important in financial sector, for example, in the stock exchange market where 10 ms of latency could potentially result in a 10% drop in revenues for a company [21]. No matter how fast you can execute a trade command, if your market data is delayed relative to competing traders, you will not achieve the expected fill rates and your revenue will drop [22]. Low latency trading has moved from executing a transaction within several seconds to milliseconds, to microseconds, and even nanoseconds. Nowadays, a millisecond improvement in network speeds offers competitive advantage for financial institutions [23].

2. Latency issue in optical networks

It is important to look at latency as consisting of different components. Latency is a time delay experienced in transmission system and it describes how long it takes for data to get from transmission side to receiver side. In a fiber optical communication systems it is essentially the length of optical fiber divided by the speed of light in fiber core, supplemented with delay induced by optical and electro optical elements plus any extra processing time required by system also called overhead [24]. Signal processing delay can be reduced by using parallel processing based on large scale integration CMOS technologies [25].

Added to the latency due to propagation in the fiber, there are other path building blocks which affect the total data transport time [17, 24]. These elements include opto-electrical conversation, switching and routing, signal regeneration, amplification, chromatic dispersion compensation, compensation of polarization mode dispersion (PMD), data packing, digital signal processing (DSP), protocols and addition forward error correction (FEC) [24]. Data transmission speed over optical metro network must be carefully chosen. If we upgrade 2.5 Gbit/s link to 10 Gbit/s link then CD compensation or amplification may be necessary, but it also will increase overall system latency. For optical lines with transmission speed more than 10 Gbit/s (e.g. 40 Gbit/s) a need for coherent detection arises. In coherent detection systems CD can be electrically compensated using DSP which also adds latency. Therefore some companies avoid using coherent detection for their low-latency network solutions [26].

From the standpoint of personal communications, effective dialogue requires latency < 200 ms, an echo needs > 80 ms to be distinguished from its source, remote music lessons require latency < 20 ms, and remote performance < 5 ms. It has been reported that in virtual environments, human beings can detect latencies as low as 10 to 20 ms. In trading industry or in tele-health every microsecond matters. But in all cases, the lower latency we can get the better system performance we can achieve [25, 27].

2.1. Single mode optical fiber

In standard single-mode fiber (SMF) major part of the light signal travels in the core while a small amount of light travels in the cladding. Optical fiber with lower group index of refraction provides an advantage in low latency applications [27].

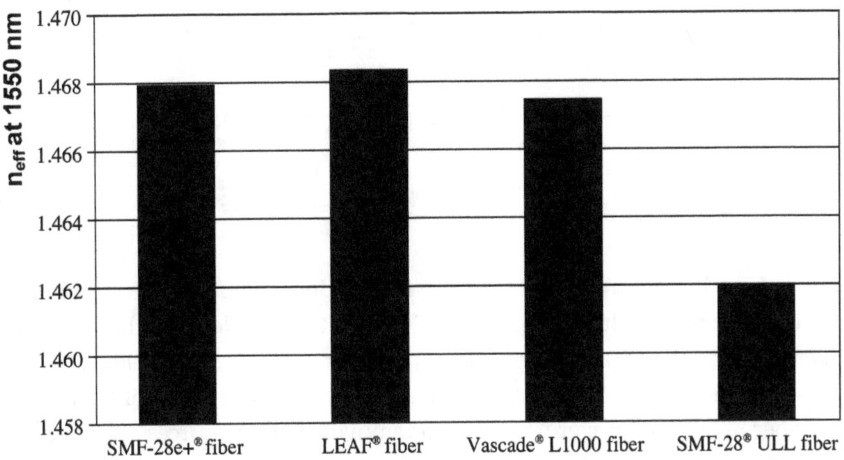

Figure 1. Effective group index of refraction impact of various commercially available Corning single mode fiber types [28].

Therefore it is useful to use a parameter "effective group index of refraction (n_{eff}) instead of "index of refraction (n)" which only defines the refractive index of core or cladding of single mode fiber. The n_{eff} parameter is a weighted average of all the indices of refraction encountered by light as it travels within the fiber, and therefore it represents the actual behavior of light within a given fiber [28]. Figure 1 illustrates the impact of profile shape on n_{eff} by comparing its values for several Corning single mode fiber products with different refractive index profiles.

It is known that speed of light in vacuum is 299792.458 km/s. Assuming ideal propagation at the speed of light in vacuum, an unavoidable latency value can be calculated as following in Equation (1):

$$\Delta_{\text{light}} = \frac{1\ \text{km}}{c} = \frac{1\ \text{km}}{299792.458\ \text{km}/\text{s}} = 3.336\ \mu s \tag{1}$$

However, due to the fiber's refractive index light travels more slowly in optical fiber than in vacuum. In standard single mode fiber defined by ITU-T G.652 recommendation the effective group index of refraction (n_{eff}), for example, can be equal to 1.4676 for transmission on 1310 nm and 1.4682 for transmission on 1550 nm wavelength [29]. By knowing n_{eff} we can express the speed of light in selected optical fiber at 1310 and 1550 nm wavelengths, see Equations (2) and (3):

$$V_{1310\text{nm}} = \frac{c}{n_{\text{eff}}} = \frac{299792.458\ \text{km}/\text{s}}{1.4676} = 204273.956\ \text{km}/\text{s} \tag{2}$$

$$V_{1550\text{nm}} = \frac{c}{n_{\text{eff}}} = \frac{299792.458\ \text{km}/\text{s}}{1.4682} = 204190.477\ \text{km}/\text{s} \tag{3}$$

By knowing speed of light in optical fiber at different wavelengths (see Equation (2) and (3)) optical delay which is caused by 1 km long optical fiber can be calculated as following:

$$\Delta_{1310nm} = \frac{1\text{ km}}{V_{1310nm}} = \frac{1\text{ km}}{204273.956\text{ km / s}} = 4.895\ \mu s \tag{4}$$

$$\Delta_{1550nm} = \frac{1\text{ km}}{V_{1550nm}} = \frac{1\text{ km}}{204190.477\text{ km / s}} = 4.897\ \mu s \tag{5}$$

As one can see from Equations (4) and (5), propagation delay value of optical signal is affected not only by the fiber type with certain n_{eff}, but also with the wavelength which is used for data transmission over fiber optical network. It is seen that optical signal delay values in single mode optical fiber is about 4.9 μs. This value is the practical lower limit of latency achievable for 1 km of fiber in length if it were possible to remove all other sources of latency caused by other elements and data processing overhead [7].

Photonic crystal fibers (PCFs) can have very low effective refractive index, and can propagate light much faster than in SMFs [25]. For example, hollow core fiber (HCF) may provide up to 31% reduced latency relative to traditional fiber optics [30, 31]. But there is a problem that attenuation in HCF fibers is much higher compared to already implemented standard single mode fibers (for SMF α=0.2 dB/km but for HCF α=3.3 dB/km at 1550 nm) [32]. It is reported even 1.2 dB/km attenuation obtained in hollow-core photonic crystal fiber [33].

2.2. Chromatic dispersion compensation

Chromatic dispersion (CD) occurs because different wavelengths of light travel at different speeds in optical fiber. CD can be compensated by dispersion compensation module (DCM) where dispersion compensating fiber (DCF) or fiber Bragg grating (FBG) is employed [34].

A typical fiber access optical network will require DCF approximately 15 to 25% of the overall fiber length. It means that use of DCF fiber adds about 15 to 25% to the latency of the fiber [17, 18]. For example, 100 km long optical metro network where standard single mode fiber (SMF) is used, can accumulate chromatic dispersion in value about 1800 ps/nm at 1550 nm wavelength [29]. For full CD compensation is needed about 22.5 km long DCF fiber spool with large negative dispersion value (typical value is-80 ps/nm/km) [35]. If we assume that light propagation speed in DCF fiber is close to speed as in SMF then total 100 km long optical network's latency with CD compensation using DCF DCM is about 0.6 ms.

Solution how to avoid need for chromatic dispersion compensation or reduce the length of necessary DCF fiber is to use optical fiber with lower CD coefficient value. For example, non-zero dispersion shifted fibers (NZ-DSFs) were developed to simplify CD compensation while making a wide band of channels available. NZ-DSF fiber parameters are defined in ITU-T G. 655 recommendation [36]. Today NZ-DSF fibers are optimized for regional and metropolitan high speed optical networks operating in the C-and L-optical bands. For C band it is defined that wavelength range is from 1530 to 1565 nm, but for L band it is from 1565 to 1625 nm [37].

For commercially available NZ-DSF fiber chromatic dispersion coefficient can be from 2.6 to 6.0 ps/nm/km in C-band and from 4.0 to 8.9 ps/nm/km in L-band. At 1550 nm region typical CD coefficient is about 4 ps/nm/km for this type of fiber. It can be seen that for G.655 NZ-DSF fiber CD coefficient is about four times lower than for standard G.652 SMF fiber [27, 38]. Since these fibers have lower dispersion than conventional single mode, simpler modules are used that add only up to 5% to the transmission time for NZ-DSF [7]. This enables a lower latency than using SMF fiber for transmission. Another solution how to minimize need for extra CD compensation or reduce it to the necessary minimum is dispersion shifted fiber (DSF) which is specified in ITU-T G.653 recommendation. This fiber is optimized for use in the 1550 nm region and has no chromatic dispersion at 1550 nm wavelength. Although, it is limited to single-wavelength operation due to non-linear four wave mixing (FWM), which causes optical signal distortions [39].

If CD is unavoidable another technology for compensation of accumulated CD is a deployment of fiber Bragg gratings (FBG). DCM with FBG can compensate several hundred kilometers of CD without any significant latency penalty and effectively remove all the additional latency that DCF-based networks add [7]. In other words, a lot of valuable microseconds can be gained by migrating from DCF DCM to FBG DCM technology in optical metro network [40]. Typical fiber length in an FBG used for dispersion compensation is about 10 meters. Therefore, normally FBG based DCM can introduce from 5 to 50 ns delay in fiber optical transmission line [27].

One of solutions how to avoid implementation of DCF DCM which introduces addition delay is coherent detection where new transmission formats like quadrature phase-shift keying (QPSK) can be used. However, it must be mentioned that it can be a poor choice from a latency perspective due to the added digital signal processing (DSP) time they require. This additional introduced delay can be up to 1 μs [24, 31].

2.3. Optical amplifiers

Another key optical component which adds additional time delay in optical transmission line is optical amplifier. Erbium doped fiber amplifiers (EDFA) is widely used in fiber optical networks. EDFA can amplify signals over a band of almost 30 to 35 nm extending from 1530 to1565 nm, which is known as the C-band fiber amplifier, and from 1565 to 1605 nm, which is known as the L-band EDFA. The great advantage of EDFAs is that they are capable of simultaneously amplifying many WDM channels and there is no need to amplify each individual channel separately. EDFAs also remove the requirement for optical-electrical-optical (OEO) conversion, which is highly beneficial from a low-latency perspective. However it must be taken into account that EDFA contains few meters erbium-doped optical fiber (Er^{3+}) which adds extra latency, although this latency amount is small compared with other latency contributors. Typical EDFA amplifier contains up to 30 m long erbium doped fiber. These 30 m of additional fiber add 147 ns (about 0.15 μs) time delay [27].

Solution how to avoid extra latency if amplification is necessary is use of Raman amplifier instead of EDFA or together (in tandem) with EDFA. This combination provides maximal signal amplification with minimal latency. Raman amplifiers use a different optical characteristic to amplify the optical signal [9, 27]. Raman amplification is realized by using stimulated Raman scattering. The Raman gain spectrum is rather broad, and the peak of the gain is centered about 13 THz (100 nm in wavelength) below the frequency of the pump signal used. Pumping a fiber using a high-power pump laser, we can provide gain to other signals, with a peak gain obtained 13 THz below the pump frequency. For example, using pumps around 1460–1480 nm wavelength provides Raman gain in the 1550–1600 nm window, which partly cover C and L bands. Accordingly, we can use the Raman's effect to provide gain at any wavelength we want to amplify. The main benefit regarding to latency is that Raman amplifier pump the optical signal without adding fiber to the signal path, therefore we can assume that Raman amplifier adds no latency [41].

2.4. Transponders and opto-electrical conversion

Any transmission line components which are performing opto-electrical conversation increase total latency. One of key elements used in opto-electrical conversation are transponders and muxponders. Transponders convert incoming signal from the client to a signal suitable for transmission over the WDM link and an incoming signal from the WDM link to a suitable signal toward the client. Muxponder basically do the same as transponder except that it has additional option to multiplex lower rate signals into a higher rate carrier (e.g. 10 Gbit/s services up to 40 Gbit/s transport) within the system in such a way saving valuable wavelengths in the optical metro network [41].

The latency of both transponders and muxponders varies depending on design, functionality, and other parameters. Muxponders typically operate in the 5 to 10 μs range per unit. The more complex transponders include additional functionality such as in-band management channels. This complexity forces the unit design and latency to be very similar to a muxponder, in the 5 to 10 μs range. If additional FEC is used in these elements then latency value can be higher [7]. Several telecommunications equipment vendors offer simpler, and lower-cost, transponders that do not have FEC or in-band management channels or these options are improved in a way to lower device delay. These modules can operate at much lower latencies from 4 ns to 30 ns. Some vendors also claim that their transponders operate with 2 ns latency which is equivalent to adding about a half meter of SMF to a fiber optical path [41].

2.5. Optical signal regeneration

For low latency optical metro networks it is very important to avoid any regeneration and focus on keeping the signal in the optical domain once it is entered the fiber. An optical-electronic-optical (OEO) conversion takes about 100 μs, depending on how much processing is required in the electrical domain [24]. Ideally a carrier would like to avoid use of FEC or full 3R (signal power, shape and length) regeneration. 3R regeneration needs OEO conversion which adds unnecessary time delay. Need for optical signal regeneration is determined by transmission data rate involved, whether dispersion compensation or amplification is required, and how many nodes the signal must pass through along the fiber optical path [26].

2.6. Forward error correction and digital signal processing

It is necessary to minimize the amount of electrical processing at both ends of fiber optical connection. FEC, if used (for example, in transponders) will increase the latency due to the extra processing time [7]. This approximate latency value can be from 15 to 150 μs based on the algorithm used, the amount of overhead, coding gain, processing time and other parameters [27].

Digital signal processing (DSP) can be used to deal with chromatic dispersion (CD), polarization mode dispersion (PMD) and remove critical optical impairments. But it must be taken into account that DSP adds extra latency to the path. It has been mentioned before that this additional introduced delay can be up to 1 μs [31].

2.7. Example of latency optimized optical network

In this section we present optimized model of 160 km long optical metro access network operating at 10 Gbit/s bitrate. The aim of this section to show how much optical network can be optimized in terms of latency by replacing conventional latency inducing optical and electrical components to low-latency components. In Figure below we propose two optical transmission networks – conventional and latency optimized, see Fig. 2(a) and Fig. 2(b).

First optical metro network (see Fig. 2(a)) consists of two transponders, DCF DCM for CD pre-compensation, two EDFA amplifiers to boost optical signal, and two 80 km long ITU-T G.652 SMF fiber spans.

Pre-compensation configuration of accumulated chromatic dispersion was chosen because our researches proved it more effective than post-compensation method [34]. Fiber optical line is chosen 160 km in length because typically fiber optical metro networks are up to 200 km long [2].

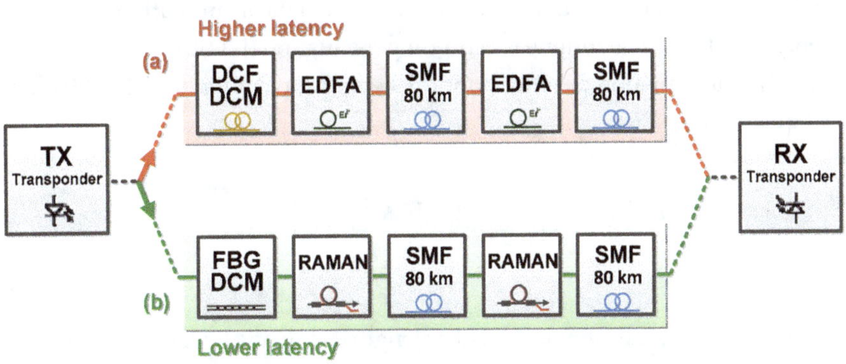

Figure 2. Optical metro network with (a) conventional elements and (b) latency optimized elements.

Second network scheme is latency optimized (see Fig. 2(b)), where we replaced conventional transponder by ultra-low latency transponders, DCF DCM replaced by FBG DCM and EDFA amplifiers are replaced by Raman amplifiers.

For total latency calculations of both optical systems we assume that 64 byte packet which adds 51.2 ns network interface delay is transmitted. Here we used typical latency parameters we described previously in this work and compared results; see from obtained data (see Table 2).

As one can see from obtained data (see Table 2), total latency amount is reduced by 21.6% or 216.24 μs. In this case the main benefit comes from optimization of chromatic dispersion compensation scheme from DCF DCM to FBG DCM and replacement of conventional transponders to low latency transponders.

Line component	Conventional network	Latency optimized network
2 x 80 km ITU-T G.652 SMF fiber spans	4.9 μs x 160 km = 784 μs	4.9 μs x 160 = 784 μs
1 CD compensation module (DCM)	DCF DCM: 25% of 784 μs = 196 μs	FBG DCM: 50 ns
2 Optical amplifiers	EDFA: 2 x 0.15 μs = 0.3 μs	RAMAN adds no delay
2 Transponders	2 x 10 μs = 20 μs	2 x 4 ns = 8 ns
2 Network interface delays	2 x 51.2 ns = 102.4 ns	2 x 51.2 ns = 102.4 ns
Total latency value	**1 ms**	**0.784 ms**

Table 2. Latency calculation for conventional and latency optimized network.

3. Measurement technique and accuracy

Our research is based on powerful and accepted mathematical simulation software OptSim. It solves complex differential nonlinear Schrödinger equation (NLSE) using split-step Fourier method (SSFM). This equation describes optical signal propagation over the fiber and can be written as Equation (6) [42]:

$$\frac{\partial}{\partial z} \cdot A + \frac{\alpha^1}{2} \cdot A + j \cdot \frac{\beta_2}{2} \cdot \frac{\partial^2}{\partial t^2} \cdot A - \frac{\beta_3}{6} \frac{\partial^3}{\partial t^3} \cdot A = j \cdot \gamma \cdot |A|^2 \cdot A \tag{6}$$

where A(t, z) is complex optical field; z is fiber length, [km]; $\alpha^1$is linear attenuation coefficient of an optical fiber, [km^{-1}]; $\beta_2$is the second order parameter of chromatic dispersion, [ps^2/nm]; $\beta_3$is the third order parameter of chromatic dispersion, [ps^3/nm]; $\gamma$is nonlinear coefficient, [W$^{-1}\cdot$km^{-1}]; t is time, [s]. NLSE takes into the account linear and nonlinear affects and they influence to optical signal distortions. The principle of split-step method is better illustrated by (6), which can be written as follows [42]:

$$\frac{\partial}{\partial z}\cdot A(t,\ z)=(\hat{D}+\hat{N})\cdot A(t,\ z) \tag{7}$$

$\hat{D}$is linear operator responsible for linear effects such as dispersion and attenuation [42]:

$$\hat{D}=-\frac{\alpha^{l}}{2}-j\cdot\frac{\beta_2}{2}\cdot\frac{\partial^2}{\partial t^2}+\frac{\beta_3}{6}\cdot\frac{\partial^3}{\partial t^3} \tag{8}$$

$\hat{N}$is nonlinear operator, which takes into account Kerr and other nonlinear effects (NOEs) [42]:

$$\hat{N}=j\cdot\gamma\cdot|A|^2\cdot A \tag{9}$$

In general split-step method is based on assumption that linear and nonlinear effects affect optical signals independently. This statement can be considered as true if we assume that all fiber length z is being divided into sufficiently small spans Δz, and only then these linear and nonlinear effects by turns are taken into account for each $\frac{\Delta z}{2}$ segment.

There are two basic algorithms for realization of SSFM: time domain split step (TDSS) and frequency domain split step (FDSS). These two algorithms differ only with an approach that is being used for calculation of linear operator$\hat{D}$. While in both cases nonlinear operator $\hat{N}$ is being calculated in time domain [42].

Operator $\hat{D}$ is being fully characterized by its impulse response h(t) and it is mathematically correct to calculate its influence to A(t, z) optical field using products of mathematical convolution. In TDSS case it can be written as follows [42]:

$$A_L[n]=A[n]*h[n]=\sum_{k=-\infty}^{\infty}A[k]\cdot h[n-k] \tag{10}$$

This algorithm calculates this convolution in time domain and precisely obtains time delay values between signals with different wavelength. In OptSim software this TDSS algorithm is realized using finite impulse reaction (FIR) filters. This sophisticated technique provides complete control of an overall mistake that may occur during all process of calculating. By contrast FDSS calculates $\hat{D}$ in frequency domain but firstly for this algorithm is necessary to calculate fast Fourier transformation (FFT) from A[n] signal samples and from h(t) impulse reaction. Then it is necessary to use invers FFT (FFT^{-1}) to convert obtained data array to time sample domain. FDSS algorithm can be mathematically described using following equation [42]:

$$A_L^{'}[n]=A[n]\otimes h[n]=FFT^{-1}\{FFT(A[n])\times FFT(h[n])\} \tag{11}$$

As one can see, then in this case circular convolution is used for obtaining signal sample array $A_L'[n]$. This array may contain fewer samples than it is necessary to obtain actual convolution products—$A_L[n]$ sample array. Hence this algorithm is easier to implement than TDSS and it requires less computation time and resources but serious errors may occur during calculation [42].

Equation (7) can be solved assuming that $\hat{D}$ and $\hat{N}$ operators are independent and fiber span Δz length is small enough (5-100 meters, depending on the simulation accuracy requirements). Then optical signal after propagation over Δz span can be described in the following manner [42]:

$$A(t,\ z+\Delta z) \cong \exp\left[\frac{\Delta z}{2}\cdot\hat{D}\right]\cdot\exp\left\{\Delta z\cdot\hat{N}\left[A\left(t,\ z+\frac{\Delta z}{2}\right)\right]\right\}\cdot exp\left(\frac{\Delta z}{2}\cdot\hat{D}\right)\cdot A(t,\ z) \qquad (12)$$

For the evaluation of system performance will be used such parameter as Q-factor and BER value. Q=7.03 (16.94 dB) corresponds to the commonly used reference for BER of 10^{-12}. Q factor and BER confidence interval magnitude depends on the total number of simulated bits N_{total} [42]:

$$dev\left[Q^*\right] \equiv \sigma_Q \cong \frac{Q}{\sqrt{2\cdot N_{total}}} \qquad (13)$$

where Q is Q-factor value that can be calculated using following Equation (9) [42]:

$$Q = \frac{|\mu_1 - \mu_0|}{\sigma_1 + \sigma_0} \qquad (14)$$

where $\mu_{1,0}$ and $\sigma_{1,0}$ are the mean and the standard deviation of the received signal, when a logical "1" and "0" is transmitted, and $\pi \approx 3.14$ [42].

$$BER = \frac{1}{2}\cdot erfc\left(\frac{Q}{\sqrt{2}}\right) \qquad (15)$$

Using Equation (13) Q-factor uncertainty range can be expressed as [42]:

$$range = 20\cdot\log_{10}\left(\frac{1+\sqrt{\frac{2}{N_{total}}}}{1-\sqrt{\frac{2}{N_{total}}}}\right) \qquad (16)$$

Q-factor uncertainty range for 1024 simulated bits that is used in our schemes is equal to 0.77 dB. For 1024 of simulated bits these intervals are:

$$Q_{for16.94dB} \in \left[16.55;17.31\right],\ \left[dB\right] \qquad (17)$$

$$\log 10\{BER_{\text{for } 10^{-12}}\} \in [-12.97; -11.04] \tag{18}$$

Q-factor and BER value 95% ($\pm 2\sigma_Q$) confidence intervals for 16.94 dB nominal value can be obtained using Equation (13) and Equation (15) assuming that we are dealing with Gaussian distribution.

4. Background of spectrum slicing technique and WDM PON models

Spectrum slicing technique is one of basic techniques available in WDM PON systems in order to reduce the cost of components and simplify the passive network architecture. There incoherent broadband light source (BLS) is sliced and equally spaced multi-wavelength channels is generated [2]. The aim of spectrum slicing is to employ a single BLS for transmission on a large number of wavelength channels (see Fig. 3).

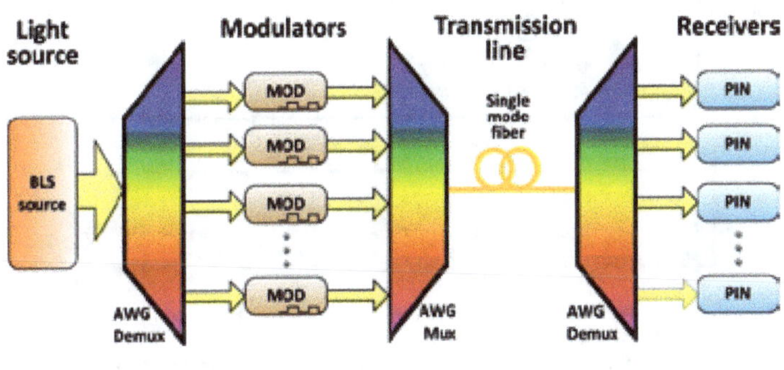

Figure 3. Operational principle of spectrum sliced multi-channel fiber optical transmission system.

At first, BLS is sliced with arrayed-waveguide grating (AWG). Afterwards, optical slices are modulated, multiplexed by second AWG and transmitted over standard single mode optical fiber line. Channels are demultiplexed by third AWG located after SMF and received by direct detection optical receiver. This receiver can be PIN photodiode or avalanche photodiode (APD). Broadband light sources like LED, SLED (superluminescent diode) or ASE can be used in spectrum sliced systems for data transmission. Transmission power available for each separate optical channel depends on the slice width. It should be considered that a larger slice will increase not only the total channel power but also increase the influence of dispersion and therefore the number of available WDM channels [2]. In our research we choose ASE as BLS for spectral slicing because it has the highest optical output power compared to other above mentioned broadband light sources.

4.1. Experimental models of WDM PON transmission system

Erbium doped fiber amplifier (EDFA) emits high power amplified spontaneous emission (ASE) noise in C band (wavelength from 1530 to 1565 nm) and L band (wavelength from 1565 to 1625 nm) if there is no signal to be amplified. This effect is used to design broadband ASE light source in this research. ASE noise generation and gain occurs along all EDFA fiber length and it depends on erbium ion (Er^{3+}) emission and absorption spectrum [14]. Broadband ASE light source realization can be done in different ways: by using one EDFA or connecting several amplifiers in cascade mode (one after another). The latter method allows achieving almost flat ASE output spectrum and higher output power because of better Er^{3+}ions usage along several amplifiers.

In our research the broadband ASE light source is constructed from two EDFAs combined in cascaded mode. The smoothest ASE output spectrum can be achieved if 400 mW output power for all pump lasers is used, where first EDFA amplifier is pumped in co-propagating direction on 1480 nm wavelength, and second EDFA is pumped in co-propagating direction on 1480 nm as well as in counter-propagating direction on 980 nm (see Fig. 4).

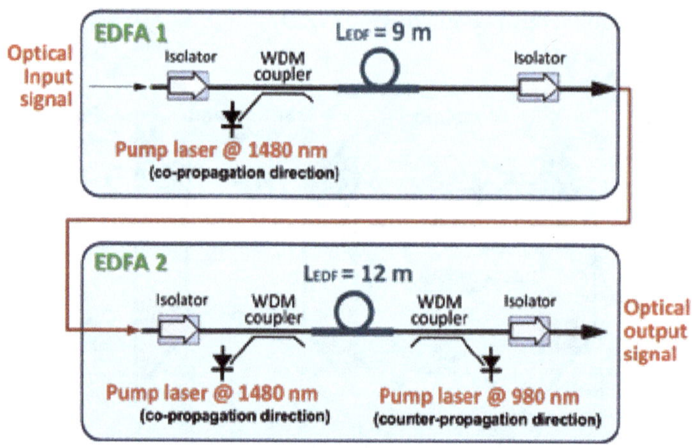

Figure 4. Detailed scheme of cascaded EDFA system with optimized parameters.

Erbium doped fiber span with length of 9 m is used for first EDFA and 12 m long erbium doped span is used for second EDFA. In this manner a broadband ASE source with almost flat spectrum and total output power on the output of cascaded EDFA system about+23 dBm (200 mW) is being constructed. The output spectrum of realized broadband ASE noise-like light source which will be spectrally sliced using AWG unit is shown in Figure below (see Fig. 5).

The highlighted area in figure shows the 1545.322 nm to 1558.983 nm wavelength range (192.3 THz to 194.0 THz frequency, centered on 1552.524 nm wavelength or 193.1 THz in frequency) which will be used for our 8-channel and 16-channel spectrally sliced WDM-PON systems.

As one can see (see Fig. 5), fluctuations of optical power level are minimal and spectrum in this employed region is almost flat.

This section describes simulation scheme of SS-WDM PON system with up to 16 channels, 2.5 Gbit/s transmission speed per channel (total system's capacity is up to 40 Gbit/s) and chromatic dispersion compensation module (DCM).

The performance of simulated scheme was evaluated by the obtained bit error ratio (BER) value of each channel in the end of the fiber optical link. Basis on ITU recommendation G.984.2 it is taken into account that BER value for fiber optical transmission systems with data rate 2.5 Gbit/s per channel must be below 10^{-10}.

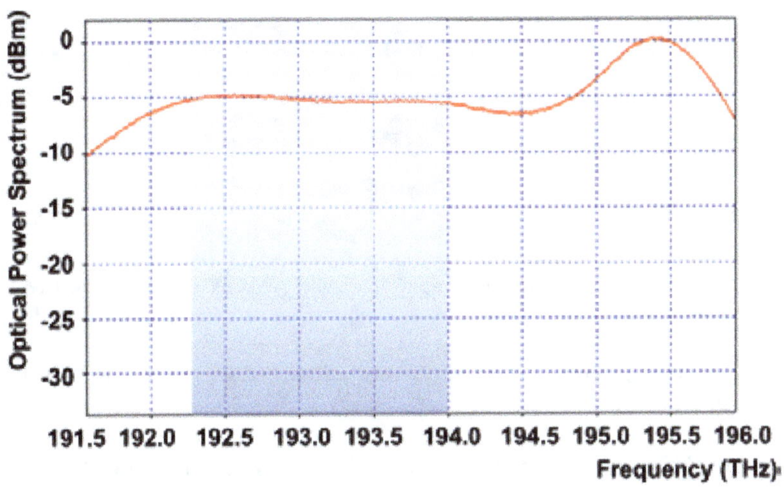

Figure 5. Output spectrum of noise-like ASE broadband light source.

As one can see in Figure 6, SS-WDM PON simulation scheme we built consists of up to 16 channels. Frequency grid is anchored to 193.1 THz and channel spacing is chosen equal to 100 GHz frequency interval. This frequency grid and interval is defined in ITU-T recommendation G.694.1. Broadband ASE light source is spectrally sliced using 8-channel or 16-channel flattop AWG filter (AWG1) with channel spacing equal to 100 GHz (0.8 nm in wavelength). Using this AWG unit we can obtain equally arranged optical channels (slices) with dense channel interval 100 GHz. Insertion losses of AWG units are simulated using additional attenuation blocks (attenuators). It is taken into account that simulated high-performance AWG multiplexers and demultiplexers are absolutely passive optical components with insertion loss up to 3 dB each. After spectrum slicing operation implemented by AWG1, optical slices are transmitted to optical line terminals (OLTs). OLTs are located at central office (CO). Each OLT consists of data source, non-return-to-zero (NRZ) driver, and external Mach-Zehnder modulator (MZM). Each MZM has 5 dB insertion losses, 20 dB extinction ratio, modulation voltage V_π of 5 Volts and maximum transmissivity offset voltage 2.5 Volts.

Generated bit sequence from data source is sent to NRZ driver where electrical NRZ pulses are formed. Afterwards formed electrical NRZ pulses are sent to MZM modulator. MZM modulates optical slices received from AWG1 and forms optical signal according to electrical drive signal. These formed optical pulses from all OLTs are coupled by AWG multiplexer

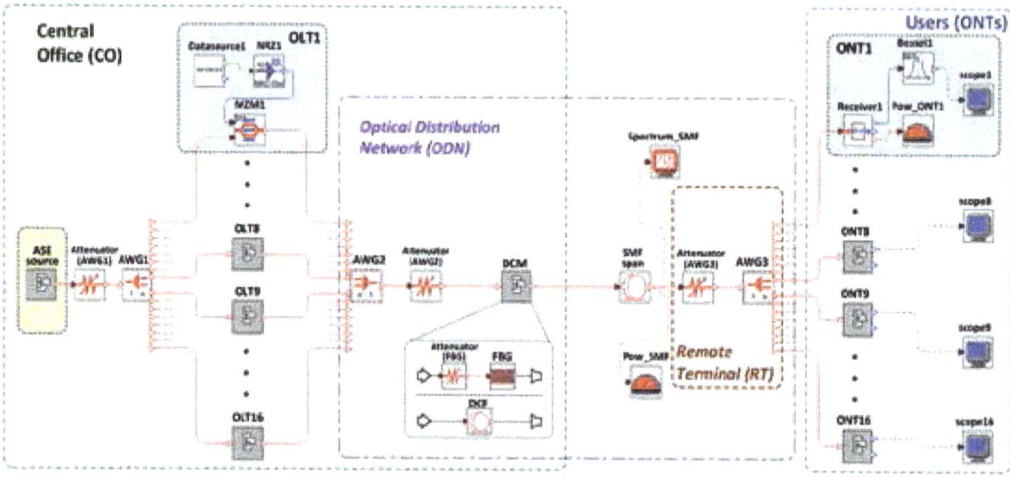

Figure 6. Simulation model of proposed high-speed up to 16-channel AWG filtered and ASE seeded SS-WDM PON system with DCM module.

(AWG2) and sent into standard optical single mode fiber (SMF) defined in ITU-T recommendation G.652 [12].

Information from OLT is transmitted to an optical network terminal (ONT) or user over the fiber optical transmission link called optical distribution network (ODN). ODN consists of AWG multiplexer, two optical attenuators, dispersion compensation module (DCM), SMF with length up to 20 km and two AWG demultiplexers. SMF fiber used in simulation setup has large effective core area A_{eff}=80 μm^2, typical attenuation α=0.2 dB/km, dispersion coefficient D=16 ps/(nm km) and dispersion slope D_{sl}=0.07 ps/(nm^2 km) at the reference wavelength λ=1550 nm.

As one can see in middle of Figure 6, two simulated CD compensation methods implemented in DCM module are FBG and DCF. In our research we use DCF fiber with A_{eff}=20 μm^2, attenuation coefficient α=0.55 dB/km, dispersion coefficient D=-80 ps/(nm km) and dispersion slope D_{sl}=0.19 ps/(nm^2 km) at the reference wavelength λ=1550 nm. For full compensation of chromatic dispersion accumulated by 10 km of SMF fiber we need about 2 km of DCF fiber. Simulated FBG is tunable in terms of both reference frequency and total dispersion value which can be compensated. Additional attenuator is used for simulation of FBG's 3 dB insertion loss. For the most accurate results we used real parameters of standard DCF fiber and tunable FBG in our simulation setup. In the end of fiber optical link 8 or 16 optical channels are separated using AWG demultiplexer (AWG3) which is located in remote terminal (RT). Receiver section includes ONT units. Each ONT consists of sensitivity receiver with PIN photodiode (sensitivity S=-25 dBm at sensitivity reference error probability BER=10^{-10}), Bessel electrical lowpass filter (3-dB electrical bandwidth B_E=1.6 GHz), optical power meter and electrical probe to evaluate the quality of received optical data signal (capture eye diagram or spectrum). Optical signal is converted to electrical signal using PIN photodiode and filtered with Bessel electrical filter for noise reduction.

4.2. Advanced model of SS-WDM PON system

This section describes simulation scheme of more advanced, bidirectional 8-channel SS-WDM PON system with symmetrical 10 Gbit/s transmission speed per channel, gain saturated SOAs before optical modulators and chromatic dispersion compensation module (DCM) in downstream direction. For cost saving purposes DCM is not implemented in upstream direction because the influence of CD on 20 km long optical transmission line, where conventional laser based system (e.g. distributed feedback laser) is used may be considered as negligible [7]. Total system's capacity in downstream as well as in upstream direction is up to 80 Gbit/s.

Simulation scheme consists of 8 channels in downstream direction (see Fig. 7(a)) and 8 channels in upstream direction (see Fig. 7(b)). Channel spacing for both directions is chosen equal to 100 GHz frequency or 0.8 nm wavelength intervals. As one can see in Figure 7(a), broadband ASE light source, is spectrally sliced using 8-channel flattop AWG filter (AWG1) with channel spacing equal to 100 GHz. Using this AWG unit we can obtain equally arranged optical slices with dense channel interval. Insertion losses of AWG units are simulated using additional attenuation blocks. It is taken into account that simulated AWG multiplexers and demultiplexers are absolutely passive optical components with insertion loss up to 3 dB per unit. After spectrum slicing operation realized by first AWG1, optical slices are transmitted to optical line terminals (OLTs). OLTs are located at central office (CO).

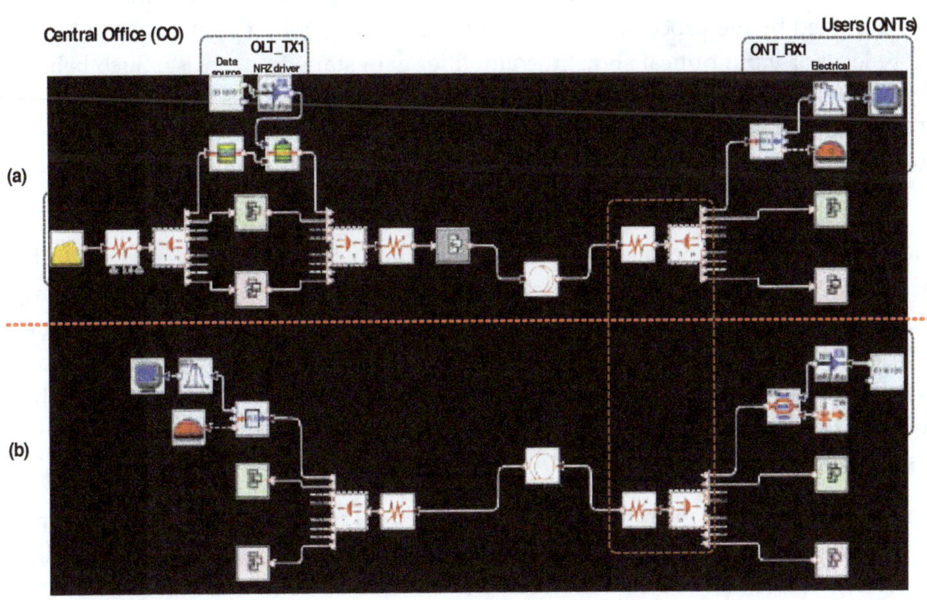

Figure 7. Setup of proposed bidirectional 8-channel SS-WDM PON system with data transmission speed 10 Gbit/s per channel. In (a) is shown downstream direction which is realized by using spectrum-slicing technique, and in (b) upstream direction where conventional laser based solution is used.

Each transmitter (Tx) of OLT consists of data source, non-return-to-zero (NRZ) driver, semiconductor optical amplifier (SOA) and polarization-insensitive electro-absorption modulator (EAM). Compared with polarization-sensitive Mach–Zehnder modulator (MZM), the use of EAM gives a 3-dB gain in optical power and signal to noise ratio (SNR) [43]. Each EAM has 3 dB insertion loss and maximum applied voltage is 5 Volts. SOA is gain saturated and amplifying incoming carrier spectral slice and used for suppression of excess intensity noise (EIN) which comes from the spontaneous–spontaneous beating between different wavelength components of the spectrally sliced incoherent broadband ASE light source [43].

Generated bit sequence from data source is sent to NRZ driver where electrical NRZ pulses are formed and sent to EAM to drive it. Each EAM modulate amplified incoming spectral slices according to electrical drive signal and forms optical pulses. These formed optical pulses from all OLT transmitters (OLT_TX1 to OLT_TX8) are coupled by AWG multiplexer (AWG2) and sent into standard 20 km long SMF fiber.

We also used additional gain saturated semiconductor optical amplifier (SOA) before modulator to suppress excess intensity noise (EIN) which originates from the ASE source [43].

In general SOAs, due to their structure can be characterized as a semiconductor lasers without feedback, as the same mechanism is used in order to amplify incident light as in laser diodes. Semiconductor optical amplifiers are the type of amplifiers in which electrical energy is applied as a pump to achieve population inversion in the active region, and amplification is achieved via the process of stimulated recombination luminescence. When the intensity of the input optical signal increases, at a certain point the whole carrier population of the upper energy level is occupied by the process of amplification. Therefore, for a specific SOA at the certain intensity level of input optical signal the amplifier gain starts to decrease. Such behavior of semiconductor amplifiers is known as gain saturation. When gain saturation occurs, characteristics of the gain become nonlinear (device gain becomes inversely proportional to the input intensity). Therefore, intensity noise suppression of input optical signal takes place when operating in such a nonlinear mode.

The main idea of intensity noise suppression is following: when intensity of input optical signal is lower, saturated SOA will provide higher optical gain coefficient than during the moment when input optical signal reaches its maximal value. Therefore, on the output of SOA amplitude difference between intensity drops and peaks will not be so large when signal is observed in time domain, and obtained optical radiation will be more similar to the output of continuous wavelength lasers, which are normally used as carriers in cases when external optical signal intensity modulation is used.

Information from OLT transmitter part is sent to receiver part of optical network terminals (ONT_RX1 to ONT_RX8) over the fiber optical transmission link referred as optical distribution network (ODN). In our case for both transmission directions ODN consists of AWG multiplexers/demultiplexers, optical attenuators, and 20 km long SMF spans. In downstream direction FBG DCM is used for chromatic dispersion compensation. It is tunable in terms of both reference frequency and total dispersion value which can be compensated. Additional attenuator is used for simulation of FBG's insertion loss which is set 3 dB. In the end of downstream fiber optical link 8 optical channels are separated by using AWG demultiplexer (AWG3) which is located in remote terminal (RT).

Each ONT receiver (ONT_RX1 to ONT_RX8) consists of sensitivity receiver with PIN photodiode (sensitivity is-19 dBm at BER=10^{-12}), Bessel electrical lowpass filter (3-dB bandwidth is 7.5 GHz), optical power meter and electrical scope to evaluate the quality of received signal (e.g. show eye pattern). Optical signal is converted to electrical signal using PIN photodiode and filtered with Bessel electrical filter for noise reduction.

As one can see in Fig. 7(b), which represents upstream optical transmission, conventional laser based system (where each user has its own individual light source) is realized. Each user's transmitter (ONT_TX1 to ONT_TX8) has its own continuous wavelength light source modulated by Mach-Zehnder modulator. Upstream transmission from user side to OLT receiver block (OLT_RX1 to OLT_RX2) is realized over separate 20 km long optical fiber span. OLT receiver structure is the same as for ONT receiver. As it mentioned before, dispersion compensation is not used in upstream transmission for cost reduction and network architecture's complexity reduction purposes.

In the simulation model it was decided to set all geometrical and material parameters of the active layer of the SOA as it was used previously in [43]. There these parameters were set as in a standard InGaAlAs travelling wave SOA with negligible residual facet reflectivity.

The only parameters which values were changed were the length of active layer and input bias current. In the proposed solution it was required to obtain gain saturation of the amplifier for input optical power of at least 5.5 dBm. Therefore, it was required to reduce the maximal optical gain that can be provided by the SOA. This was achieved by shortening the length of active layer of the SOA from 750 μm to 100 μm, in such a way significantly reducing the population of available carriers in the active layer, hence also lowering the value of amplifier's maximal gain. It was required to set appropriate value of input bias current to obtain such level of population inversion in active layer, which would provide BER values below the 10^{-12} threshold for detected signals in all ONTs.

Best results were obtained if pumping current of 370 mA was used for our SOAs located in OLTs transmitters. As mentioned in introduction part, when intensity of input optical signal is lower, saturated SOA will provide higher optical gain coefficient than during the moment when input optical signal reaches its maximal value. Instantaneous optical power at input and output of the SOA is represented in Figures 8(a) and 8(b) respectively.

It can be clearly seen that after processing through the SOA intensity fluctuations of the spectrally sliced optical flow have been reduced significantly. Unfortunately not all of the intensity noise was suppressed. The provided solution of intensity noise suppression applies only to the intensity noise which frequency is low in respect to the spontaneous carrier lifetime. The SOA carrier changes are rate limited by their lifetimes, and carrier changes cannot keep up with the high frequency intensity changes. Suppression of intensity noise, which frequency is high in respect to the carrier lifetime, requires more detailed investigation and a different approach for the SOA configuration [20].

Spectrum at the input and output of the SOA amplifier can be observed in Figure 9. By comparing both curves, it can be seen that spectrum of spectral slice (i.e. spectrum of spectral sliced carriers) at the output of the SOA is clearly broadened. Also, upper part (peak) of the

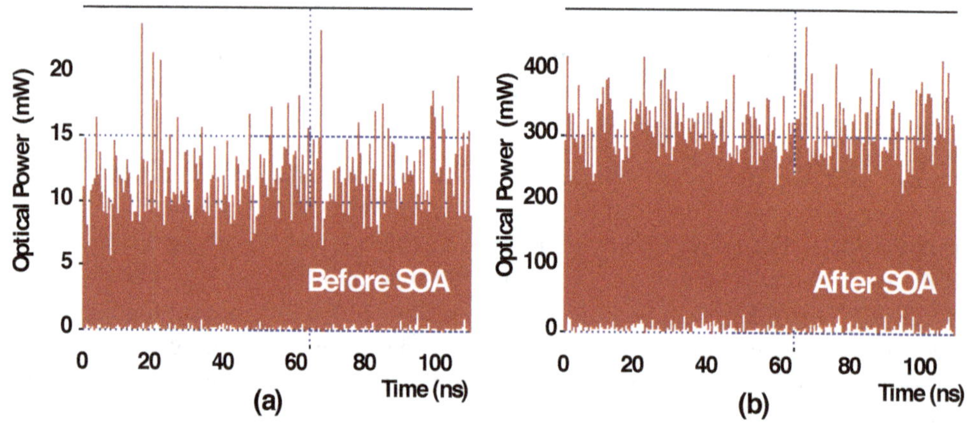

Figure 8. Time diagram of optical signal: (a) before SOA and (b) on the output of SOA.

spectral slice obtained a visible slope with the intensity peak shifted towards to lower frequency.

The spectral broadening of spectral slices, each containing multiple carriers, is caused by occurrence of two non-linear effects which became well seen as the SOA began to operate in the gain saturated mode: intra-channel four wave mixing (FWM) and self-phase modulation (SPM) of the carriers. Appearance of slope at the peak of spectral-sliced carriers is related to SPM, as phase modulation also has the effect of shifting the peak power toward longer wavelengths (lower frequencies) as the light travels through the SOA [43].

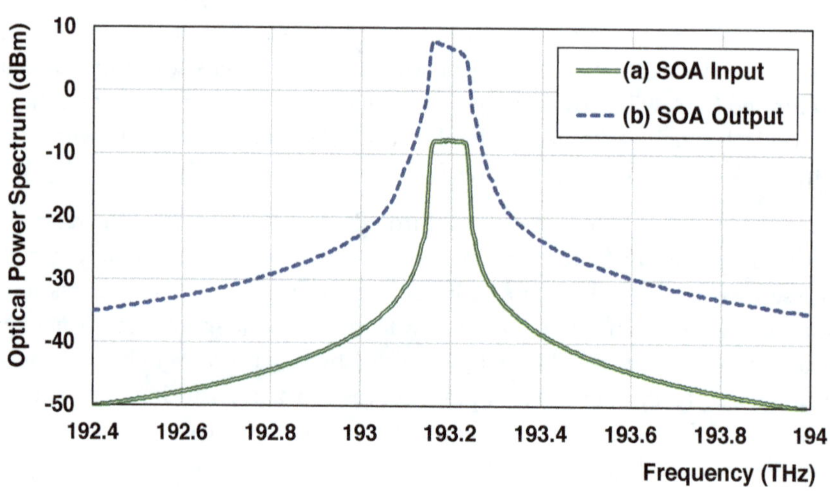

Figure 9. Optical spectra of spectral slices filtered by AWG1 demultiplexer: (a) before SOA and (b) on the output of SOA after amplification.

In Figure 10 it is shown spectrum of downstream optical signals on the output of all OLT transmitters (before multiplexing by AWG2 unit) and spectrum in the input of ONTs (after demultiplexing by AWG3).

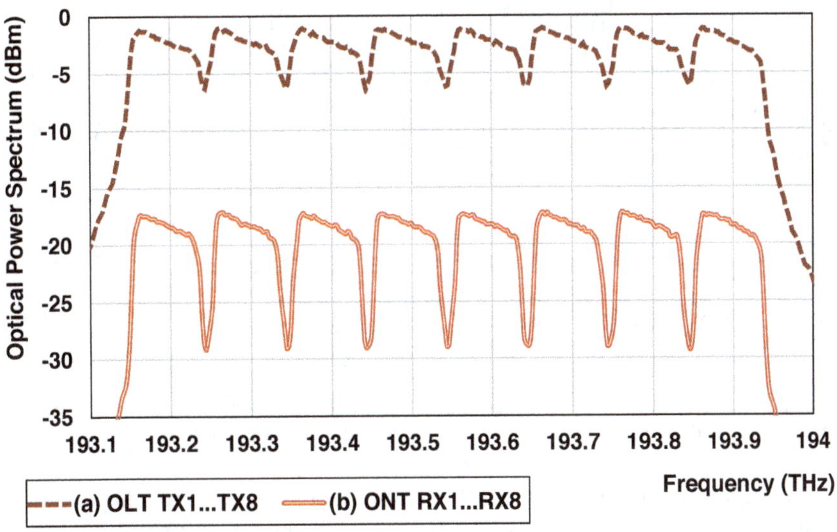

Figure 10. Optical power spectra of transmitted signals for spectrum-sliced WDM-PON downstream direction: (a) on the output of OLT transmitters and (b) on the input of ONT receivers (users) after 20 km transmission.

We found that optimal 3-dB bandwidth value of flat-top type AWG units for maximal quality of received signal and acceptably low crosstalk between channels must be about 90 GHz for spectrum sliced downstream and 50 GHz for laser based upstream system [7]. In case of ASE spectral slicing flat-top type AWG units shows good filtering performance because of its amplitude transfer function which provides good WDM channel separation at the same time passing sufficiently high optical power. The larger is width of spectral slice the better performance we can obtain. However, arising crosstalk between channels must be taken into account in this case [7]. It means that there is a tradeoff between optical filter bandwidth and crosstalk between spectrum sliced channels which can result in system's performance limitations.

In Figure 11 is shown spectrum of optical signals on the output of laser based ONT transmitters and in the input of OLT receivers after demultiplexing by AWG4 unit. As one can see all 8 upstream channel are well separated and any major spectral distortions is not observed. It must be mentioned, that for CD compensation and performance as well as network reach improvement of proposed 8-channel bidirectional SS-WDM PON system fiber Bragg grating dispersion compensation module (FBG DCM) was used. It is seen in Figure below (see Fig. 12(a)) that without CD compensation system performance is poor and successful data transmission is not possible.

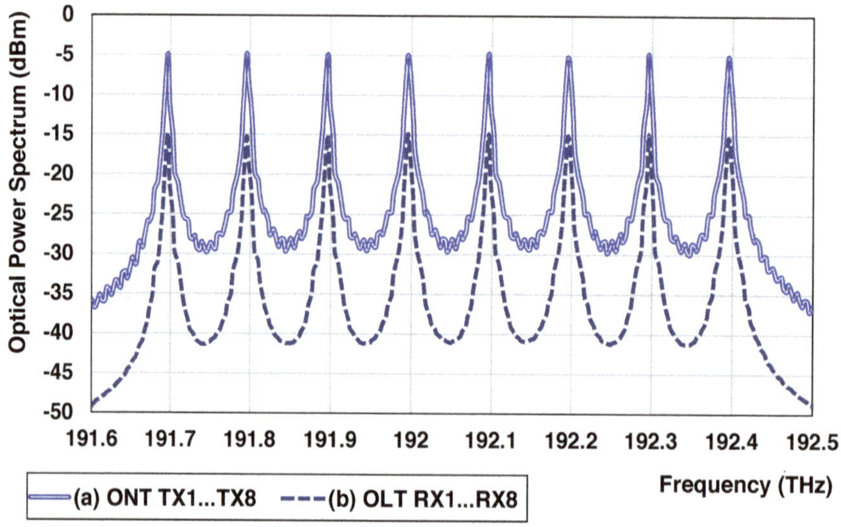

Figure 11. Optical power spectra of transmitted signals for laser based WDM-PON upstream direction: (a) on the output of ONT transmitters and (b) on the input of OLT receivers after 20 km transmission.

Theoretical value of accumulated CD for 20 km SMF fiber span is about 320 ps/nm. We found that optimal CD amount which must be compensated by FBG DCM for downstream 20 km SMF span is 310 ps/nm. Usage of gain saturated SOA for noise suppression was another key element to build spectrum sliced WDM-PON system operating below desired BER threshold. As it is shown in Figure 12(b), without gain saturated SOA downstream signal performance already after transmitter (back-to-back (B2B) measurement) is seriously affected by noise and BER is well above defined threshold (BER<10^{-12}).

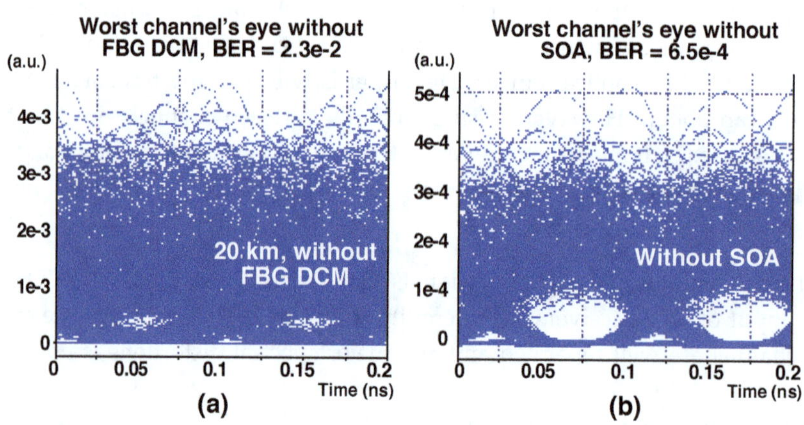

Figure 12. Received eye diagrams of downstream signals for bidirectional 8-channel spectrum sliced WDM-PON system: (a) after 20 km without FBG DCM for CD compensation and (b) without use of SOA for noise suppression (B2B measurement).

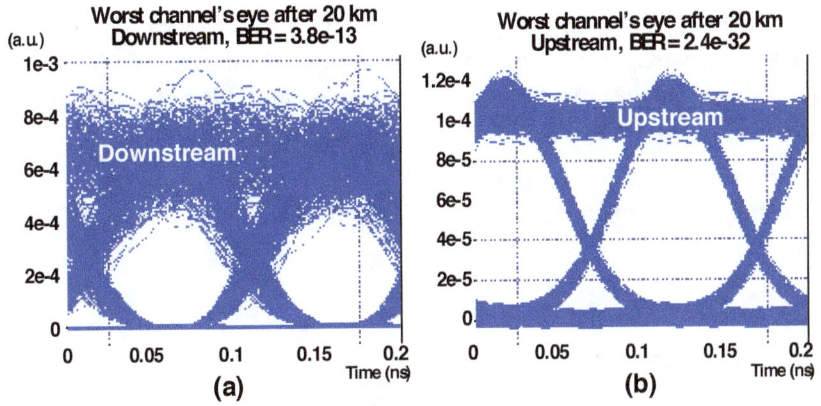

Figure 13. Received eye diagrams for bidirectional 8-channel WDM-PON system after 20 km transmission: (a) spectrum sliced downstream using SOA before modulator as well as FBG DCM for CD compensation, and (b) upstream signal eye pattern without additional DCM.

When SOA amplifier and FBG DCM were used for performance improvement of downstream direction, quality of obtained signal in receiver was sufficient and BER was below minimal threshold (see Fig. 13(a)). It is clearly seen that noise at logical "1" level is suppressed and eye opening became noticeably larger. It is due to the fact that gain saturated SOA is amplifying low intensity signal more than comparatively high signal and compress intensity fluctuations existing in input light signal, which in our case is spectrally sliced ASE. In Figure 13(b) we see, that eye diagram of upstream signal is quite good and eye is wide open. Some signal distortion due to CD is also seen, but for transmission distance of 20 km it can be considered as negligible.

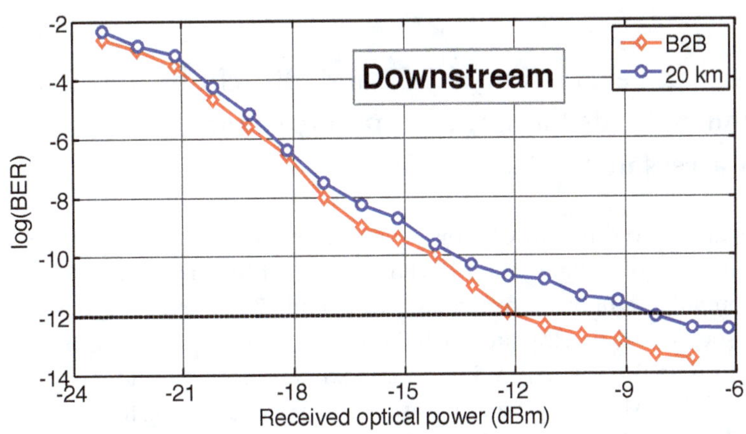

Figure 14. Measured BER performance of spectrum sliced downstream light signals in B2B configuration and after transmission through 20 km SMF fiber span.

For evaluation of downstream spectrum sliced system performance BER curve was measured (see Fig. 14). It can be observed, that power penalty to receive optical signal for 8-channel spectrum sliced downstream with desired BER<10^{-12} after 20 km transmission is 3.7 dB. This penalty is introduced by the cross talk effects, dispersion and due to the noise-like nature of broadband ASE light source.

In upstream transmission direction (see Fig. 15) minimal received power to obtain BER<10^{-12} must be more than-18.9 dBm for B2B configuration and-17.6 dBm for 20 km SMF transmission line.

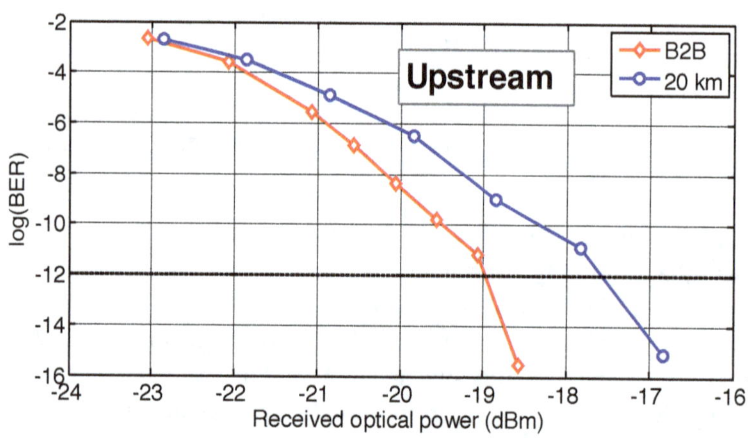

Figure 15. Measured BER performance of separate laser based upstream light signals in B2B configuration and after transmission through 20 km SMF fiber span.

As one can see in Figure 15, power penalty is about 1.3 dB, which can be explained by the influence of CD which broadens optical pulses as they travel along the optical fiber span.

5. Comparison and evaluation of DCF and FBG chromatic dispersion compensation methods for usage in up to 16 channel spectrum sliced WDM-PON access networks

There are compared two different CD compensation methods for improvement of maximal reach and performance of 8-channel and 16-channel AWG filtered SS-WDM PON system with flattened ASE broadband light source. In Figures 16 and 17 it is shown optical spectra on the output of ASE source and spectra after each flat-top AWG unit. We found that optimal 3-dB bandwidth value of flat-top type AWG unit for maximal system performance must be about 90 GHz for both SS-WDM PON systems. In case of ASE spectral slicing flat-top type AWG unit shows good filtering performance and high OSNR because of its amplitude transfer function that provides good WDM channel separation at the same time passing sufficient high optical power [12].

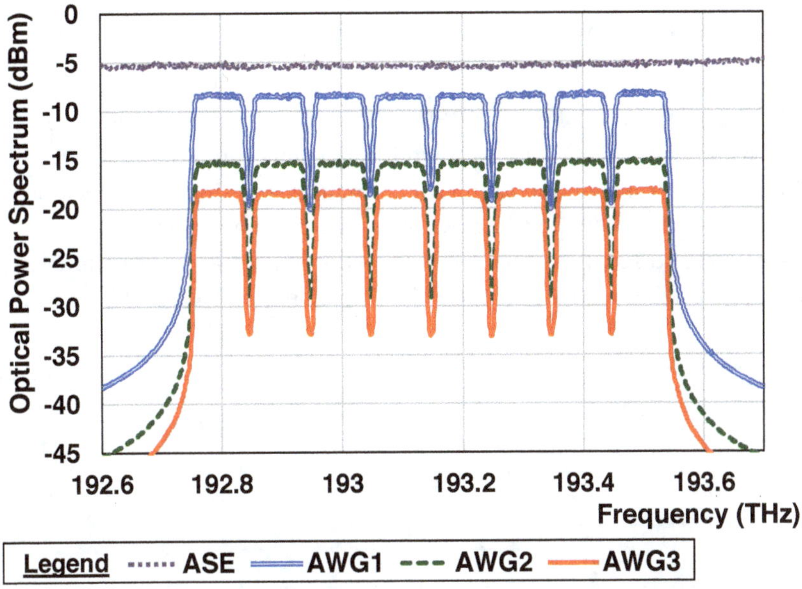

Figure 16. Optical output power spectrum of ASE source and 8-channel SS-WDM PON system after each stage of flat-top type AWG unit.

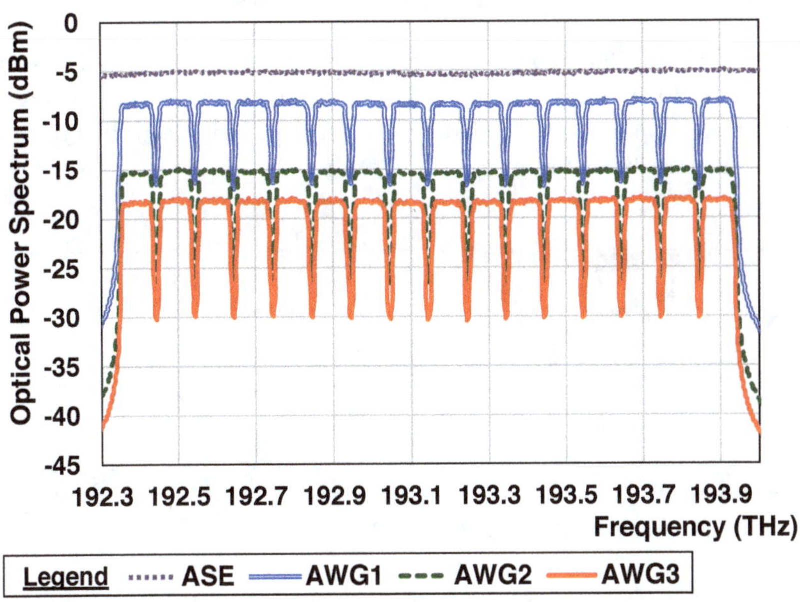

Figure 17. Optical output power spectrum of ASE source and 16-channel SS-WDM PON system after each stage of flat-top type AWG unit.

First, we studied both SS-WDM PON systems in back to back (B2B) configuration (without DCM and SMF fiber span in ODN) which was the reference to which we compared all other measurements.

In Figures 18(a) and 19(a) we show the BER versus optical received power for 2.5 Gbit/s 8-channel and 16-channel SS-WDM PON downlink transmission in B2B configuration as well as after 10 km 20 km transmission without chromatic dispersion compensation. In Figures 18(b) and 18(c), as well as 19(b) and 19(c) we show the BER versus optical received power for 2.5 Gbit/s 8-channel and 16-channel SS-WDM PON downlink transmission after 10 km and 20 km transmission realizing CD compensation by implementing DCF and FBG respectively.

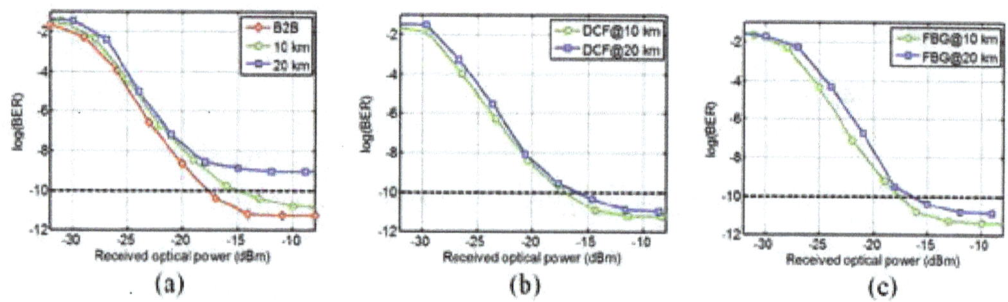

Figure 18. BER versus received optical power at the receiver for 8-channel SS-WDM PON system (a) without CD compensation, (b) using DCF for CD compensation, (c) using FBG for CD compensation.

As one can see in Figure 18(a), power penalty to receive optical signal for 8-channel system with BER<10^{-10} after 10 km transmission is 2.7 dB. The minimal BER value of received signal after 20 km transmission is above defined BER threshold and it means that qualitative data transmission is not possible in this case.

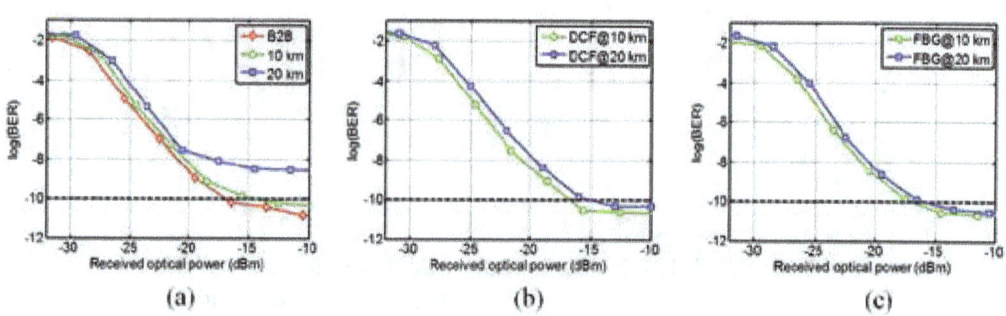

Figure 19. BER versus received optical power at the receiver for 16-channel SS-WDM PON system (a) without CD compensation, (b) using DCF for CD compensation, (c) using FBG for CD compensation.

In Figure 20 are shown eye diagrams of received signal in B2B configuration (BER<10^{-10}) and after 20 km long SMF span (here BER>10^{-10}) without CD compensation.

Theoretical value of accumulated CD for 10 km SMF fiber span is about 160 ps/nm and 320 ps/nm for 20 km span. We found that optimal CD compensation amount that must be compensated by FBG for 10 km span is 125 ps/nm for both 8-channel and 16-channel systems. For 20 km SMF span we must compensate 290 ps/nm in 8-channel system and 280 ps/nm in 16-channel system. In order to make optimal CD compensation in 8-channel system we used 2.5 km DCF fiber before 10 km long SMF fiber span, and 4.7 km DCF fiber before 20 km span. For 16-channel system it was used 2 km long DCF before 10 km span and 4.5 km DCF before 20 km SMF span. In case of FBG it was used incomplete CD compensation, but in case of DCF the optical line was overcompensated.

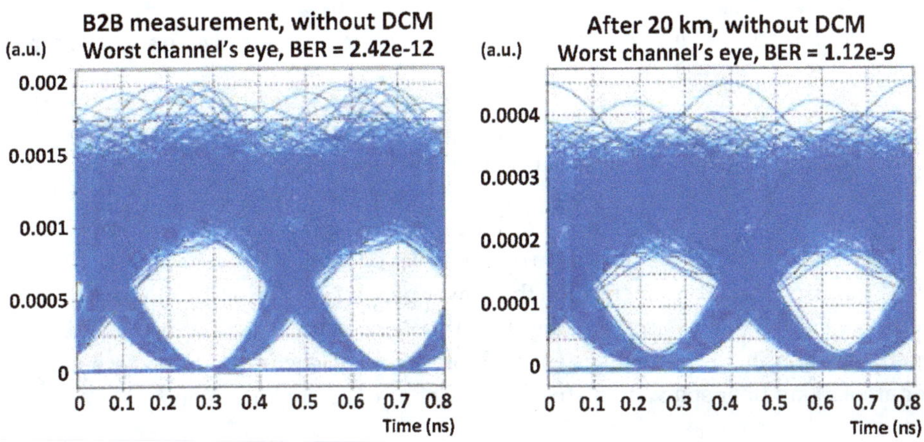

Figure 20. Eye diagrams and BER values of received signals from worst channels of 8-channel SS-WDM-PON scheme without DCM module for measured B2B signal and signal after 20 km transmission.

Usage of DCF for accumulated CD compensation provides extension of network reach from 10 km to 20 km as well as improvement of network performance for both line lengths, (see Fig. 18(b)). In this figure it is seen that, for a BER of 10^{-10} using DCF for CD compensation, the power penalty to pass from 10 km transmission to 20 km transmission is 1.3 dB. This penalty is introduced by the cross talk effects, dispersion and due to the noise-like nature of broadband ASE light source. The power penalty for BER of 10^{-10} to pass from 10 km to 20 km is 1.1 dB in case when FBG is used for compensation of accumulated CD for reach extension and performance improvement of proposed 8-channel SS-WDM PON system, (see Fig. 18(c)). The comparison between DCF and FBG in terms of performance improvement shows that FBG provides higher performance improvement.

After investigation of 8-channel SS-WDM PON system the simulation model of 16-channel system was realized. Results show that the overall performance of this system is slightly lower than 8-channel system using the same ASE source. As one can see in Figure 19(a), power penalty to receive optical signal after 10 km with defined BER is 2.8 dB. Performance of transmission system is too low to provide data transmission over 20 km SMF fiber span with BER$<10^{-10}$ (see Fig. 21).

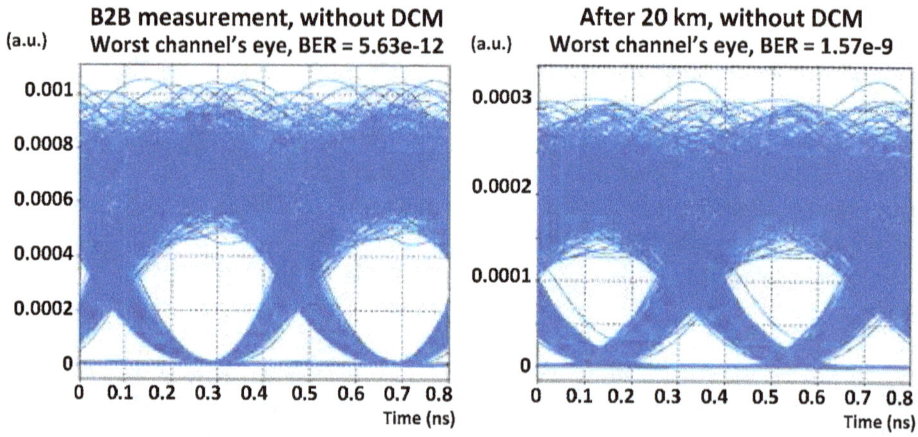

Figure 21. Eye diagrams and BER values of received signals from worst channels of 16-channel SS-WDM-PON scheme without DCM module for measured B2B signal and signal after 20 km transmission.

It is shown (see Fig. 19(b)) that usage of DCF fiber for CD compensation improve access system's performance and it is capable to operate over 20 km SMF span with BER<10^{-10}. In case when DCF is used for CD compensation the power penalty for BER of 10^{-10} to pass from 10 km to 20 km is 1.5 dB. If FBG is used (see Fig. 19(c)) power penalty is 1 dB. When FBG is used for CD compensation in 16-channel system BER of 10^{-10} is reached at-17 dBm received optical power after 10 km fiber span and at-16 dBm power after 20 km long span.

When DCF fiber is used for CD compensation minimal received power must be-16.7 dBm for 10 km SMF fiber span and-15.2 dBm for 20 km fiber span or transmission line. As it is seen from obtained results performance between FBG and DCF for CD compensation and system efficiency improvement of 8-channel and 16-channel SS-WDM PON systems are close. However, FBG provides better CD compensation performance. Instead of DCF fiber it can be used at higher optical powers without inducing nonlinear optical effects and additional signal distortion which can occur due to small effective core area of a DCF fiber.

6. Conclusion

We report on causes of latency in fiber optical networks and solutions how to minimize it. Latency is a critical parameter in a wide set of applications like financial transactions, telemedicine, cloud services and other real-time applications which requires optical transmission line with the smallest available time delay. It was shown that latency can be greatly reduced by replacing conventional network components with low-latency components and by optimizing optical network path trying to keep it short as possible. Future telecommunication networks must be cost and energy efficient, in the same time supporting high bandwidths and low latency. We showed the realization of cost effective up to 16-channel dense WDM-PON system where spectrum sliced spectrally uniform ASE source is used as a light source. It is shown the design of this broadband spectrally-uniform ASE source with+23 dBm output power in system's operating wavelength range by using two EDFA amplifiers connected in cascade mode. We also demonstrated excess intensity noise (EIN) suppression by using additional SOA in transmitter section as well as experimentally compared two different chromatic compensation methods, namely dispersion compensation fiber (DCF) and fiber Bragg grating (FBG) for implementation of signal performance improvement in high-speed spectrum sliced optical access systems.

Author details

Vjaceslavs Bobrovs*, Jurgis Porins, Sandis Spolitis and Girts Ivanovs

Riga Technical University, Institute of Telecommunications, Latvia

References

[1] Spolitis S., Ivanovs G. Extending the reach of DWDM-PON access network using chromatic dispersion compensation. In: Proceedings of 2011 IEEE Swedish Communication Technologies Workshop, SWE –CTW 2011, Stockholm, Sweden, pp. 29-33, 2011.

[2] El-Sahn Z.A., Mathlouthi W., Fathallah H., LaRochelle S. and Rusch L.A. Dense SS-WDM over legacy PONs: smooth upgrade of existing FTTH networks. In: Journal of Lightwave Technology, Vol. 28, No. 10, 1485-1495, May15, 2010.

[3] Cisco Inc. Cisco visual networking index: Forecast and methodology, 2012–2017. (2013).

[4] Yuang M.C., Hsu D.Z., Tien P.L., Chen H.Y., Wei C.C., Chen S.H., and Chen J. An energy and cost efficient WDM/OFDMA PON system: design and demonstration. In: Journal of Lightwave Technology, vol.31, issue.16, pp.2809-2816, 2013.

[5] Ruffini M., Doyle L., Payne D., Mehta D., O'Sullivan B. and Quesada L. Deployment case studies of an energy efficient protected LR-PON architecture. In: Proceedings of 16th International Conference on Optical Network Design and Modeling (ONDM), pp.1-6, 2012.

[6] Newaz S.H.S., Laghari K., Myoung L.G., Crespi N., and SeongGon C. Energy efficient shared WDM-PON. In: Proceedings of 14th International Conference on Advanced Communication Technology (ICACT), pp.1017-1020, 2012.

[7] Spolitis S., Bobrovs V., Ivanovs G. Realization of combined chromatic dispersion compensation methods in high speed WDM optical transmission systems. In: Journal of Electronics and Electrical Engineering, vol. 10, issue 116, pp. 33-38, 2011.

[8] Wong M. A., Zulkifli N., Idrus S. M. Analysis of the next generation green optical access networks. In: Proceedings of the 18th Asia-Pacific Conference on Communications (APCC 2012), pp. 498-503, 2012.

[9] Bobrovs V., Udalcovs A., Spolitis S., Ozolins O., Ivanovs G. Mixed chromatic dispersion compensation methods for combined HDWDM systems. In: Proceedings of International Conference on Broadband and Wireless Computing, Communication and Applications (BWCCA), pp. 313–319, 2011.

[10] Choi B.H., Lee S.S. The effect of AWG-filtering on a bidirectional WDM-PON link with spectrum-sliced signals and wavelength-reused signals. In: Optics Communications, vol. 284, issue 24, pp. 5692–5696, 2011.

[11] ITU-T Recommendation G.984.2. Gigabit-capable passive optical networks (GPON): Physical media depend (PMD) layer specification, 2003.

[12] Spolitis S., Bobrovs V., Ivanovs G. Investigation of high-speed AWG filtered spectrum-sliced WDM PON system. In: Proceedings of 2012 8th International Symposium on Communication Systems, Network and Digital Signal Processing, CSNDSP 2012, Poznan, Poland, 2012.

[13] Kay R. Pragmatic Network Latency Engineering Fundamental Facts and Analysis. cPacket Networks White Paper, 2009.

[14] O3b Networks Ltd. What is Network Latency and Why Does It Matter? 2011 http://www.o3bnetworks.com/media/40980/white%20paper_latency%20matters.pdf

[15] Lightower Fiber Networks. Low Latency in Every Industry. White paper. 2013. www.lightower.com/wp-content/uploads/2011/10/Lightower-Whitepaper-Low-Latency-in-Every-Industry.pdf

[16] Fujitsu. Low-latency networks for trading infrastructure. Whitepaper. 2011, www.fujitsu.com/downloads/TEL/fnc/whitepapers/LowlatencyTrading.pdf

[17] Nordell S., Network latency – how low can you go? Lightwave Magazine, November 1, 2012. http://www.lightwaveonline.com/articles/print/volume-29/issue-6/feature/network-latency-how-low-can-you-go.html

[18] Stephen H., Breaking Barriers to Low Latency. Lightwave Magazine, June 1, 2010. www.lightwaveonline.com/articles/print/volume-27/issue-6/features/breaking-barriers-to-latency-tradeoffs.html

[19] Theodoras J. ADVA Optical Networking, Ultra Low-Latency Financial Networking, 2012. http://www.advaoptical.com/~/media/Resources/White%20Papers/WP_Low-latency.ashx

[20] New Paradigm Resources Group (NPRG). The Low-Latency Services Market, 2010.

[21] http://www.nprg.com/Media/PDF/130-low-latency-services

[22] MRV Communications Inc. Wall street's dream – trading at the speed of light, 2013

[23] http://www.interlinkweb.com/systemics/email/MRV-AN-LowLatencyWallSt.pdf

[24] Corvil Insight Inc., "Low-Latency Market Data", White Paper, 2010. www.cisco.com/web/strategy/docs/finance/corvil_Latency_mkt_data.pdf

[25] Lam S., Low latency drives HK data center services for financials, 2013 www.computerworld.com/s/article/9237877/Low_latency_drives_HK_data_center_services_for_financials

[26] Mazzarese D. Minimizing latency in long-haul networks. Lightwave Magazine, October 1, 2011 www.ofsoptics.com/press_room/media-pdfs/Minimizing-Latency-Lightwave-9-2011.pdf

[27] Kawanishi T., Kanno A., Yoshida Y., Kitayama K., Impact of wave propagation delay on latency in optical communication systems. In: Proc. SPIE Photonics West, Optical Metro Networks and Short-Haul Systems V, December 26, 2012.

[28] Stephen H. Breaking Barriers to Low Latency. Lightwave Magazine, June 1, 2010. www.lightwaveonline.com/articles/print/volume-27/issue-6/features/breaking-barriers-to-latency-tradeoffs.html

[29] Jay A. John. Low Signal Latency in Optical Fiber Networks. Corning Optical Fiber. In: Proceedings of the 60th IWCS Conference, 2011.

[30] Corning Inc. Explanation of the Sources of Variation in Optical Fiber Effective Group Index of Refraction Values, White paper, 2012.

[31] Corning Inc. Corning SMF-28e+optical fiber specification datasheet, 2013.

[32] Poletti F., Wheeler N.V., Petrovich M.N., Baddela N., Numkam Fokoua E., Hayes J.R., Gray D.R., Li Z., Slavik R. and Richardson D.J. Towards high-capacity fibre-optic communications at the speed of light in vacuum. In: Nature Photonics 7, 279–284, 2013.

[33] Freiberger M., Templeton D., Mercado E. Low latency optical services. In: Optical Fiber Communication Conference and Exposition (OFC/NFOEC), 2012 and the National Fiber Optic Engineers Conference, vol., no., pp.1,3, 4-8 March 2012.

[34] Jung, Y., Sleiffer V.A.J.M., Baddela N., Petrovich M.N., Hayes J.R., Wheeler N.V., Gray D.R., Numkam Fokoua E., Wooler J.P., Wong N.H., Parmigiani F., Alam S.U., Surof J., Kuschnerov, M., Veljanovski V., de Waardt H., Poletti, F. and Richardson D.J. First demonstration of a broadband 37-cell hollow core photonic bandgap fiber and its application to high capacity mode division multiplexing. In: Optical Fiber

Communication Conference and Exposition and the National Fiber Optic Engineers Conference (OFC/NFOEC), 2013, pp.1,3, 17-21 March, 2013.

[35] Roberts P.J., Couny F., Sabert H., Mangan B.J., Williams D.P., Farr L., Mason M.W., Tomlinson A., T. Birks A., Knight J.C., and Russell P. St.J. Ultimate low loss of hollow-core photonic crystal fibres. In: Opt. Express 13, 236-244, 2005.

[36] Bobrovs, V., Spolitis, S., and Ivanovs, G. Comparison of chromatic dispersion compensation techniques for WDM-PON solution. In: Proc. of 2nd Baltic Congress on Future Internet Communications, pp.64,67, 25-27 April 2012.

[37] Bobrovs V., Spolitis S., Ivanovs G. and Gavars P. Performance Improvement of High Speed Spectrum-Sliced Dense WDM-PON System. In: Proc. of IX International Symposium on Telecommunications, pp.1-6, 25-27 October, 2012.

[38] ITU-T Recommendation G.655. Characteristics of a non-zero dispersion-shifted single-mode optical fibre and cable, 2009.

[39] Chomycz B. Planning Fiber Optic Networks. – McGraw-Hill, 401 pp, 2009.

[40] OFS. Technical specification of TrueWave RS Low Water Peak (LWP) Fiber, 2013.

[41] ITU-T Recommendation G.653. Characteristics of a dispersion-shifted, single-mode optical fibre and cable, 2010.

[42] Fredrik S., Bengt J., and David M. FBG dispersion compensation improves amplification designs and strategies", Lightwave Magazine, August 1, 2008.www.lightwaveonline.com/articles/print/volume-25/issue-8/applications/fbg-dispersion-compensation-improves-amplification-designs-and-strategies-54888677.html

[43] Agrawal G. Fiber – Optic Communication Systems. Third Edition. – USA: John Wiley and Sons, 561 pp, 2002.

[44] Bobrovs, V., Ivanovs G. Investigation of Mixed Data Rate and Format Transmission in WDM Networks, Electronics and Electrical Engineering. Vol. 4(84), pp. 63-66, 2008.

[45] Al-Qazwini Z., Kim H. Ultranarrow Spectrum-Sliced Incoherent Light Source for 10-Gb/s WDM PON. In: IEEE Journal of Lightwave Technology, vol.30, issue 19, pp. 3157-3163, Oct. 2012.

7

Application of Complex Wavelet Packet Transform (CWPT) in Coherent Optical OFDM (CO-OFDM) Communication Systems

Y. Ben-Ezra and B.I. Lembrikov

1. Introduction

Coherent optical orthogonal frequency division multiplexing (CO-OFDM) is a modulation format that attracted wide interest due to its high spectral efficiency (SE) and robustness against chromatic dispersion (CD) and polarization mode dispersion (PMD). CO-OFDM communication systems with coherent detection combine high SE and high receiver sensitivity [1], [2]. CO-OFDM transmission at 1 Tb/s can be realized [1]. OFDM is implemented using fast Fourier transform (FFT) which results in inter carrier interference (ICI) and inter symbol interference (ISI) [1], [2]. Usually, in order to avoid ICI and ISI, a so-called cyclic prefix (CP) is inserted into OFDM symbols. The addition of CP requires an increase of a bandwidth and sampling rate of analog-to-digital converter (ADC) and digital-to-analog converter (DAC) decreasing SE [1], [2]. The need for CP can be avoided if the wavelet packet transform (WPT) is used in CO-OFDM systems instead of Discrete Fourier Transform (DFT) and inverse DFT (IDFT) [2]. In such a case, a signal can be expanded in an orthogonal set of so-called wavelets [1], [2]. WPTs provide orthogonality between OFDM subcarriers based on the wavelets instead of sinusoids. Wavelets have finite length. For this reason, wavelet transforms (WTs) have both frequency and time localization. It has been shown that WPT-OFDM single-polarization system can mitigate CD of 3380 ps/nm at bit rate of 112 Gb/s [1].

An alternative real-time multiplexing technique characterized by a high SE is the Nyquist wavelength division multiplexing (N-WDM) [3]. N-WDM is made up of temporal sinc-pulses [3]. Electrically generated Nyquist pulses are shaped with finite duration impulse response (FIR) filters, have nearly rectangular spectra, and are transmitted in independent WDM channels [3]. For the higher order filters the spectrum approaches to a rectangle, and the inter channel guard bands can be kept small without introducing ICI [3]. N-WDM has been investigated both theoretically and experimentally [3], [4]. For the polarization-multiplexed quadrature phase-shift keying (PM-QPSK) transmitter based on two QPSK integrated modulators in nonreturn-to-zero (NRZ) modulation regime, and standard single mode fiber (SSMF) with the typical values of parameters such as bit rate of 111 Gb/s per channel, loss 0.22dB/km, dispersion 16.7 ps/nm/km, noise-figure 5 dB, BER=$4x10^{-3}$ *at* the maximum distance of 2300 km, performance of CO-OFDM and N-WDM techniques appeared to be similar [3],[4].

Recently, dual-polarization transmission has been investigated as a promising technique for future high bit rate communication systems [1], [5]. In such a case, the optical fiber two-input two-output channel model is described by a two-element Jones vector for any OFDM symbol on a subcarrier [6]. In Fourier transform (FT) based optical OFDM (FTO-OFDM), individual subcarriers are single-sideband, and the phase dispersion caused by CD and PMD can be compensated. In WPT-OFDM systems, on the contrary, the modulated signals are double-sideband. Unlike CD, PMD does not possess the phase symmetry [5]. Consequently, the Jones matrixes for the positive and negative sidebands are not equal, the dispersions of the two sidebands are different, and their addition does not represent the initial real WT basis. As a result, the orthogonality is broken, and inter-packet-interference occurs [7].

We proposed a novel type of CO-OFDM based on the dual-tree complex WPT (DT-CWPT) recently developed by N. Kingsbury [7]. DT-CWPT utilizes two wavelet filter banks (FBs) [8].

1. A first wavelet FB is used in the discrete WT (DWT).
2. A second wavelet FB is constructed in such a way that its impulse responses are approximately the discrete Hilbert transforms of the first wavelet FB.

The first and the second FBs are the real and imaginary parts of the DT-CWPT, respectively. We have shown that PMD can be compensated by digital signal processing using a DT-CWPT. SE and the transmission distance are substantially increased. Numerical simulations show that the 1 Tb/s single-channel CO-OFDM transmission with high SE of 7.88 bit/s/Hz over the distance of 1800 km can be realized.

The proposed Chapter is constructed as follows. In Section 2, we discuss the fundamentals of CO-OFDM systems; in Section 3, the WPT based OFDM systems are considered; PMD influence on the CO-OFDM dual-polarization transmission is reviewed in Section 4. In Sections 5 and 6, the structure and the properties of CO-OFDM system based on DT-CWT and DT-CWPT are described, respectively. In Section 7, the original results are presented for the DT-CWPT based single-channel CO-OFDM transmission system including a laser-diode and Mach-Zehnder external modulator. We describe the methods of the CD, PMD, and nonlinearity mitigation and present the numerical simulations results. Conclusions are presented in Section 8.

2. Fundamentals of CO-OFDM systems

In this Section, we consider the fundamentals of CO-OFDM communication systems OFDM is a kind of multicarrier modulation (MCM) where the data information is carried over many lower rate subcarriers [9]. The block diagram of an OFDM system is shown in Fig. 1 [9].

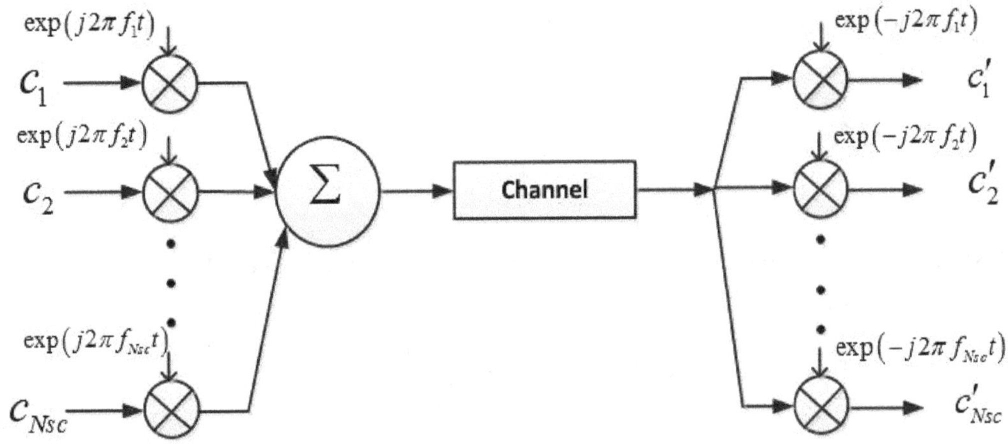

Figure 1. A diagram of OFDM system.

The fundamental advantages of OFDM are its robustness against channel dispersion and the ease of phase and channel estimation in a time-varying environment; the essential disadvantages of OFDM are high peak-to-average power ratio (PAPR) and sensitivity to frequency and phase noise [9]. The MCM transmitted signal is represented as follows [9].

$$s(t) = \sum_{i=-\infty}^{\infty} \sum_{k=1}^{N_{sc}} c_{ki} s_k (t - iT_s) \tag{1}$$

$$s_k(t) = \Pi(t) \exp(j2\pi f_k t) \tag{2}$$

$$\Pi(t) = \begin{Bmatrix} 1, & (0 < t \le T_s) \\ 0, & (t \le 0, t > T_s) \end{Bmatrix} \tag{3}$$

where c_{ki} is the i-th information symbol at the k-th subcarrier, s_k is the waveform for the k-th subcarrier, N_{sc} is the number of subcarriers, f_k is the frequency of the k-th subcarrier, T_s is the symbol period, and $\Pi(t)$ is the pulse shaping function. The detected information symbol c'_{ki} is given by [9].

$$c'_{ki} = \frac{1}{T_s} \int_0^{T_s} r(t - iT_s) \exp(-j2\pi f_k t) dt \tag{4}$$

where $r(t)$ is the received time domain signal. In the case of the classical MCM nonoverlapped bandlimited signals are used which results in the excessive bandwidth and decrease of SE. In the case of OFDM, the overlapped orthogonal signal sets are determined by the following condition for any two orthogonal subcarriers [9].

$$\frac{1}{T_s} \int_0^{T_s} s_k s_l^* dt = \delta_{kl} \tag{5}$$

Substituting expression (2) into (5) and carrying out the integration we obtain.

$$\exp\left[j\pi(f_k - f_l)T_s\right] \frac{\sin\left[\pi(f_k - f_l)T_s\right]}{\left[\pi(f_k - f_l)T_s\right]} = \delta_{kl} \tag{6}$$

The orthogonality condition (6) is satisfied for the following relationship between the subcarrier frequencies

$$f_k - f_l = m\frac{1}{T_s}; m = 1, 2, ... \tag{7}$$

In such a case, the orthogonal subcarrier sets can be recovered with the matched filters without ICI even for the strong signal overlapping [9]. OFDM modulation and demodulation can be implemented by using IDFT/DFT where the m-th samples s_m and r_m of the MCM transmitted and received signals $s(t)$ and $r(t)$, respectively, are given by [9].

$$s_m = \sum_{k=1}^{N} c_k \exp\left[j2\pi \frac{(k-1)(m-1)}{N}\right] = F^{-1}\{c_k\}; c'_k = F\{r_m\} \tag{8}$$

Here $F\{r_m\}$ is the Fourier transform,

$m \in [1, N]$, $f_k = (k-1)/T_s; k \in [k_{\min} + 1, k_{\min} + N]$, $k_{\min}$ is an arbitrary integer, and T_s/N is the sampling interval. Expressions (8) show that the discrete value of the transmitted signal $s(t)$ is an N-point IDFT of the information symbol c_k, and the received information symbol c'_k is an N-point DFT of the received sampled signal r_m [9]. The PAPR of the OFDM signal is given by the following expression [9]:

$$PAPR = \max\left\{|s(t)|^2\right\} / E\left\{|s(t)|^2\right\}, t \in [0, T_s] \quad (9)$$

where $\max\{|s(t)|^2\}$ and $E\{|s(t)|^2\}$ are the maximum value and the expectance of $|s(t)|^2$.

In order to implement IDFT/DFT DAC and ADC are needed. IDFT/DFT is characterized by two important advantages [9].

1. The computation complexity measured with the number of complex multiplications is reduced to $(N/2)\log_2(N)$ which is almost linear with respect to the subcarriers number N.
2. The OFDM system architecture is relatively simple. It does not require complex radio frequency (RF) oscillators and filters.

OFDM is characterized by the ISI and ICI as it was mentioned above. These phenomena are caused by a large number of subcarriers [9]. CP, i.e. the cyclic extension of the OFDM waveform into the guard interval (GI) Δ_G, has been proposed in order to prevent ISI and ICI [9]. If the GI is long enough to contain the intersymbol transition, then the remaining part of the OFDM symbol satisfies the orthogonality conditions (6), (7) and receiver cross-talk occurs only within GI [10]. The addition of CP requires an increase of a bandwidth and sampling rate of ADC and DAC. CP is an easily recognizable feature of an OFDM system which results in the signal vulnerability to interception by surveillance receiver [11]. Hence, the elimination of CP reduces the probability of interception and improves SE [11]. CP can be inserted implicitly into the definition of the MCM transmitted signal $s(t)$ (1)-(3) by modification of equation (3) which takes the form [9].

$$\Pi(t) = \begin{cases} 1, & (-\Delta_G < t \le T_s) \\ 0, & (t \le -\Delta_G, t > T_s) \end{cases} \quad (10)$$

OFDM is characterized by the dispersion robustness, high SE, possibility of linear and nonlinear impairment mitigation [9]. For this reason, OFDM appeared to be an attractive long-haul transmission format for the optical communication systems [9], [12], [13].

The optical communication systems can be divided into two groups. The first group is based on optical wireless, multimode fiber (MMF) systems and plastic optical fiber (POF) systems where the OFDM signal is represented by the intensity of the optical signal; the second group is based on the single mode fiber (SMF) techniques where the OFDM signal is represented by the optical field [13]. The first group uses the direct detection technique, while the second group uses the coherent detection technique [9]. The optical OFDM systems based on the coherent detection are called CO-OFDM [9], [12]. A generic CO-OFDM system consists of five functional blocks: the RF OFDM transmitter, the RF-to-optical (RTO) up-converter, the optical channel, the optical-to-RF (OTR) down-converter, and the RF OFDM receiver with the coherent detector [12]. CO-OFDM manifests higher receiver sensitivity, SE, and robustness against polarization dispersion [9]. CO-OFDM can achieve high SE by overlapping subcarrier spectrum and at the same time avoiding interference by using coherent detection and signal set orthogonality [9], [12]. CO-OFDM communication systems are characterized by the linearity in RTO up-conversion and optical-to-RF (OTR) down-conversion [9], [12]. The electrical bandwidth requirements for the CO-OFDM transceiver can be greatly reduced by using direct up/down conversion which results in the low cost of the high-speed electronic circuits [9], [12]. SE η of the CO-OFDM system is given by [9], [12].

$$\eta = \frac{2R}{B_{OFDM}} \approx 2\alpha; \alpha = \frac{t_s}{T_s} \tag{11}$$

where $R = N_{sc} / T_s$ is the total symbol rate, $B_{OFDM} = (2/T_s) + (N_{sc} - 1)/t_s$ is the OFDM bandwidth, $N_{sc} \gg 1$, the factor of 2 accounts for two light wave polarizations in the fiber, and t_s is the observation period.

CO-OFDM communication systems possess the following advantages as compared to the intensity modulation with direct detection (IM/DD) systems [14].

1. The shot-noise limited receiver can be achieved with a sufficient local oscillator (LO) power.
2. The frequency resolution at the intermediate frequency (IF) or baseband stage is high enough in order to separate close wavelength-division multiplexed (WDM) channels in the electric domain.
3. The phase detection improves the receiver sensitivity compared with IM/DD systems.
4. The advanced multilevel modulation formats can be introduced into optical communications by using phase modulation.

CO-OFDM communication systems can achieve high SE transmission through higher order modulation such as 64-QAM in single polarization and 16-QAM in dual polarization due to the DACs at the transmitter [9]. As a result, CO-OFDM techniques can be applied to the optical long-haul transmission at 100 Gb/s and beyond [9]. Recently, the 1.15 Tb/s no-guard-interval CO-OFDM (NGI-CO-OFDM) DP-QPSK superchannel transmission over the distance of 4520 km with SE=3.75b/s/Hz with BERs below the stringent forward error correction (FEC) limit has been demonstrated experimentally [15]. The pure silica core fiber and the erbium doped fiber amplifier (EDFA) have been used [15].

3. WPT based CO-OFDM systems

Conventional CO-OFDM systems based on DFT and IDFT can be replaced with WPT based CO-OFDM [1]. The sinusoidal functions are infinitely long in the time domain while wavelets have finite length being localized in time and in frequency domains [1]. Wavelet signal analysis can be a base for an effective computational algorithm which is faster and simpler than FFT [16]. Wavelets have been used in optical communications for time-frequency multiplexing and ultrafast image transmission [16]. A signal may be expanded in an orthogonal set of wavelet packets (WPs) as the basis functions, each channel occupies a wavelet packet (WP), and IDWPT/ DWPT are used at the transmitter and receiver, respectively [1], [16]. The theory and possible applications of continuous wavelet transform (CWT) and DWT are presented in [1], [16]-[19].

In particular, CO-OFDM system SE can be improved by avoiding CP. The problem of the CP elimination can be solved by using WPT-OFDM [16]. A mitigation of CD of 3380 ps/nm at 112 Gb/s without CP in the WPT-OFDM system has been demonstrated [1]. Typically, WPT-OFDM optical communication systems are single-polarization. However, the dual-polarization transmission is a promising technique for 1 Tb/s Ethernet transport [1]. Here we present some definitions and relationships essential for DWT applications.

CWT $W_T(a, \tau)$ of a given function $f(t)$ with respect to a mother wavelet (MW) $\psi(t)$ is defined as follows [17], [18]:

$$W_T(a,\tau)=\frac{1}{\sqrt{|a|}}\int_{-\infty}^{\infty}\psi^*\left(\frac{t-\tau}{a}\right)f(t)dt \tag{12}$$

where the real numbers a and τ are the scaling and shifting, or translation parameters, respectively, and asterisk means complex conjugation. Note that in many practically important cases MW $\psi(t)$ is real. The functions $\psi^{a,\tau}(s)=|a|^{-1/2}\psi((s-\tau)/a)$ are called wavelets [18]. The set of wavelets is orthogonal and can be used as a basis instead of sinusoidal functions [16]. It is possible to localize the events described by $f(t)$ in time and frequency domains simultaneously by means of WT choosing the appropriate values of the parameters a and τ [17].

DWT is given by [17], [18]

$$W_T^{m,n}(a,\tau)=a_0^{-m/2}\int_{-\infty}^{\infty}\psi^*\left(a_0^{-m}t-n\tau_0\right)f(t)dt \tag{13}$$

where $m,\ n\in Z$, Z is the set of all integers, and the constants $a_0>1,\ \tau_0>1$. Comparison of eqs. (12) and (13) shows that $a=a_0^m$ and $\tau=n\tau_0 a_0^m$ [17]. The orthogonal wavelet series expansions can be successfully used in DSP and multiplexing when the scaling and translation parameters are discrete [17]. In such a case, a signal $s(t) \in V_0$ can be represented by a smooth approximation at resolution 2^M, obtained by combining translated versions of the basic scaling function $\phi(t)$, and M details at the dyadic scales $a=2^l$, $(l=1, 2, ..., M-1)$ obtained by combining shifted and dilated versions of the MW $\psi(t)$ as follows [17].

$$s(t)=\sum_k 2^{-M/2} c_M[k]\phi(2^{-M}t-k\Delta\tau)+\sum_{l=1}^{M}\sum_k 2^{-l/2} d_l[k]\psi(2^{-l}t-k\Delta\tau) \tag{14}$$

Here a subspace $V_0 \in L^2(R)$, $L^2(R)$ is a the linear vector space of square integrable functions, $2^{-l/2}\phi(2^{-l}t-k\Delta\tau)$ and $2^{-l/2}\psi(2^{-l}t-k\Delta\tau)$ are the orthonormal bases for the subspaces $V_l \in L^2(R)$ and $W_l \in L^2(R)$, respectively, $V_l \perp W_l$, $(l, k) \in Z$, $c_l[k]$ and $d_l[k]$ are the scaling and detail coefficients, respectively, at resolution 2^l, $\Delta\tau$ is the time interval coinciding with the inverse of the free spectral range (FSR).

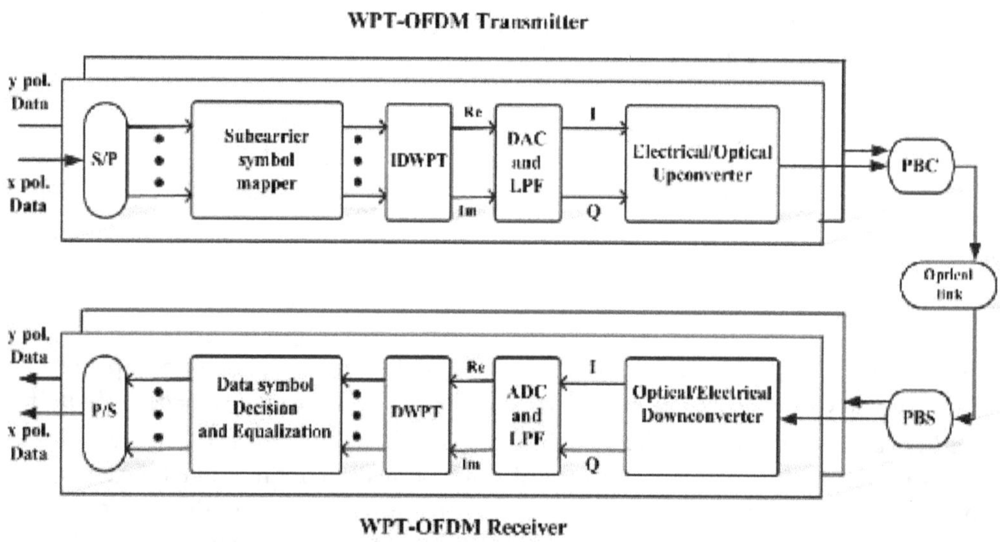

Figure 2. Block diagram of WPT-OFDM communication system. PBC/PBS: polarization beam Combiner/Splitter.

The block diagram of the WPT-OFDM communication system is shown in Fig. 2 [1]. In WPT-OFDM system each channel occupies a wavelet packet which plays a role of a subcarrier in wavelet domain. At the transmitter, IDWPT reconstructs the time domain signal from wavelet packets [1]. At the receiver, DWPT is used for the decomposition of the time domain signal into wavelet packets. The decomposition is carried out by means of successive low-pass and high-pass filtering in the time domain [1]. For WPT-OFDM, the basis function wavelets are finite in time, and the inter-symbol orthogonality in WT is caused by the shift orthogonal property of the waveforms [1], [17], [18]. In WT, symbols are overlapped in time domain [1].

As a result, the symbol duration increases. The symbol duration increase, in turn, provides the CD tolerance and eliminates the need for CP [1].

Real-valued wavelets are used for the real signal processing [1]. However for fiber optic channels, the up-conversion to the optical domain generates two spectral sidebands: a positive sideband and a negative one [1], [5]. In such a case, PMD influence on WT-OFDM is essential. It will be considered in Section 4.

4. PMD influence on the CO-OFDM communication system performance

In this Section we consider PMD in CO-OFDM communication systems and its influence on their performance [1], [5], [6], [9]. CO-OFDM transmission system can be defined as 2×2 multiple-input multiple-output (MIMO)-OFDM channel in the presence of polarization dispersion effects [7]. SMF supports two degenerate modes of orthogonal polarization [9]. However, the degeneracy may be broken due to the asymmetry in fiber geometry caused by the manufacture process or mechanical stress [9]. In such a case, a fiber birefrigence $B = |n_x - n_y|$ occurs where $n_{x,y}$ are the refractive indices for x and y polarization, respectively [9]. An optical signal with the wavelength λ propagating through the SMF is coupled from one polarization to the other with the coupling period $L_B = \lambda / B$ [9]. Typical values of these parameters are $B \approx 10^{-7}$, $L_B \approx 10m$ at $\lambda = 1.5\mu m$ [9]. Then, the transmitted OFDM time-domain signal $s(t)$ (1) and the i-th information symbol at the k-th subcarrier c_{ki} (2) should be replaced with the Jones vectors for the two polarizations given by [6], [9].

$$s(t) = \begin{pmatrix} s_x \\ s_y \end{pmatrix}; \vec{c}_{ki} = \begin{pmatrix} c_{ki}^x \\ c_{ki}^y \end{pmatrix} \tag{15}$$

GI Δ_G in such a case must be long enough in order to handle the fiber CD and PMD. It takes the form [6], [9].

$$\Delta_G \geq \frac{c}{f^2} |D_t| N_{sc} \Delta f + (DGD)_{\max} \tag{16}$$

where c is the speed of light, f is the optical carrier frequency, D_t is the total accumulated CD in units of ps/pm, Δf is the subcarrier channel spacing, and $(DGD)_{\max}$ is the maximum budgeted differential-group-delay (DGD) which is typically chosen to be of 3.5 times of the mean PMD in order to satisfy inequality (16). The Jones vector of the received symbol $\vec{c'}_{ki} = (c'^{x}_{ki}, c'^{y}_{ki})^t$ with a sufficiently long symbol period for the k-th subcarrier in the i-th OFDM symbol is given by [6], [9].

$$\vec{c'}_{ki} = \exp(j\phi_i)\exp(j\Phi_D(f_k))T_k\vec{c}_{ki} + \vec{n}_{ki} \tag{17}$$

where $\vec{n}_{ki}=(n_{ki}^x, n_{ki}^y)^t$ is the noise including two polarization components, ϕ_i is the OFDM symbol phase noise, or common phase error (CPE) caused by the phase noises from the lasers and RF LOs at the transmitter and receiver, $\Phi_D(f_k)=\pi c D_t f_k^2 / f^2$ is the phase dispersion caused by the fiber CD, $T(k)$ is the Jones matrix for the fiber link given by [6], [9].

$$T(k) = \prod_{l=1}^{N} \exp\left\{\left(-\frac{1}{2}j\vec{\beta}_l f_k - \frac{1}{2}\vec{\alpha}_l\right)\bullet\vec{\sigma}\right\} \tag{18}$$

N is the number of PMD and polarization dependent loss (PDL) cascading elements, $\vec{\beta}_l$ is the l-th element birefringence vector, $\vec{\alpha}_l$ is the l-th element PDL vector, and $\vec{\sigma}$ is the Pauli matrix vector.

The CO-OFDM system architecture in the fiber optic channel can be divided into four groups related to the number of the transmitters and receivers used in the polarization dimension [6], [9].

1. Single-input single-output (SISO) where one optical OFDM transmitter and one optical OFDM receiver are used for CO-OFDM transmission.The SISO configuration is susceptible to the polarization mode coupling in a fiber. It requires a polarization controller (PC) before the receiver in order to align the input signal polarization with the LO polarization. However, in the case of large PMD, PC cannot prevent the polarization rotation between subcarriers. As a result, SISO CO-OFDM is susceptible to polarization-induced fading and should not be implemented in applications.

2. Single-input two-output (SITO) where one optical OFDM transmitter and two optical OFDM receivers are used, one for each polarization. In such a system, PC is not needed. SITO CO-OFDM system is resilient to PMD when the polarization diversity receiver is used. The PMD in the fiber link may even improve the system margin against PDL-induced fading.

3. Two-input single-output (TISO) where two optical OFDM transmitters and one optical OFDM receiver are used. This time, each transmitter is used for one specified polarization. The PC is not needed at the transmit end. The polarization diversity transmitter can achieve PMD resilience. In the TISO scheme, the same information symbol is repeated in two consecutive symbols which results in the electrical and optical efficiency reducing by half and the two times increase of the optical signal-to-noise ratio (OSNR) requirement as compared to SITO scheme.

4. Two-input two-output (TITO) where a polarization diveristy optical OFDM transmitter and a polarization diveristy optical OFDM receiver are used. In such a scheme, the transmitted OFDM information symbol c_{ik} is considered as polarization modulation or

polarization multiplexing. Hence, the channel capacity is doubled as compared to the SITO scheme. The PMD rotates subcarrier polarization and does not influence the channel capacity doubling. In the case of the TITO scheme, the polarization diversity receiver permits to avoid the polarization tracking at the receiver.

Comparison of the properties of the different CO-MIMO-OFDM architectures shows that SITO-and TITO-OFDM are the preferred configurations [9].

Consider the PMD influence on the transmission in a 2x2 MIMO CO-OFDM system. In such a case, the Jones matrix (18) for the fiber link on the k-th subcarrier in the first-order PMD approximation is given by [1], [5].

$$T(k) = M^{-1}\begin{bmatrix} \exp(-j\pi f_k \tau_{link}) & 0 \\ 0 & \exp(j\pi f_k \tau_{link}) \end{bmatrix} M \tag{19}$$

where τ_{link} is the DGD of the link, and the matrix M has the form.

$$M = \begin{bmatrix} \cos\left(\frac{\theta}{2}\right)\exp\left(-j\frac{\varphi}{2}\right) & -\sin\left(\frac{\theta}{2}\right)\exp\left(-j\frac{\varphi}{2}\right) \\ \sin\left(\frac{\theta}{2}\right)\exp\left(j\frac{\varphi}{2}\right) & \cos\left(\frac{\theta}{2}\right)\exp\left(j\frac{\varphi}{2}\right) \end{bmatrix} \tag{20}$$

Here θ, φ are the polar and azimuth angle of the principle state of polarization. In FTO-OFDM systems, the individual subcarrier is single-sideband by nature, and $\Phi_D(f_k)$, $T(k)$ can be estimated and compensated [5]. In WT based OFDM systems the modulated signals are double-sideband as it was mentioned above. CD influence on such a system is insignificant since the two sidebands have the same phase dispersion: $\Phi_D(f_k)=\Phi_D(-f_k)$. On the contrary, PMD does not have phase symmetry: $T(k)\neq T(-k)$. As a result, at the receiver two sidebands experience two different dispersions, and their addition does not reproduce the real wavelet basis, the orthogonality is broken, and the inter-packet-interference occurs [1], [5]. The WPT-OFDM systems are more susceptible to PMD then conventional FT-OFDM systems [1], [5].

5. Dual-Tree Complex Wavelet Transform (DT-CWT)

DWT has two main disadvantages [7], [8].

1. It is not shift-invariant. For this reason, the small shift in the input signal can cause major variations in the energy distribution between DWT coefficients at different scales.
2. It is characterized by a low directional selectivity for the diagonal elements since the wavelet filters are separable and real.

The DT-CWT has been proposed recently [7], [8]. DT-CWT consists of two WTs operating in parallel on an input signal [8]. In order to construct a DT-CWT, each sub band should be repeatedly decomposed using low-pass/high-pass PR FBs [8]. The response of each branch of the second WP FB is the discrete Hilbert transform of the corresponding branch of the first WP FB [8]. These WTs correspond to two FBs. The block diagram of DT-CWT is shown in Fig. 3 [8].

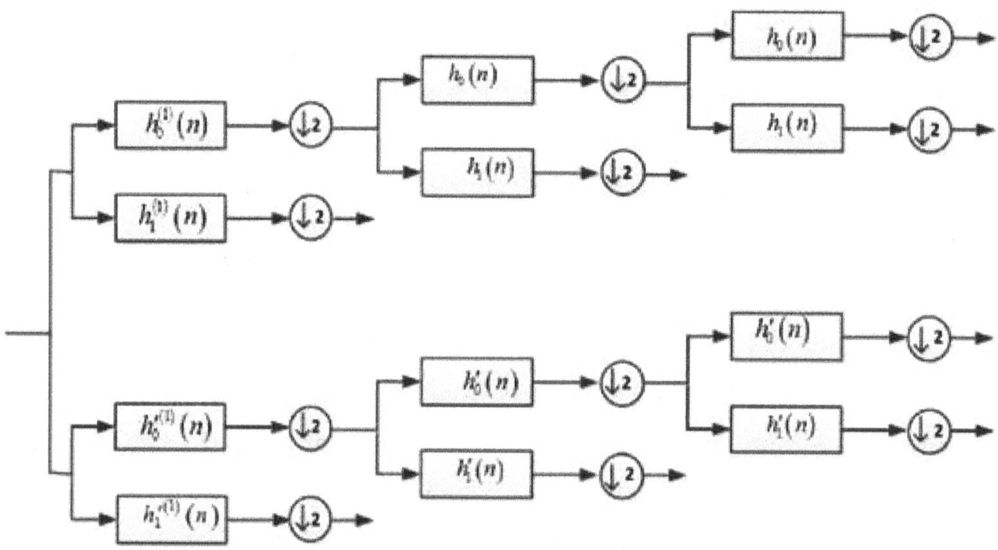

Figure 3. The DT-CWT consists of two wavelet FBs operating in parallel.

The first wavelet FB is determined by DWTs, the second FB contains the discrete Hilbert transforms of the first wavelet FB [8]. The first FB is the real part and the second FB is the imaginary part of the CWPT [8].

The wavelet function $\psi(t)$ associated with the first FB and the corresponding scaling function $\phi(t)$ have the form [8].

$$\psi(t)=\sqrt{2}\sum_{n} h_1(n)\phi(2t-n);\phi(t)=\sqrt{2}\sum_{n} h_0(n)\phi(2t-n) \tag{21}$$

For an orthonormal wavelet basis, the LPFs and high-pass filters (HPFs) have the following relationships $H_1(e^{j\omega})=-e^{-j\omega d}H_0^*(e^{j(\omega-\pi)})$, or equivalently $h_1(n)=(-1)^n h_0(d-n)$ where d is an odd integer [8]. The wavelet function $\psi'(t)$ and the filters $h'_0(n)$, $h_1'(n)$ for the DT-CWPT imaginary part are defined in a similar way [6]. The filters $\{h_0(n), h_1(n)\}$ and $\{h'_0(n), h'_1(n)\}$ are assumed to be finite impulse response (FIR) conjugate quadrature filter (CQF) pairs [8]. In the case of the ideal DT-CWT we have.

$$\psi_c(t) = \mathrm{H}_{Hilbert}\{\psi(t)\} \tag{22}$$

where $H_{Hilbert}\{\psi(t)\}$ is the Hilbert transform of $\psi(t)$ and the complex wavelet $\psi(t)=\psi(t)+j\psi'(t)$ is approximately analytic [20]. In order to satisfy the perfect reconstruction (PR) conditions, the filters are designed in such way that the two low pass filters (LPFs) should satisfy the half-sample delay condition: $h'_0(n)\approx h_0(n-0.5)$ [20]. If each wavelet $\psi(t)$, $\psi'(t)$ are orthogonal to their integer translates, then the Hilbert relation between them is satisfied under the condition [8]:

$$H_0'(\exp(j\omega)) = \exp(-j0.5\omega)H_0(\exp(j\omega)), |\omega| < \pi \tag{23}$$

Then, for the HPFs we have [8]

$$H_1'(\exp(j\omega)) = -j\,\mathrm{sgn}(\omega)\exp(j0.5\omega)H_1(\exp(j\omega)), |\omega| < \pi \tag{24}$$

Here sgn is the signum function, $H'_{0,1}(\exp(j\omega))$, $H_{0,1}(\exp(j\omega))$ are the z-transforms of $h'_{0,1}(n)$, $h_{0,1}(n)$ on the unit circle, respectively [8]. It can be shown that the first stage of the DT-CWT must be different from the following stages [8]. The equivalent response $H^{(k)}(\exp(j\omega))$ of the kth stage of the first FB terminated with $H_1(\exp(j\omega))$ for $k>1$ is given by:

$$H^{(k)}(\exp(j\omega)) = H_1\left(\exp\left(j2^{k-1}\omega\right)\right)\prod_{m=0}^{k-2} H_0\left(\exp\left(j2^m\omega\right)\right), |\omega| < \pi \tag{25}$$

The equivalent response $H^{'(k)}(\exp(j\omega))$ of the second FB's corresponding branch can be obtained by replacing H_i with H'_i given by [8]:

$$H'^{(k)}(\exp(j\omega)) = -j\,\mathrm{sgn}(\omega)\exp(j0.5\omega)H^{(k)}(\exp(j\omega)), |\omega| < \pi \tag{26}$$

which is equivalent to the relationship

$H^{'(k)}(\exp(j\omega)) = -\exp(j0.5\omega)H_{Hilbert}\{H^{(k)}(\exp(j\omega))\}$ [8]

DT-CWT has the following properties [7].

1. Approximately shift-invariance.
2. Good directional selectivity in 2 dimensions.
3. Perfect reconstruction (PR) using short linear-phase filters.

4. Limited redundancy.

5. Efficient order-N computation.

6. Dual-Tree Complex Wavelet Packet Transform (DT-CWPT)

DT-CWT can be generalized to DT-CWPT by using the bases of discrete wavelet packet transforms (DWPTs). DT-CWPT is approximately shift-invariant, provides a geometrically oriented signal analysis in multiple dimensions and permits the transformation of a double-sideband spectrum into an approximately single-sideband one [7], [8]. The invert DT-CWPT is provided by the inversion of its real and imaginary parts [8].

The construction of the DT-CWPT requires the repeated decomposition of the sub bands by using low-pass/high-pass PR FBs [8]. The PR FBs should be chosen in such a way that the response of each branch of the second wavelet packet FB is the discrete Hilbert transform of the corresponding branch of the first wavelet packet FB [8]. Under this condition, each sub band of the DT-CWPT will be analytic [8]. Taking into account that the discrete Hilbert transform is a linear time-invariant (LTI) system we can write the following relationships for the discrete Hilbert transform pair of the filters $g(n)$, $h(n)$ [8].

$$G(\exp(j\omega)) = j\,\mathrm{sgn}(\omega)H(\exp(j\omega)), |\omega| < \pi \tag{27}$$

$$G(\exp(j\omega))C(\exp(j\omega)) = j\,\mathrm{sgn}(\omega)H(\exp(j\omega))C(\exp(j\omega)), |\omega| < \pi \tag{28}$$

Then, the convolutions $g(n) * c(n)$, $h(n) * c(n)$ are also a discrete Hilbert transform pair [8].

It has been shown that a DT-CWPT consisting of two wavelet packet FBs operating in parallel can be produced where some filters in the second wavelet packet FB are the same as those in the first wavelet packet FB [8].

The first of these two wavelet FBs for a four-stage DT-CWPT is shown in Fig. 4 [8]. The second wavelet packet FB is obtained by replacing the first stage filters $h_i^{(1)}(n)$ with $h_i^{(1)}(n-1)$ and by replacing $h_i(n)$ with $h_i'(n)$ for $i \in \{0, 1\}$ [8]. The filters F_i are unchanged in the second wavelet packet FB [8].

In order to demonstrate the development of the M-band DT-CWT based on the Hilbert pairs of M-band wavelets, consider the construction of the four-band DT-CWT (M=4) from a given two-band DT-CWT which is defined by a two-channel orthonormal FB $\{h_0^{(2)}(n), h_1^{(2)}(n)\}$ associated with the scaling function $\phi^{(2)}(t)$ and the wavelet $\psi^{(2)}(t)$ [8]. Their Fourier transforms (FTs) have the form [8].

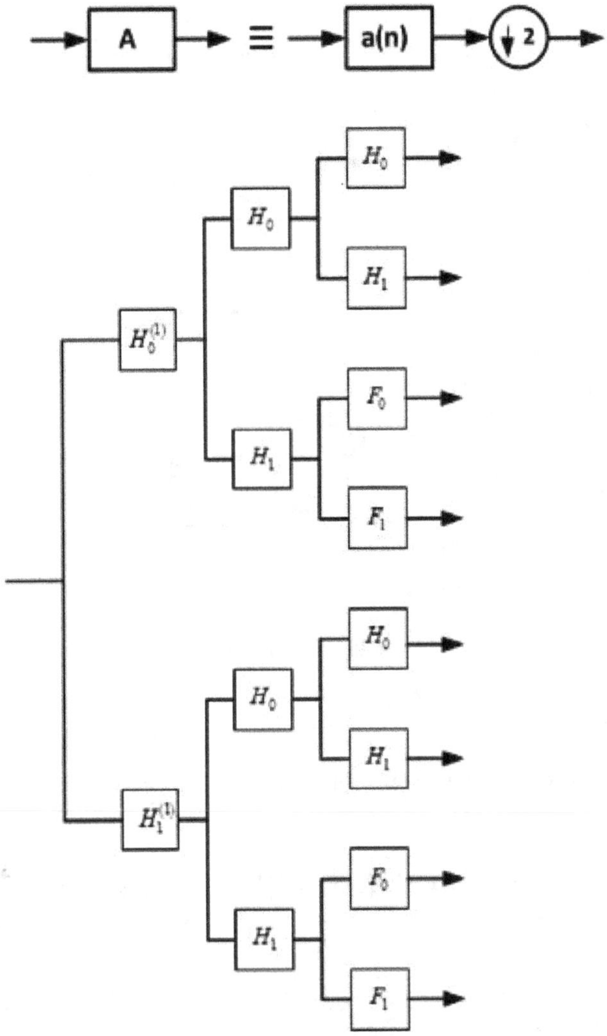

Figure 4. First wavelet packet FB of a four-stage DT-CWPT.

$$\Phi^{(2)}(\omega)=\prod_{l=1}^{\infty}\left[\frac{1}{\sqrt{2}}H_0^{(2)}\left(\frac{\omega}{2^l}\right)\right];\Psi^{(2)}(\omega)=\frac{1}{\sqrt{2}}H_1^{(2)}\left(\frac{\omega}{2}\right)\Phi^{(2)}\left(\frac{\omega}{2}\right) \tag{29}$$

A two-band DT-CWT can be created if we have a second two-channel orthonormal FB $\{h'^{(2)}_0(n), h'^{(2)}_1(n)\}$ with the associated scaling function $\phi'^{(2)}(t)$ and the wavelet $\psi'^{(2)}(t)=H_{Hilbert}\{\psi^{(2)}(t)\}$, where the complex wavelet $\psi^{(2)}(t)+j\psi'^{(2)}(t)$ is analytic [8]. In order to construct a four-band CWT, another two-channel orthonormal FB $\{f_0(n), f_1(n)\}$ is necessary [8]. The corresponding four-channel orthonormal FB, i.e a DWPT is shown in Fig. 5 [8].

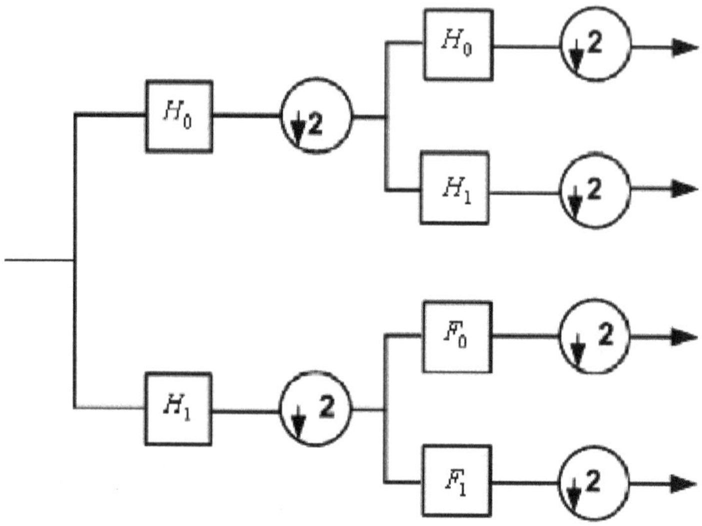

Figure 5. Discrete wavelet packet transform (DWPT).

A second WPT such that the wavelets associated with the two WPTs form the Hilbert transform pairs is given by [8].

$$H_0^{(4)}\left(\exp(j\omega)\right) = H_0^{(2)}\left(\exp(j\omega)\right)H_0^{(2)}\left(\exp(j2\omega)\right) \tag{30}$$

$$H_1^{(4)}\left(\exp(j\omega)\right) = H_0^{(2)}\left(\exp(j\omega)\right)H_1^{(2)}\left(\exp(j2\omega)\right) \tag{31}$$

$$H_2^{(4)}\left(\exp(j\omega)\right) = H_1^{(2)}\left(\exp(j\omega)\right)F_0\left(\exp(j2\omega)\right) \tag{32}$$

$$H_3^{(4)}\left(\exp(j\omega)\right) = H_1^{(2)}\left(\exp(j\omega)\right)F_1\left(\exp(j2\omega)\right) \tag{33}$$

It can be shown that the FB determined by eqs. (30)-(33) is equivalent to the FB shown in Fig. 5 [8].

7. Simulation results for the single channel DT-CWPT based CO-OFDM system

We have carried out the numerical simulations based on DT-CWPT for the single-channel CO-OFDM transmission system including the optical front elements such as a laser-diode, and

Mach-Zehnder external modulator [21]. The nonlinear effects are neglected since Hilbert transform is LTI system [8]. We used the 16 quadrature amplitude modulation (QAM) with $128=2^7$ sub bands, 7 decomposition levels, bit rate of 100 Gb/s and the bandwidth of 25 GHz. The typical values of transmission process are following: the input optical signal power is equal to 10 mW; CP is 1%; the phase noise frequency in the case of the coherent detection is equal to 500 KHz; the overall noise amplification level of the link is 50 dB; the pre-amplifier gain at the receiver input is 25 dB; the optical fiber dispersion constant is $17\, ps/(nm \cdot km)$; the PMD constant is 0.1 ps/km$^{1/2}$. The 8 bit DAC and ADC have been used. In order to mitigate the group velocity dispersion and PMD we used the Least Mean Square (LMS) and Viterbi-Viterbi Digital Signal Processing (DSP) algorithms [22]. We used in our simulations the complex wavelets based on the Q-shift filters [7], [8], [23]. The design of pairs of wavelet bases where the wavelets form a Hilbert transform pair was also discussed in detail in [24], [25]. The numerical simulation results are presented in Figs. 6-11. The QAM-16 modulated signal constellations for the transmission distance of 100 km, 700 km and 1700 km in the case of the DT-CWPT based CO-OFDM system are shown in Figs.6-8, respectively.

The QAM-16 modulated signal constellations for the transmission distance of 1200 km and 1300 km in the case of the conventional CO-OFDM system are shown in Figs.9-10, respectively. In both cases CP=1% has been used.

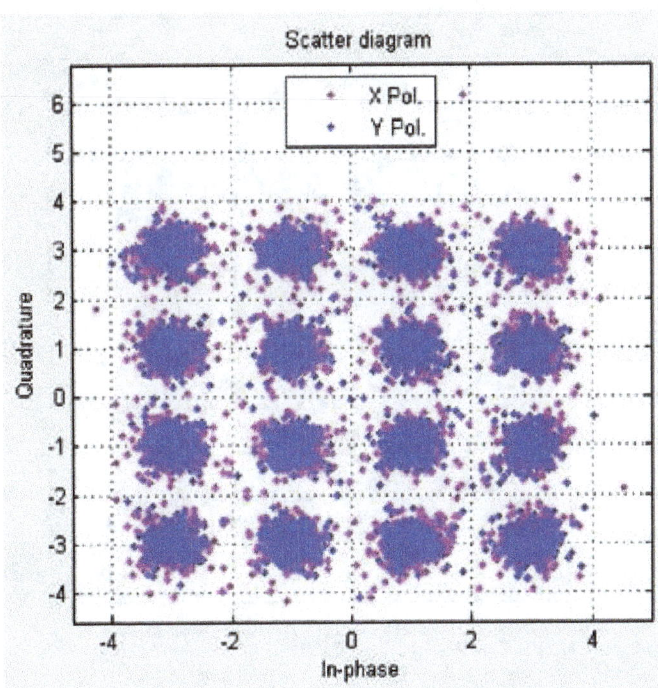

Figure 6. The QAM-16 modulated signal constellation for the transmission distance of 100 km, CP 1%, BER=$10^{-2.56}$, DT-CWPT based CO-OFDM system.

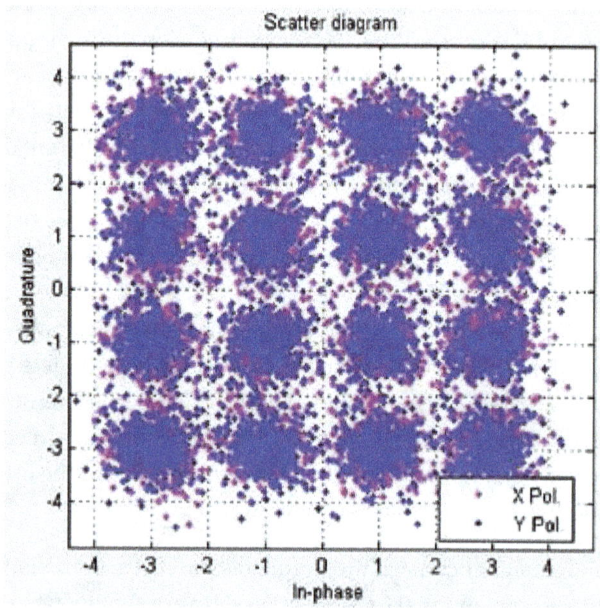

Figure 7. The QAM-16 modulated signal constellation for the transmission distance of 700 km, CP 1%, BER=$10^{-2.13}$, DT-CWPT based CO-OFDM system.

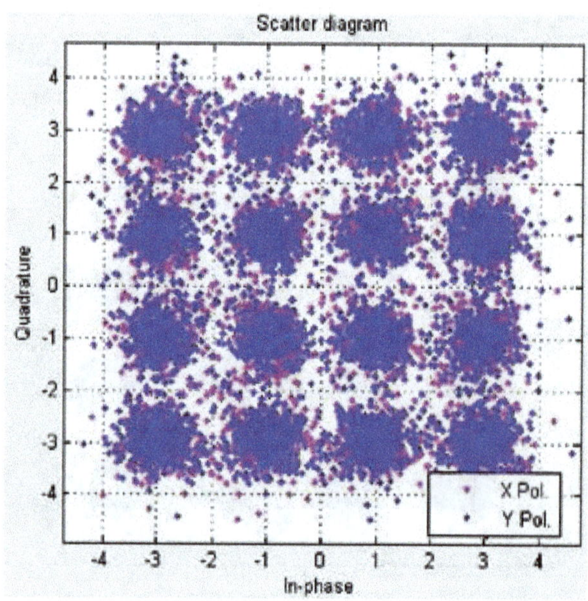

Figure 8. The QAM-16 modulated signal constellation for the transmission distance of 1700 km, CP 1%, BER=$10^{-2.08}$, DT-CWPT based CO-OFDM system.

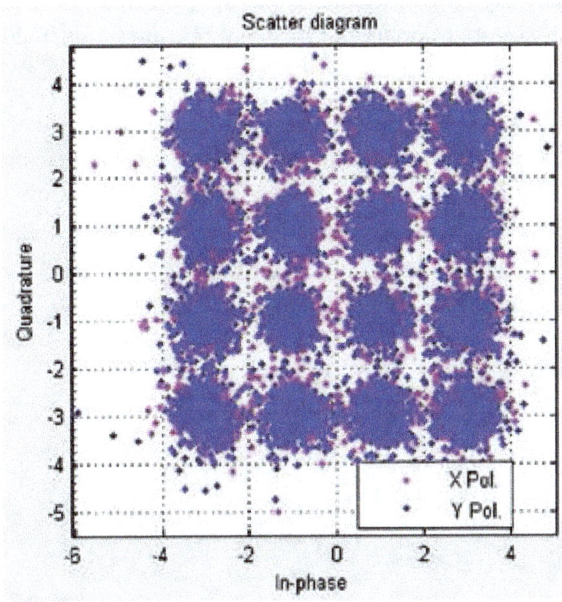

Figure 9. The QAM 16 modulated signal constellation for the transmission distance of 1200 km, BER=$10^{-2.08}$, CP 1% Conventional CO-OFDM system.

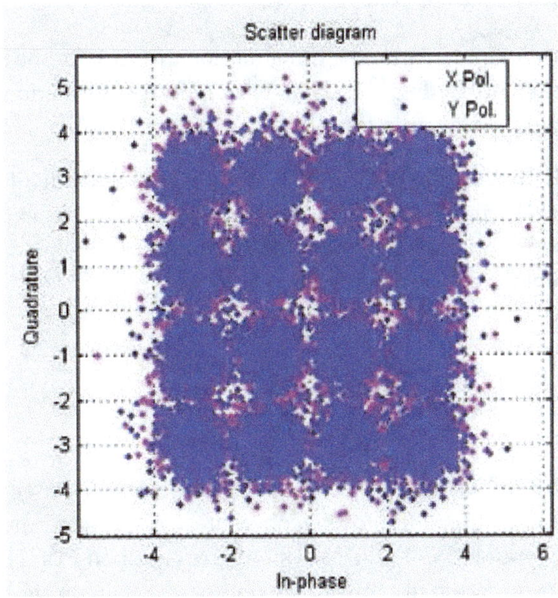

Figure 10. The QAM 16 modulated signal constellation for the transmission distance of 1300 km, BER=$10^{-1.83}$, CP 1% Conventional CO-OFDM system.

The QAM-16 modulated signal constellations for the transmission distance of 1200 km and 1300 km in the case of the conventional CO-OFDM system are shown in Figs.9-10, respectively. In both cases CP=1% has been used.

The comparison of Figs. 8 and 10 clearly shows that the performance of the DT-CWPT based OFDM communication system is better than the performance of the conventional CO-OFDM communication system. Indeed, for the same modulation format and CP, $BER = 10^{-1.83} = 1.48 \times 10^{-2}$ at the distance of 1300 km for the conventional CO-OFDM as compared to $BER = 10^{-2.08} = 8.32 \times 10^{-3}$ at the distance of 1700 km for the DT-CWPT based OFDM system.

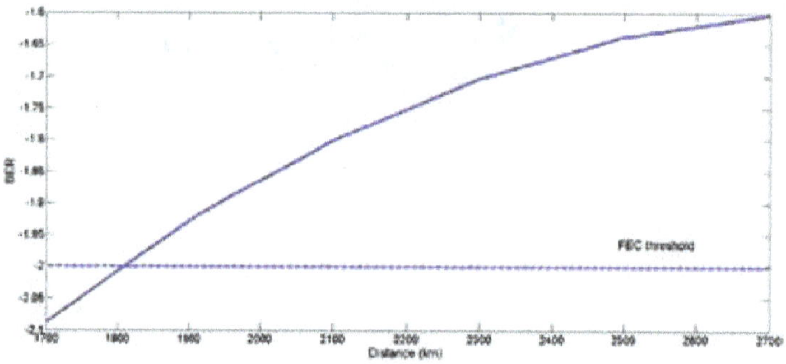

Figure 11. The BER dependence on the transmission distance for the DT-CWPT based CO-OFDM system.

The bit error rate (BER) dependence on the transmission distance for the DT-CWPT based CO-OFDM system is presented in Fig. 11. The forward error correction (FEC) threshold of BER=10^{-2} corresponds to the transmission distance of about 1800 km.

We have evaluated SE for the dual polarization signal, sampling rate of 26.6 GHz, and QAM 16 modulation format. The maximum available SE in such a case is 8bit/s/Hz. Assuming the FEC penalty of 20% per band, the CP penalty of 1.5%, and channel and timing estimation of 5%, we obtain the actual $SE = 8 \times 0.735 = 5.84 bit / s / Hz$.

8. Conclusions

CO-OFDM communication systems are characterized by high SE, possibility of the advanced modulation formats applications due to the coherent detection, and the high receiver sensitivity [1], [9], [13]. Conventional CO-OFDM systems are based on FTs [1], [9]. In such systems, low rate subcarriers are orthogonally transformed into time domain signals [1]. The condition of subcarrier orthogonality prevents spectral overlapping and ICI [1], [9]. The basis functions of the conventional CO-OFDM system are the infinitely long in time sinusoids. The 1-Tb/s CO-OFDM single-channel signal transmission over 600 km SSMF with SE of 3.3 bit/s/Hz and without Raman amplification and dispersion compensation has been reported [26]. In CO-

OFDM systems, the CD and PMD impairments can be compensated by choosing an appropriate length of CP [1], [9]. However, the CP insertion decreases the CO-OFDM communication system SE [1], [9].

Recently, the CO-OFDM systems based on WPTs have been proposed [1], [2]. Wavelets used as the basis functions in these advanced systems have finite length in the time domain. For this reason, WPT-OFDM systems do not need CP which results in a higher SE [1], [2]. The WPT-OFDM can mitigate a CD of 3380 ps/nm at 112 Gb/s rate without CP [1]. However, the high performance has been demonstrated for a CO-OFDM system based on a single-polarization [5]. The most promising technique for achieving high SE and maximizing power efficiency is coherent detection with polarization multiplexing where the in-phase (I) and quadrature (Q) signals are used in two field polarizations [27]. Information can be encoded in all available degrees of freedom, while the the compensation of transmission impairments can be provided by DSP [22], [27]. Advanced FEC coding can be also implemented [27]. Taking into account the importance of the polarization multiplexing, it is necessary to investigate the PMD influence on the dual-polarization transmission [5], [6].

In WPT-OFDM communication systems the modulated signals are double-sideband [5]. It has been shown that the CD influence on WPT-OFDM is compensated automatically due to the phase symmetry of the both sidebands while the PMD does not possess the phase symmetry [5]. As a result, the two sidebands have two different dispersions, their addition does not reproduce the real wavelet basis, orthogonality condition is broken, and inter-packet-interference occurs [5].

In this chapter, we considered the fundamentals of WPT based CO-OFDM communication systems and the PMD influence on such systems. It appeared to be that DT-CWPT can generate the single-sideband wavelets in frequency domain [5], [7], [8], [20], [21], [23]-[25]. We briefly discussed the properties of complex wavelets. The DT-CWPT consists of two real wavelet FBs where the second wavelet FB is the Hilbert transform of the first one and it represents an imaginary part of the CWT. The DT-CWPT approach to CO-OFDM communication systems mitigates the PMD impact on the system perfomance.

We have carried out the numerical simulations for the DT-CWPT based CO-OFDM system and QAM-16 modulation format. The constellation and BER dependence on the distance clearly show that the transmission over the distance of 1800 km with high actual **SE** of 5.85 bit/s/Hz taking into account the FEC and CP penalties, channel and timing estimation can be achieved due to the PMD mitigation provided by the DT-CWPT based CO-OFDM system.

We compared the performance of the conventional CO-OFDM and the DT-CWPT based CO-OFDM communication systems. The performance of the DT-CWPT based OFDM communication system is better than the performnace of the conventional CO-OFDM communication system: for the same modulation format and CP, BER at the distance of 1300 km for the conventional CO-OFDM is larger than BER at the distance of 1700 km for the DT-CWPT based OFDM system. We used in our simulations the already developed complex wavelets [23]-[25]. In future theoretical investigations, we suppose to find more appropriate wavelet bases using the so-called best-basis algorithms [8].

Author details

Y. Ben-Ezra and B.I. Lembrikov

Department of Electrical Engineering, Holon Institute of Technology, Holon, Israel

References

[1] Li An, Shieh W., and Tucker R.S. Wavelet packet transform-based OFDM for optical communications. Journal of Lightwave Technology, 2010; 28(24), 3519-3528.

[2] Ben Ezra Y., Lembrikov B.I., Zadok Avi, Ran Halifa R., and Brodeski D. All-optical Signal Processing for High Spectral Efficiency (SE) Optical Communication. In: Narottam Das (ed.) Optical Communication. Rijeka: InTech; 2012. p343-366.

[3] Schmogrow R., Bouziane R., Meyer M., Milder P.A., Schindler P.C., Killey R.I., Bayvel P., Koos C., Freude W., and Leuthold J. Real-time OFDM or Nyquist pulse generation – which performs better with limited resources? Optics Express 2012; 20(26) B543-B551.

[4] Bosco G., Carena A., Curri V., Poggiolini P., and Forghieri F. Performance limits of Nyquist-WDM and CO-OFDM in high-speed PM-QPSK systems. IEEE Photonic Technology Letters, 2010; 22(15) 1129-1131.

[5] Li An, Shieh W., and Tucker R.S. Impact of polarization-mode dispersion on wavelet transform based optical OFDM systems, In: proceedings of National Fiber Optic Engineers Conference, San-Diego, California, USA March 21-25, 2010, JThA5, pp.1-3.

[6] Shieh W., Yi X., Ma Y., and Tang Y. Theoretical and experimental study on PMD-supported transmission using polarization diversity in coherent optical OFDM systems. Optics Express 2007; 15(16) 9936-9947.

[7] Kingsbury N. Complex wavelets for shift invariant analysis and filtering of signals Journal of Applied and Computational Analysis 2001 10 (3) 234-253.

[8] Bayram I. and Selesnick I.W. On the dual-tree wavelet packet and M-band transforms. IEEE Transactions on Signal Processing 2008; 56(6) 2298-2310.

[9] Shieh W, Djordjevic I. Orthogonal Frequency Division Multiplexing for Optical Communications. London: Academic Press; 2010.

[10] Hillerkuss, D.et al. Simple all-optical FFT scheme enabling Tbit/s real-time signal processing, Optics Express, April 2010; 18(9) 9324-9340.

[11] Wang X.; Ho P., and Wu Y. Robust Channel Estimation and ISI Cancellation for OFDM Systems with Suppressed Features, IEEE Journal on Selected Areas in Communications 2005; 23(5) 963-972.

[12] Shieh W., Bao H., and Tang Y. Coherent optical OFDM: theory and design. Optics Express 2008; 16(2) 841-859.

[13] Armstrong J. OFDM for Optical Communications, IEEE Journal of Lightwave Technology, February 2009; 27(3) 189-204.

[14] Kikuchi K. Coherent optical communication systems, In: Kaminov, I. P.; Li, T. & Willner, A. E. (Eds.) Optical Fiber Telecommunications VB: Systems and Networks, Academic Press, Amsterdam, London, New York: Academic Press; 2008. p91-129.

[15] Da Silva E., Pataca D.M., Ranzini S. M., de Carvalho L.H.H., Juriollo A.A., da Silva M.L., Oliveira J.C.R.F. Transmission of 1.15 Tb/s NGI-CO-OFDM DP-QPSK superchannel over 4520 km of PSCF with EDFA-only amplification. Journal of Microwaves, Optoelectronics and Electromagnetic Applications, 2013; 12(SI-2) 96-103.

[16] Cincotti G., Moreolo M.S. and Neri A. Optical Wavelet Signals Processing and Multiplexing, EURASIP Journal on Applied Signal Processing, 2005; 10, 1574-1583.

[17] Rao R.M.& Bopardikar A. S. Wavelet Transforms. Introduction to Theory and Applications. Reading, Massachusetts: Addison-Wesley; 1998.

[18] Daubechies I. Ten Lectures on Wavelets. Philadelphia, Pennsylvania: Society for Industrial and Applied Mathematics; 2006.

[19] Sarkar K.T., Salazar-Palma M., Wicks M.C. Wavelet Applications in Engineering Electromagnetics. Boston: Artech House; 2002.

[20] Nerma M. H. M., Kamel N. S., Jeoti V. An OFDM based on dual tree complex wavelet transform (DT-CWT), Signal Processing: An International Journal (SPIJ), 2009; 3 (2) 14-26.

[21] Ben-Ezra Y., Brodeski D., Lembrikov B.I. High Spectral Efficiency OFDM Based on Complex Wavelet Packets. In: ICTON 2014: Proceedings of the 16th International Conference on Transparent Optical Networks, 6-10 July 2014, Graz, Austria, We.A1.4 p1-3.

[22] Kuschnerov M., Hauske F. N., Piyawanno K., Spinnler, B., Alfiad, M.S. Napoli, A., and Lankl, B. DSP for coherent single-carrier receivers, Journal of Lightwave Technology, 2009; 27(16) 3614-3622.

[23] Kingsbury N. Design of Q-shift complex wavelets for image processing using frequency domain energy minimization. In: Image Processing 2003. ICIP 2003. Proceedings of International Conference on Image Processing, 14-18 September 2003, Barcelona, Catalonia, Spain, 1, 1-1013-16.

[24] Kingsbury N. A dual-tree complex wavelet transform with improved orthogonality and symmetry properties. In: Image Processing 2000. ICIP 2000. Proceedings of International Conference on Image Processing, 10-13 September 2000, Vancouver, BC, Canada, 375-378.

[25] Selesnik I. W. Hilbert transform pairs of wavelet bases, IEEE Signal Processing Letters, 2001; 8(6) 170-173.

[26] Ma Y., Yang Q., Tang Y., Chen S., and Shieh W. 1-Tb/s single-channel coherent optical OFDM transmission over 600-km SSMF fiber with subwavelength bandwidth access. Optics Express 2009; 17(11) 9421-9427.

[27] Ip, Ezra, Lau, A.P.T., Barros, D.J.F., Kahn, J.M. Coherent detection in optical fiber systems. Optics Express, January 2008; 16(2) 753-791.

8

Nanostructured Detector Technology for Optical Sensing Applications

Ashok K. Sood, Nibir K. Dhar, Dennis L. Polla,
Madan Dubey and Priyalal Wijewarnasuriya

1. Introduction

This Chapter covers recent advances in nanostructured based detector technology, materials and devices for optical sensing applications. The authors have many years of experience working nanotechnologies that include a variety of semiconductors and other advanced materials such as GaN, ZnO, Si/SiGe, CNT and Graphene for optical sensing applications.

Optical sensing technology is critical for defense and commercial applications including optical communication. Advances in optoelectronics materials in the UV, Visible and Infrared, using nanostructures, and use of novel materials such as CNT and Graphene have opened doors for new approaches to apply device design methodology that are expected to offer enhanced performance and low cost optical sensors in a wide range of applications.

We will cover the UV band (200-400 nm) and address some of the recent advances in nano-structures growth and characterization using GaN/AlGaN, ZnO/MgZnO based technologies and their applications. We will also discuss nanostructure based Si/SiGe technologies (400-1700 nm) that will cover various bands of interest in visible-near infrared for detection and optical communication applications. The chapter will also discuss some of the theoretical and experimental results in these detector technologies.

Recent advancements in design and development of CNT based detection technologies have shown promise for optical sensor applications. We will present theoretical and experimental results on these device and their potential applications in various bands of interest.

2. UV, Visible and infrared spectrum and bands of interest

The Ultraviolet spectrum has been of interest for a variety of sensors for defense and commercial applications. The UV band is from 250-400 nanometers as shown in the figure 1. This band can be further divided into UVA and UVB bands. Each of these bands has applications for sensors, detectors and LED applications.

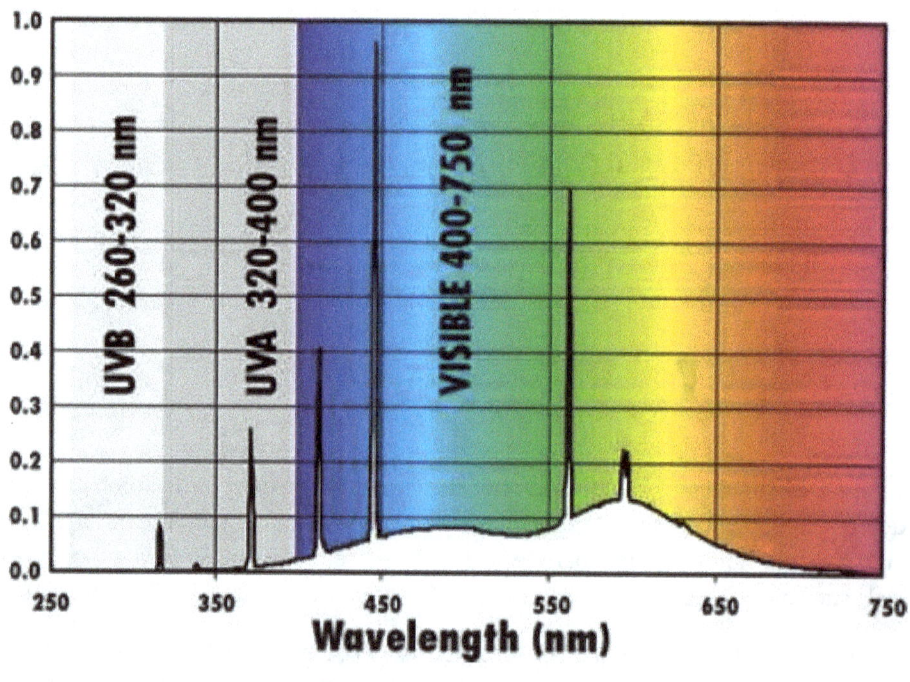

Figure 1. Overview of UV and Visible Spectral Band [1]

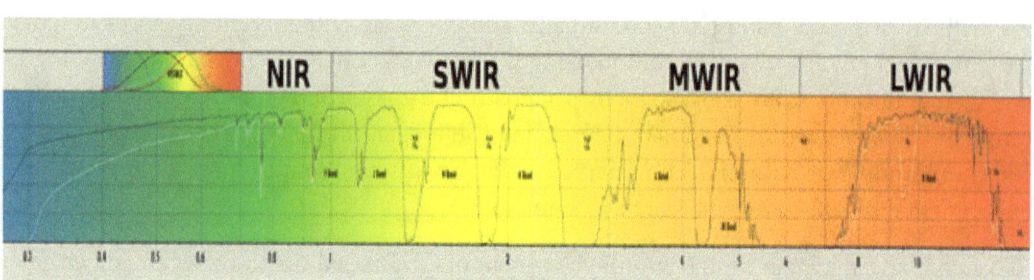

Figure 2. Definition of IR Spectral Band [1].

The word "infrared" refers to a broad portion of the electromagnetic spectrum that spans a wavelength range from 1.0 um to beyond 30 um everything between visible light and microwave radiation. Much of the infrared spectrum is not useful for ground- or sea-based imaging because the radiation is blocked by the atmosphere. The remaining portions of the spectrum are often called "atmospheric transmission windows," and define the infrared bands that are usable on Earth. The infrared spectrum is loosely segmented into near infrared (NIR, 0.8-1.1um), short wave infrared (SWIR, 0.9-2.5um), mid wave infrared (MWIR, 3-5um), long wave infrared (LWIR, 8-14um), very long wave infrared (VLWIR, 12- 25um) and far infrared (FIR, > 25um), as shown in Figure 2. The MWIR- LWIR wavebands are important for the imaging of objects that emit thermal radiation, while the NIR-SWIR bands are good for imaging scenes that reflect light, similar to visible light.

Since NIR and SWIR are so near to the visible bands, their behavior is similar to the more familiar visible light. Energy in these bands must be reflected from the scene in order to produce good imagery, which means that there must be some external illumination source. Both NIR and SWIR imaging systems can take advantage of sunlight, moonlight, starlight, and an atmospheric phenomenon called "nightglow," but typically require some type of artificial illumination at night. In lieu of photon starved scenes, arrays of infrared Light Emitting Diodes (LEDs) can provide a very cost effective solution for short-range illumination. However, achieving good performance at distances of over hundreds of meters requires more directed illumination, such as a focused beam from a laser or specialized spotlight, although special consideration of eye-safety issues is required.

3. Ultraviolet nanostructured detector array development

3.1. Applications of UV imaging technology

Imagery for identification of targets at various distances uses visible cameras, image intensifiers, shortwave IR cameras and long wave uncooled cameras. Each have distinct advantages and disadvantages and are each useful under specific sets of conditions such as light level, thermal conditions, and level of atmospheric obscuration. The shortest wavelength is desired for spatial resolution which allows for small pixels and large formats. [2- 6]

Visible cameras, if adequate light level is present, can provide high resolution, but for long range identification even under moonlit and starlit illuminations, long integration times and large optics are required and dust, smoke and fog easily defeat a single visible camera. Image intensifiers and SWIR cameras are useful in many conditions as the SWIR penetrates fog easily but requires fairly clear night skies for the upper atmospheric airglow light source, and image intensifiers require a certain level of celestial (starlight, moonlight) or light pollution irradiance. Both the SWIR and image intensifiers are limited by the diffraction resolution of the NIR to SWIR wavelengths [5-6].

For optimal resolution, the visible or ultraviolet spectrum is preferable; however, active (laser) illumination is required for long-range night imaging. Covert UV illumination is preferred over the visible and the atmosphere transmits fairly well at the longer UV wavelengths. The covert active system for high-resolution identification modeled in this paper consists of a UV laser source and a silicon CCD, AlGaN or AlGaN APD focal plane array with pixels as small as 4 microns that are spectrally tuned for the solar-blind region of the UV spectrum. The solar-blind region is optimal as virtually all of the solar radiation is absorbed at the higher altitudes leaving a pitch dark terrain even in bright day, yet for sea-level path lengths of 1 km and shorter; the UV atmospheric transmittance is still acceptable.

This combination is ideal for exploitation by a UV illuminator and UV FPA sensor. Current UV lasers can provide either continuous or pulsed energy at levels detectable by solar-blind UV detectors under relatively small optics and at 30 Hz frame rates, providing real-time high-resolution (on the order of 1 cm at 1 km) imagery. At these illumination levels and target ranges, both standard PN, PIN and APD UV detectors and silicon CCD's can be used for target identification. The model has been developed and used to include the combined effects of detector and electronics, atmospheric transmittance and UV background radiance, target size, range and reflectance, and UV laser attributes to simulate and predict both CW and pulsed laser imaging performance and to assist in the design of this prototype system [6].

3.1.1. *Model development for passive and active UV systems design*

The general equations for SNR prediction for laser illumination and APD are derived.

$$SNR = \frac{G e_{Lret}}{[F^2 G^2 (e_{Lret} + e_{bk} + e_{dk}) + (\tilde{e}_{n,amp})^2]^{1/2}} \tag{1}$$

Where G is the APD gain, F is the excess noise, the noise electron terms are the laser return shot noise, the scene noise, the dark current noise and the amp noise

Two special and frequently occurring cases are (2) for the laser power noise limited case and (3) for the amp noise limited case:

$$SNR = \frac{[e_{Lret}]^{1/2}}{F} \tag{2}$$

$$SNR = \frac{G e_{Lret}}{(\tilde{e}_{n,amp})} \tag{3}$$

The laser return in electrons for cw assuming lambertian reflection is:

$$e_{Lret,ti}=[t_i\eta\tau_o][P_{Lcw}\frac{\lambda}{hc}]\tau_a\frac{\Omega_{pix}}{\Omega_L}\rho_{tar}\tau_a^2\frac{A_o}{\pi R^2}=[t_i\eta\tau_o][P_{Lcw}\frac{\lambda}{hc}]\tau_a\frac{4A_{pix}}{\pi\theta_{fdiv}^2 f^2}\rho_{tar}\tau_a^2\frac{A_o}{\pi R^2} \tag{4}$$

Or when separated into detector/optics, atmosphere, laser and target attributes:

$$e_{Lret,ti}=[t_i\eta\tau_o\frac{A_{pix}A_o}{\pi f^2}]\tau_a^2\frac{[\Phi_{Lcw}]}{\theta_{fdiv}^2}\frac{4\rho_{tar}}{\pi R^2} \tag{5}$$

If we allow for frame summing:

$$e_{Lret,ti+}=N_{fs}[t_i\eta\tau_o\frac{A_{pix}A_o}{\pi f^2}]\tau_a^2\frac{[\Phi_{Lcw}]}{\theta_{fdiv}^2}\frac{4\rho_{tar}}{\pi R^2} \tag{6}$$

For pulse laser operation and using t_{bin} which equals t_{pulse} and the number of bins per frame $N_{bins/f}$:

$$e_{Limage}=[N_{bins/f}\eta\tau_o\frac{A_{pix}A_o}{\pi f^2}]\tau_a^2\frac{P_{Lpulse}(\lambda/hc)}{\theta_{fdiv}^2}\frac{4\rho_{tar}}{\pi R^2} \tag{7}$$

3.1.2. Systems performance metrics for UV systems design

To model the sensor and system performance, we have assumed the pixel size for a high sensitivity, detector size of 5-20 microns for the UV detector array. The fill factor of 70% is assumed typical for these small pixels. Typical quantum efficiencies have been assumed to be in the 70% range for the PIN diode and APD [5-6]. The model uses as default, an amp noise of 15 electrons per frame time, a dark current of 1e-15 amps for a 5 micron pixel or 4 nA/cm^2 or 200 electrons or about 14 noise electrons, and scene noise is effectively zero in the solar-blind region.

The model from the MODTRAN runs shown in figure 3, the daytime irradiance in the UV is insignificant in the solar-blind region. The drop-off from 0.30 microns to 0.26 microns illustrates the requirement for a UV detector with spectral response is in the solar-blind region. Figure 4 shows the UV spectral radiance at midday and the comparative laser illumination of the target at 1 km for a 6 milliradian beam divergence for powers of 1 mW and 10 mW. The left plot in the figure shows that the transmittance improves with longer UV wavelengths for all three levels of aerosols and is sufficient for 1 km lengths in our solar-blind region.

To achieve high-resolution day-night imaging and identification of targets, the following conditions and requirements must be met. While linear detection (no APD and no laser illumination) is fine for muzzle flashes and images of nearby combatants illuminated by live fire (a millisecond event), laser illumination is required for cold targets (facial recognition, profile recognition). A continuous laser and 33 msec integrations are adequate if enough laser power is available. If not, a pulsed laser with nanosecond integrations and APD detectors are required to reduce atmosphere scatter and improve detector sensitivity.

SPECTRAL BANDS		
lam hi	0.267	um
lam lo	0.265	um
lam mid	0.266	um
del lam	0.002	um
DETECTOR / FPA		
format	**256**	
dpix	**5**	um
	2.50E-07	cm2
ti cont	**33.33**	msec
t bin	**20**	nsec
t quench	**3000**	nsec
gain	**1**	
gain apd	**1**	
Fm noise	**1**	
amp noise	**5.80E+10**	
tau opt	**0.90**	
qe	**0.70**	
eta inj	**1.00**	
fill factor	**0.75**	
I dark	**5.00E-16**	a
J dark	2.00E-09	a/cm²
I surface	**1.00E-21**	a
Rload	**1.00E+06**	ohm
resp frame	1.57E-02	
resp bin	9.45E-09	
	0.63	
OPTICS		
dopt	**20**	cm
Aopt	314.16	cm2
focal length	**40**	cm
fnum	2.00	

LASER			
lambda	**0.266**	um	
P laser cont	0.10	W	
Pd laser cont	3.18E-06	W/cm2	
Ph laser cont	1.34E+17	pho/sec	
Ph laser cont frame	4.46E+15	pho/frame	
E laser pulse	**5.00E-06**	joule/pulse	j/frame
Ph laser pulse	6.69E+12	pho/pulse	3.33E-03
Ph laser bin	6.69E+12	pho/bin	watts
div full	**4**	mrad	0.100
t pulse	20	nsec	
w pulse	6	meters	

15 … uW/cm2

SCENE / TARGET TIMING			
t transit	3.33E-06 sec		lin overfill
	3333 nsec		1.25
	3.33 usec		FPA FOV (ft)
N pulses/fr	9999 max poss		5.25

bin and pulse rates	det based	mission based	det/mssion min	min in kHz	used (Hz)	used (kHz)
sample rate (max) Hz	3.31E+05	3.00E+05	3.00E+05	300.0		
N pulses/frame	11036	9999	9999	300.0	666.6	0.6666
N bins/frame	11036	9999	9999	500.0	666.6	0.6666
Dlaser tar (cm, cm²)	200	3.14E+04	1			
diameter laser at target	6.56 ft		q factor			

ELECTRONS AND NOISE from laser, scene, dark current and amp						647.5	
	electrons frame full int	electrons for bins in frame	**electrons per bin**	noise e frame full int	noise e for bins in frame	noise e per bin	
e laser ret (s)	9713.1	647.5	9.71E-01	98.55	25.45	9.86E-01	98.55
e laser ret cont	647.5	*na*	*na*	25.45	*na*	*na*	25.45
elect scene (b)	0	9.73E-17	1.46E-19	0.00	0.00	3.82E-10	0.00
e dark (d)	104	4.16E-02	6.24E-05	10.20	0.20	7.90E-03	10.20
e surface	2.08E-04	8.32E-08	1.25E-10	0.01	0.00	1.12E-05	
e kT amp				10.00	0.77	7.74E-03	
				14.28	0.80		

MODE	signal	noise s+b+d	noise b+d	SNR s+b+d	SNR b+d	SNR s+b+d	SNR b+d	
DDLM cont	**647.5**	**29.2**	**14.3**	**22.2**	**45.3**	**44.4**	**90.7**	CW
DDLM bins sum	**647.5**	**25.4**	**0.3**	**25.4**	**2267.2**	**50.9**	**4534.3**	pulsed

Figure 3. UV Sensor Model for evaluating UV Sensor Performance [6]

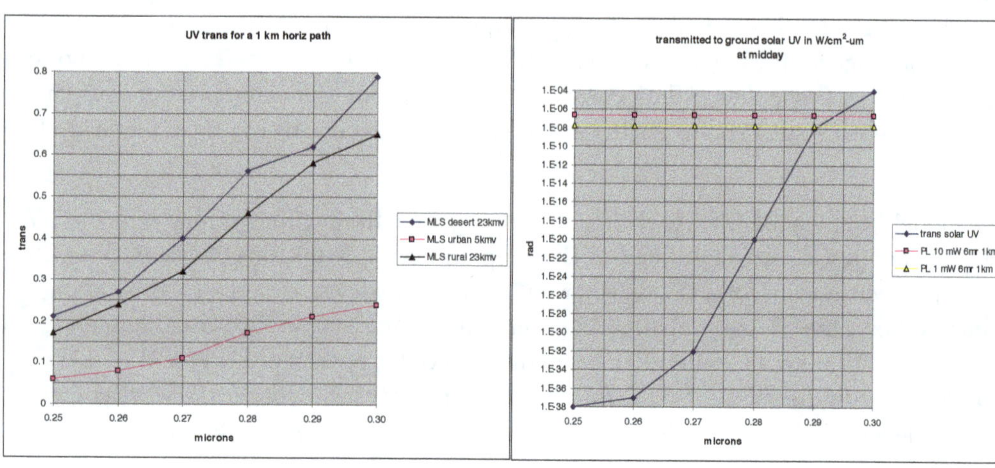

Figure 4. UV transmittance vs. wavelength for three aerosol levels (left) and UV radiance at sea level during midday and laser irradiance on the target at 1 km (6 mradian beam) from a 1mW and 10mW UV laser (right) [6]

3.2. ZnO / MgZnO nanostructures for UV applications

Zinc oxide (ZnO) is a unique wide bandgap biocompatible material system exhibiting both semiconducting and piezoelectric properties that has a diverse group of growth morphologies. Bulk ZnO has a bandgap of 3.37 eV that corresponds to emissions in the ultraviolet (UV) spectral band [7]. Highly ordered vertical arrays of ZnO nanowires (NWs) have been grown on substrates including silicon, SiO_2, GaN, and sapphire using a metal organic chemical vapor deposition (MOCVD) growth process [7]. The structural and optical properties of the grown vertically aligned ZnO NW arrays have been characterized by scanning electron microscopy (SEM), X-ray diffraction (XRD), and photoluminescence (PL) measurements [7-10]. Compared to conventional UV sensors, detectors based on ZnO NWs offer high UV sensitivity and low visible sensitivity, and are expected to exhibit low noise, high quantum efficiency, extended lifetimes, and have low power requirements [11-12]. The Photoresponse switching properties of NW array based sensing devices have been measured with intermittent exposure to UV radiation, where the devices were found to switch between low and high conductivity states at time intervals on the order of a few seconds. Envisioned applications for such sensors/FPAs potentially include defense and commercial applications [13].

Zinc oxide is a versatile functional material that provides a biocompatible material system with a unique wide direct energy band gap and exhibits both semiconducting and piezoelectric properties. ZnO is transparent to visible light and can be made highly conductive by doping. Bulk ZnO has a bandgap of 3.37 eV that includes emissions in the solar blind ultraviolet (UV) spectral band (~240-280 nm), making it suitable for UV detector applications [7]. Over this wavelength range, solar radiation is completely absorbed by the ozone layer of the earth's atmosphere, so the background solar radiation at the earth's surface is essentially zero. This enhances the capability of UV sensors in missile warning systems to detect targets such as missile plumes and flames emitting in this region.

ZnO is the basis for the one of the richest families of nanostructures among all materials taking into accounts both structure and properties. ZnO growth morphologies have been demonstrated for nanowires, nanobelts, nanocages, nanocombs, nanosprings, nanorings, and nanohelixes [7]. The development of ZnO nanowire (NW) based UV detectors offers high UV sensitivity and low visible sensitivity for missile warning related applications. Demonstration of devices using single ZnO NW strands has been widely reported in literature [7-16]. However, the development of reliable 2D arrays of aligned ZnO NWs has proven more challenging. The demonstration of reliable 2D arrays requires (1) correlation of growth process and growth parameters with the material quality of ZnO NWs, (2) correlation of the electrical and optical performance with growth parameters and fabrication processes, and (3) addressing system design challenges [17-18].

With conventional NW growth methods including electrochemical deposition, hydrothermal synthesis, and molecular beam epitaxy (MBE), it is generally difficult to scale up and control NW growth. Electrochemical deposition is well suited for large scale production but does not allow control over the NW orientation. Hydrothermal synthesis is a low temperature and low-cost process that allows growth of NWs on flexible substrates without metal catalysts, but the direction and morphology of the NWs cannot be well-controlled with this method [8-10]. The

MBE method allows monitoring of the structural quality during NW growth; however, this type of synthesis often requires use of metal catalysts as a seed layer [10], which introduces undesired defects to the structure, decreasing the crystal quality [12-16]. Chemical vapor deposition (CVD) also requires catalysts at the NW tips, and using this method the tips of the grown NWs were observed to be flat, with vertical alignment.

3.3. Characterization of ZnO NWs arrays grown on the various substrates

The samples were characterized by scanning electron microscopy (SEM) utilizing a Quanta FEG 250 system, and X-ray diffraction (XRD) using Bruker D-8 Advance X-ray diffractometer with a wavelength of 1.5406 Å corresponding to the Cu Kα line. In addition, photoluminescence (PL) measurements were performed at room temperature using a Linconix HeCd UV laser emitting at a wavelength of 325 nm. A Si detector in conjunction with at lock-in amplifier and chopper were used to measure the PL from the beam reflected off the sample at the output over the desired wavelength range [18-20].

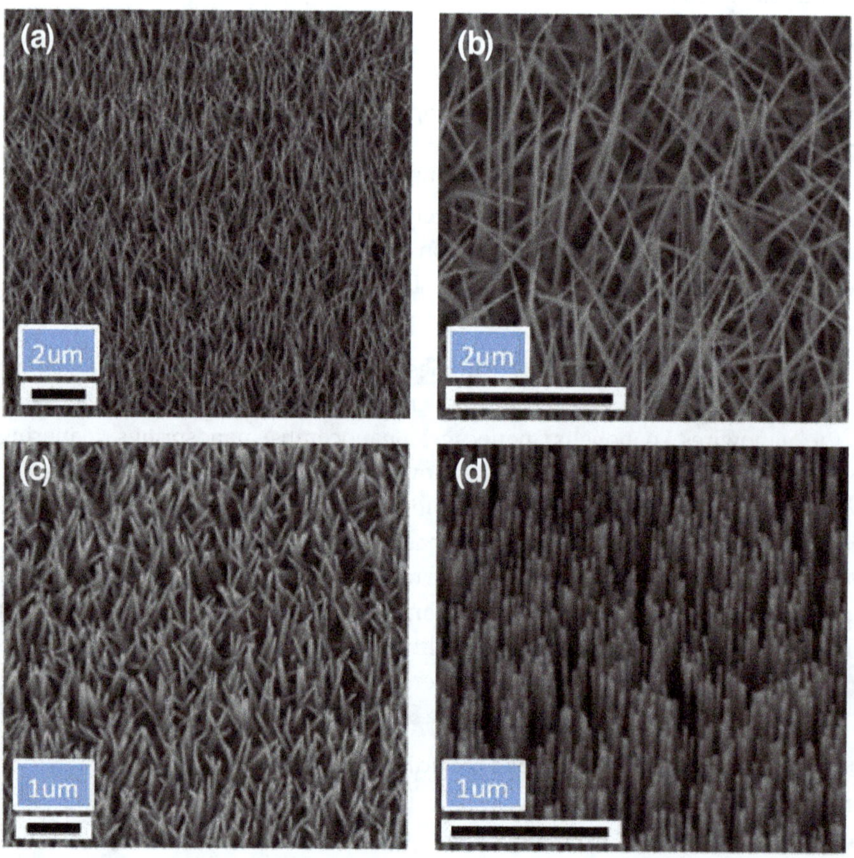

Figure 5. Scanning electron microscope (SEM) images of NWs grown on the various substrates taken at room temperature, showing NWs grown on (a) ZnO/sapphire; (b) ZnO/SiO_2/p-Si; (c) ZnO/p-Si; and (d) ZnO/GaN/sapphire.[20]

SEM was performed to explore the NWs morphology. Figure 5 show the synthesized ZnO NWs on the various substrates, which can be generally seen to have uniform distribution density. The ZnO NWs grown on sapphire [Figure 5(a)] had approximate diameters of 50-70 nm and lengths in the range of 1-2 μm. NWs grown on SiO_2 [Figure 5(b)] had diameters of 150-200 nm and lengths of 1-2 μm, and were the least vertically oriented and associated with a relatively high lattice mismatch. NWs grown on the Si (111) substrate [Figure 5(c)] had a slightly random orientation, also having diameters in the range of 150-200 nm and lengths from 1-2 μm. Finally, the NWs grown on GaN [Figure 5(d)] showed strong vertical orientation, with diameters of 20-40 nm and lengths of 0.7-1.0 μm [20].

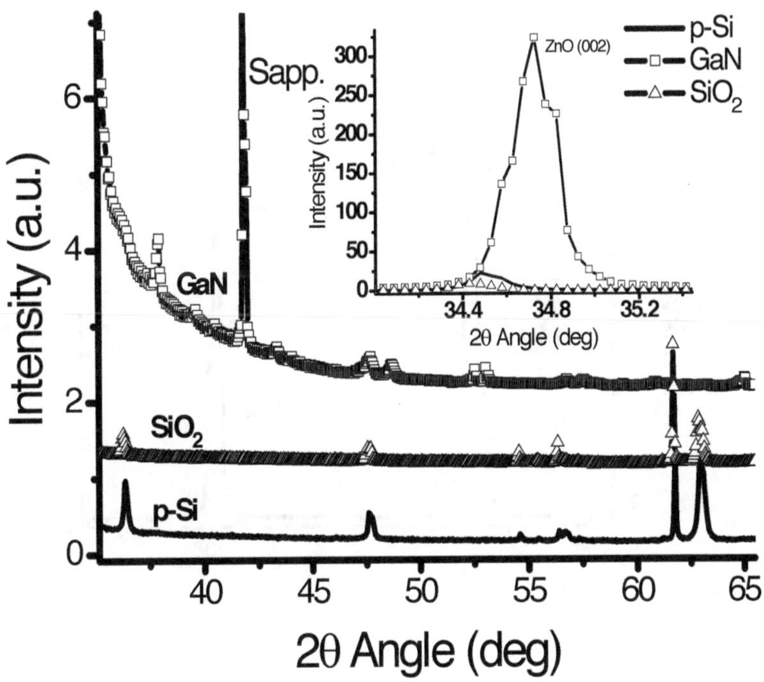

Figure 6. X-ray diffraction (XRD) of ZnO NWs grown using MOCVD on p-Si (solid), GaN/sapphire (square) and SiO_2 (triangle). The inset shows the ZnO peak associated with ZnO oriented along (002) and GaN [20].

Figure 6 shows the XRD pattern for the ZnO NWs grown on p-Si, GaN, and SiO_2 substrates [10]. The inset of Figure 2 shows dominant peaks related to ZnO (002). The peak at 34º (2θ) for ZnO grown on p-Si and SiO_2 substrates incorporated the overlapping of ZnO NWs (002) and

ZnO thin film (002). An additional diffraction peak associated with GaN was present for the GaN/sapphire substrate. ZnO NWs oriented along the (002) direction had full-widths at half maxima (FWHM) and c-lattice constants of 0.0498 (θ) and 5.1982 Å at 34.48° (2θ) for p-Si, 0.0497(θ) and 5.1838 Å at 34.58° (2θ) for GaN, 0.0865(θ) and 5.1624° at 34.38º (2θ) for SiO_2, and 0.0830°(θ) and 5.2011 Å at 34.46º (2θ) for sapphire.

The quality of the ZnO epilayers utilized as seed layers to grow ZnO NWs was also characterized. The ZnO thin films were oriented along (002) and had a maximum at 34.58º with FWHM of 0.0697 (θ) for p-Si, maximum of 34.58º with FWHM of 0.0684 (θ) for GaN, and maximum of 34.43º with FWHM of 0.0557 (θ) for SiO_2. Additional shallow diffraction peaks were observed for NWs grown on p-Si and SiO_2, which are attributed to ZnO (100, 101, 102 and 110) as can be seen from Figure 6. As shown in Figure 7, for ZnO NW growth on sapphire major peaks were observed for ZnO (002) at 34.46° (2θ) and Al_2O at 37.91° (2θ), with a minor peak for ZnO (101) at 36.34° (2θ).

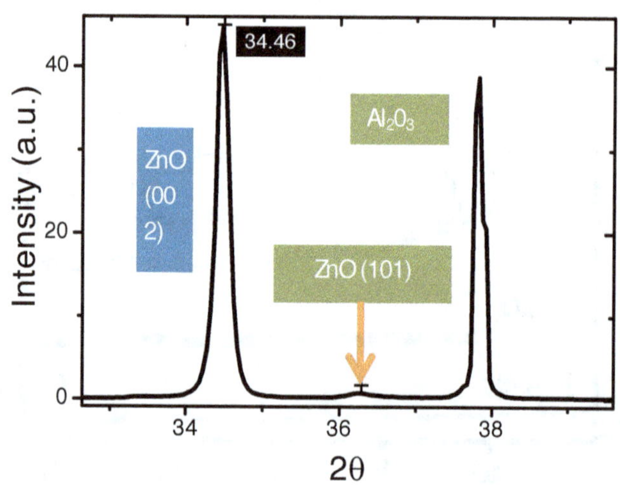

Figure 7. XRD of ZnO NWs grown using MOCVD on sapphire [20]

3.4. Photoluminescence (PL) measurements

Figure 8 shows the PL spectra for ZnO NWs grown on p-Si, GaN, and SiO_2 substrates [10]. The room temperature PL measurements were performed using a ~280 nm light source. Single peaks located at 380 nm having a FWHM of 14.69 nm and at 378 nm having a FWHM of 15 nm were observed for p-Si and SiO_2 substrates, respectively, corresponding to the recombination of excitons through an exciton-exciton collision process [18-20].

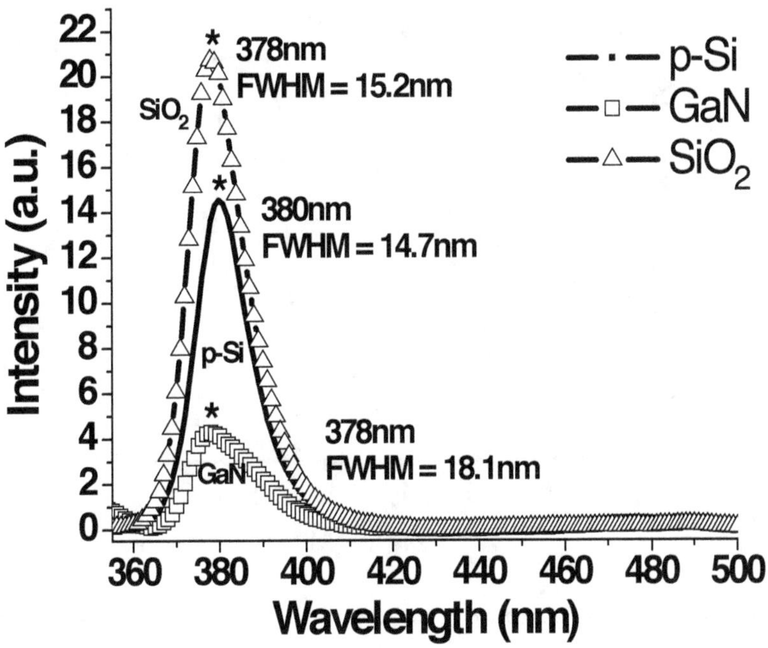

Figure 8. Photoluminescence (PL) of ZnO NWs grown on p-Si (100) (solid) with a single peak at 380 nm, GaN (square) with a stronger peak at 378 and SiO_2 (triangle) with a single peak at 378 nm [20].

No defects related to Zn or O vacancies were observed, which can be attributed to the confinement of defects at the ZnO thin film/substrate interface. For the ZnO NWs grown on GaN, a predominant peak with a FWHM of 18.18 nm was observed at 378 nm. High stress was evident for ZnO NWs grown on GaN, which can be observed in Figure 2; this can contribute to the broadening of the peak in comparison to p-Si and SiO_2. Shallow peaks identified at 474 nm and 490 nm through Lorentzian decomposition are attributed to oxygen interstitial and oxygen vacancies, respectively [20].

A UV LED lamp acquired from Sensor Electronic Technology Inc. was used to characterize the UV Photoresponse of the ZnO NW arrays [20]. The lamp comprises eight separate AlGaN based UV LEDs in a TO-3 package spanning the 240-370 nm wavelength range, with a customized power supply capable of independently monitoring and controlling the current of all or any of the LEDs. The Photoresponse was determined by first applying voltage between indium contacts on the front and back sides of a Si NW sample and measuring the resulting current in the dark, and then repeating this procedure while the sample was exposed to radiation from a UV LED at a specific wavelength.

Figure 9 shows the on-off switching characteristics of a ZnO vertical array NW device when exposed to radiation at 370 nm. This device was found to switch between low and high

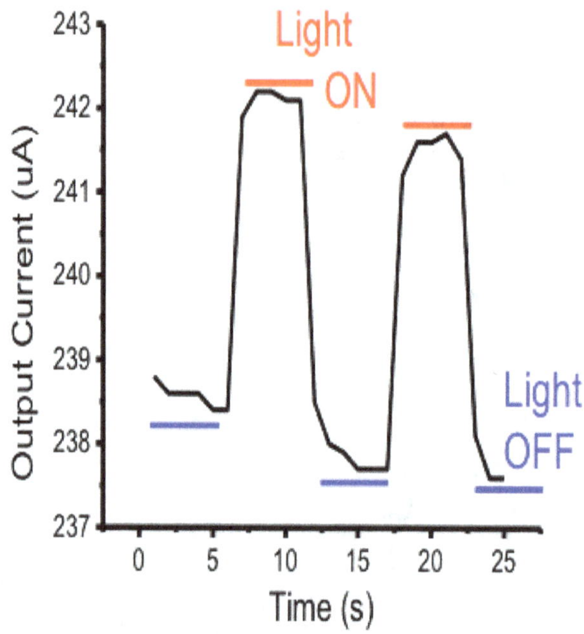

Figure 9. Switching Photoresponse characteristics of ZnO NW device when UV LED source at ~370 nm turned on and off over approximately 10 s intervals.[20]

conductivity states in approximately 3 s, a faster response than most reported thus far for ZnO NW based UV detectors.

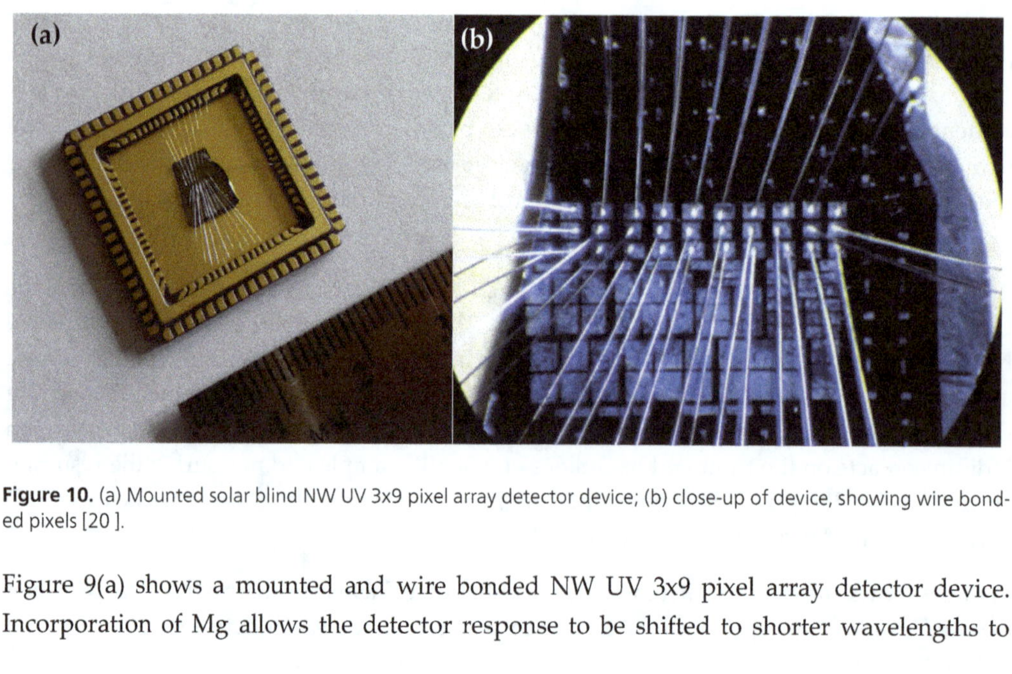

Figure 10. (a) Mounted solar blind NW UV 3x9 pixel array detector device; (b) close-up of device, showing wire bonded pixels [20].

Figure 9(a) shows a mounted and wire bonded NW UV 3x9 pixel array detector device. Incorporation of Mg allows the detector response to be shifted to shorter wavelengths to provide detection in the solar blind region. This device was tested by applying a bias between the top contacts on the pixels, which are apparent in Figure 9(b), and the back contact.

ZnO nanowires based arrays offer high sensitivity and have potential application in UV imaging systems. ZnO nanowire array based UV detectors have no moving parts, high quantum efficiency, extended lifetimes, low noise, low power requirements, and offer high sensitivity.

ZnO nanowires have also been evaluated for providing remote power for the stand alone sensors. This type of application has been extensively studied by Professor Z.L. Wang and his team at Georgia Tech [21, 22]. They have shown that ZnO nanowires can be used as nano-generators for providing remote power using the Piezo-electric effect. Photovoltaic cells or solar cells are a popular renewable energy technology, relying on approaches such as inorganic p-n junctions, organic thin films, and organic-inorganic heterojunction. However, a solar cell works only under sufficient light illumination, which depends on the location the devices will be deployed, as well as the time of the day and the weather.

Considering that mechanical energy is widely available in our living environment, They have demonstrated [21] the first hybrid cell for concurrently harvesting solar and mechanical energy through simply integrating a dye-sensitized solar cell (DSSC) and a piezoelectric nanogenerator on the two sides of a common substrate. After this, in order to solve the encapsulation problem from liquid electrolyte leakage in the first back-to-back integrated HC, early in 2011, Xu and Wang improved the prototype design of the HC and developed a compact solid state solar cell. This innovative design convoluted the roles played by the NW array to simultaneously perform their functionality in a nanogenerator and a DSSC. The design and the performance are shown in figure 11.

Based on these demonstrations of HCs for concurrently harvesting solar and mechanical energies, they have. reported an optical fiber-based three-dimensional (3D) hybrid cell, consisting of a dye-sensitized solar cell for harvesting solar energy and a nanogenerator for harvesting mechanical energy; these are fabricated coaxially around a single fiber as a core–shell structure (Figure 11). The optical fiber, which is flexible and allows remote transmission of light, serves as the substrate for the 3D DSSC for enhancing the electron transport property and the surface area, and making it suitable for solar power generation at remote/concealed locations. The inner layer of the HC is the DSSC portion, which is based on a radically grown ZnO NW array on an optical fiber with ITO as the bottom electrode. The dye-sensitized ZnO NW array was encapsulated by a stainless steel capillary tube with a Pt-coated inner wall as the photo- anode for the DSSC. The stainless steel tube also serves as the bottom electrode for the outer layer of the nanogenerator, with densely packed ZnO NWs grown on its outer wall.

Another exciting application of ZnO nanowires is designing, fabricating, and integrating arrays of nanodevices into a functional system are key to transferring nanoscale science into applicable nanotechnology as shown in Figure 12.

Recent work [22] on three-dimensional (3D) circuitry integration of piezotronic transistors based on vertical zinc oxide nanowires as an active taxel-addressable pressure/force sensor matrix for tactile imaging. Using the piezoelectric polarization charges created at a metal-semiconductor interface under strain to gate/modulate the transport process of local charge carriers, we designed independently addressable two-terminal transistor arrays, which convert mechanical stimuli applied to the devices into local electronic controlling signals.

The device matrix can achieve shape-adaptive high-resolution tactile imaging and self-powered, multidimensional active sensing. The 3D piezotronic transistor array may have applications in human-electronics interfacing, smart skin, and micro- and nano-electromechanical systems.

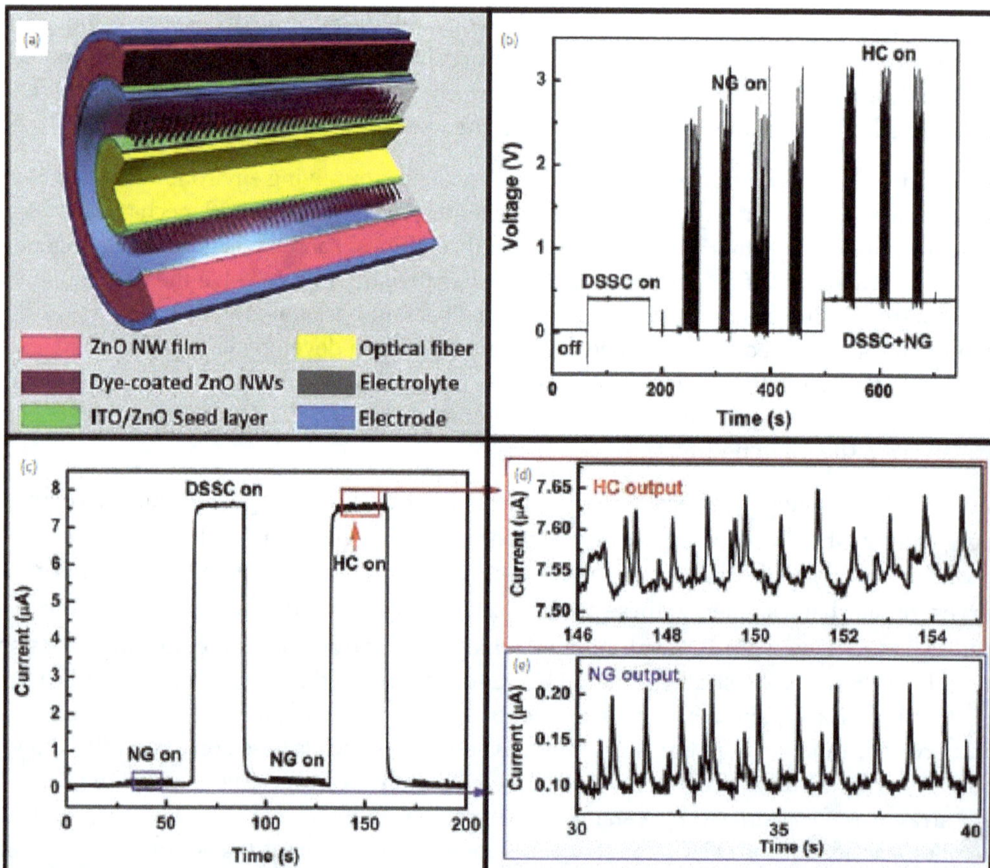

Figure 11. Design and performance of a 3D optical fiber based hybrid cell (HC) consisting of a dye-sensitized solar cell (DSSC) and a nanogenerator (NG) for harvesting solar and mechanical energy. (a) The 3D HC is composed of an optical fiber based DSSC with capillary tube as counter electrode and a NG on top. (b) Open-circuit voltage (VOC) of the HC when the NG and the DSSC are connected in series, where VOC(HC)= VOC(DSSC)+ VOC(NG). (c) Short-circuit current (ISC) of the HC when the NG and the DSSC are connected in parallel. (d) and (e) Enlarged view of ISC(HC) and ISC(NG), clearly showing that ISC(NG) is 0.13 μA, the ISC(DSSC) is 7.52 μA, and the ISC(HC) is about 7.65 μA, nearly the sum of the output of the solar cell. [21].

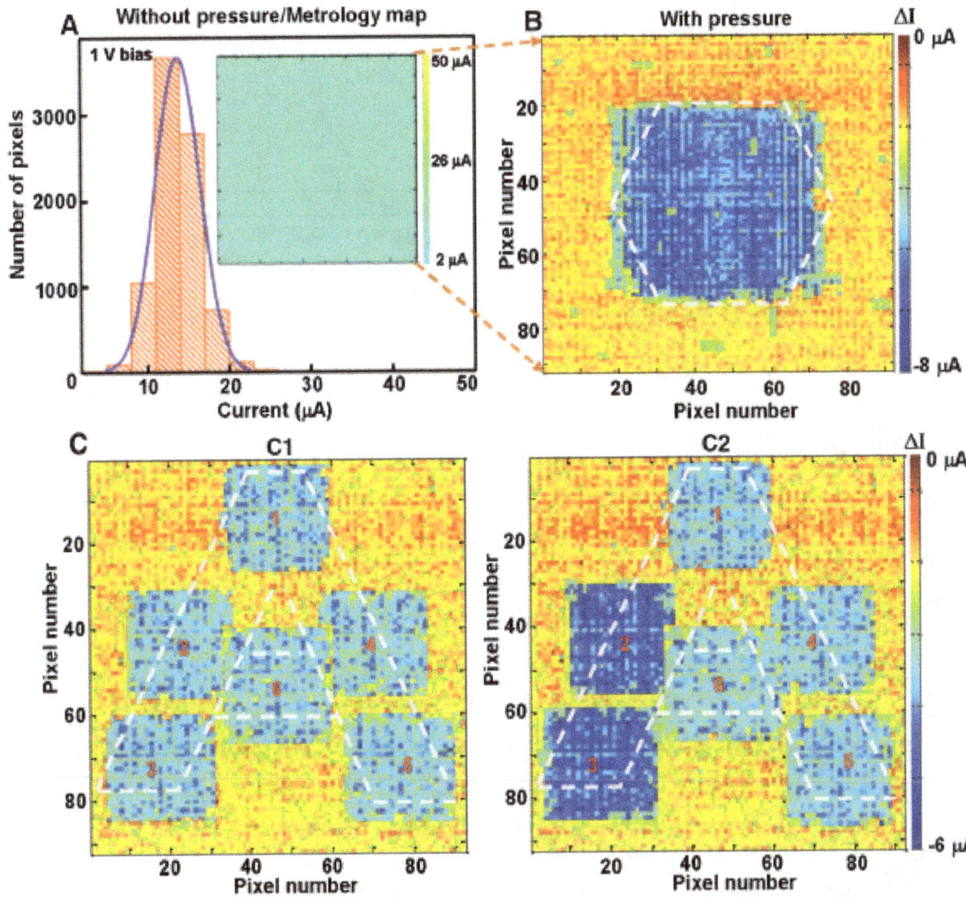

Figure 12. Tactile imaging and multidimensional sensing by the fully integrated 92 × 92 SGVPT array. (A) Metrology mapping (inset) and statistical investigation of the fully integrated SGVPT array without applying stress. (B) Current response contour plot illustrating the capability of SGVPT array for imaging the spatial profile of applied stress. Color scale represents the current differences for each taxel before and after applying the normal stress. The physical shape of the applied stress is highlighted by the white dashed lines. (C) Multidimensional sensing by an SGVPT array exhibits the potential of realizing applications such as personal signature recognition with maximum security and unique identity. The shape of a "written" letter A is highlighted by the white dashed lines. [22].

4. Development of GaN UVAPD for ultraviolet sensor applications

High resolution imaging in UV bans has a lot of applications in Defense and Commercial applications. The shortest wavelength is desired for spatial resolution which allows for small pixels and large formats. UVAPD's have been demonstrated as discrete devices demonstrating gain. The next frontier is to develop UV APD arrays with high gain to demonstrate high resolution imaging. We will discuss model that can predict sensor performance in the UV band using APD's with various gain and other parameters for a desired UV band of interest. SNR's can be modeled from illuminated targets at various distances with high resolution under standard atmospheres in the UV band and the solar blind region using detector arrays with unity gain and with high gain APD's [23-26].

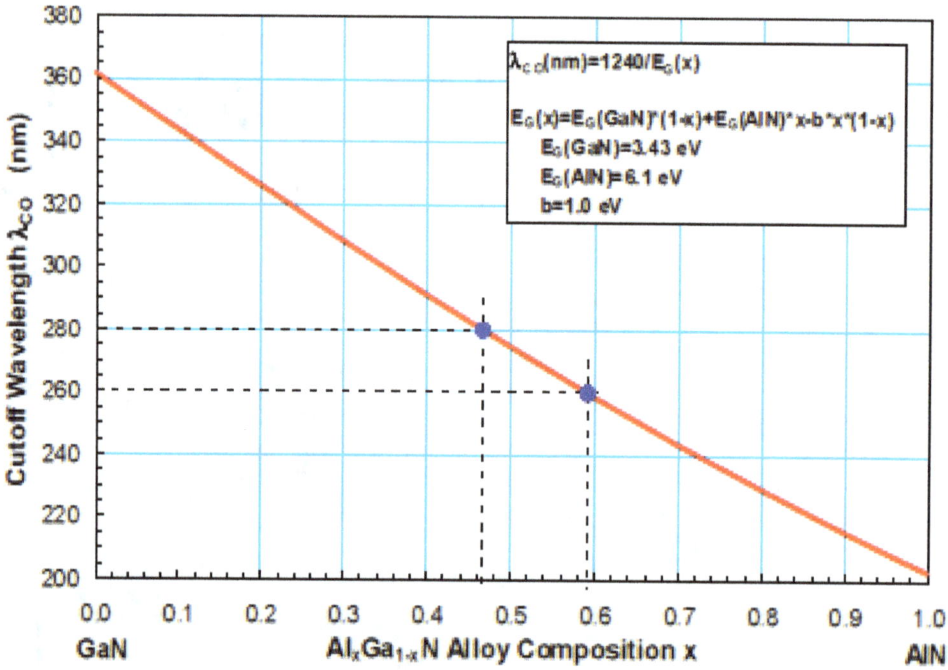

Figure 13. Relationship between alloy composition of AlGaN and the corresponding spectral cutoff for the UV detector arrays [23].

Figure 13 presents the relationship between the alloy composition of Gallium and Aluminum in $Al_xGa_{1-x}N$ that determines the cut-off wavelength of the UV detector for p-i-n [23-24] and also for UV APD's. Deep Ultra Violet (DUV) will require addition of larger composition of Aluminum in $Al_xGa_{1-x}N$. [25].

5. GaN /AlGaN UV APD growth

Figure 14 presents the High-Temperature MOCVD system by Aixtron. This new reactor design and capability has the ability to grow high quality GaN and AlGaN material with doping for GaN/AlGaN UVAPD applications [26].

Figure 15 presents the device structure of a back-side illuminated AlGaN UV APD. The substrate in this device structure is double side polished AlN substrate. The use of AlN substrate allows the UV APD device structure to be back-side illuminated and can be integrated with silicon CMOS electronics. Figure 16 presents the Reciprocal Space mapping of AlGaN on AlN substrate and Sapphire substrate. The data for sapphire substrate shows increased strain and mosaicity compared with AlN substrate.

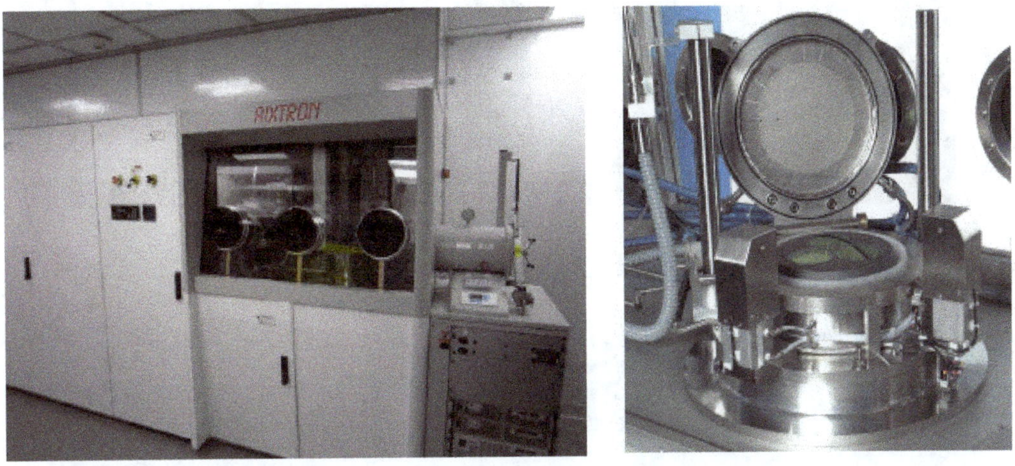

Figure 14. Photograph of New-generation AIXTRON CCS 3x2 " high temperature III-Nitride 3x2 MOCVD growth Chamber open for loading wafers showing close-coupled showerhead [26]

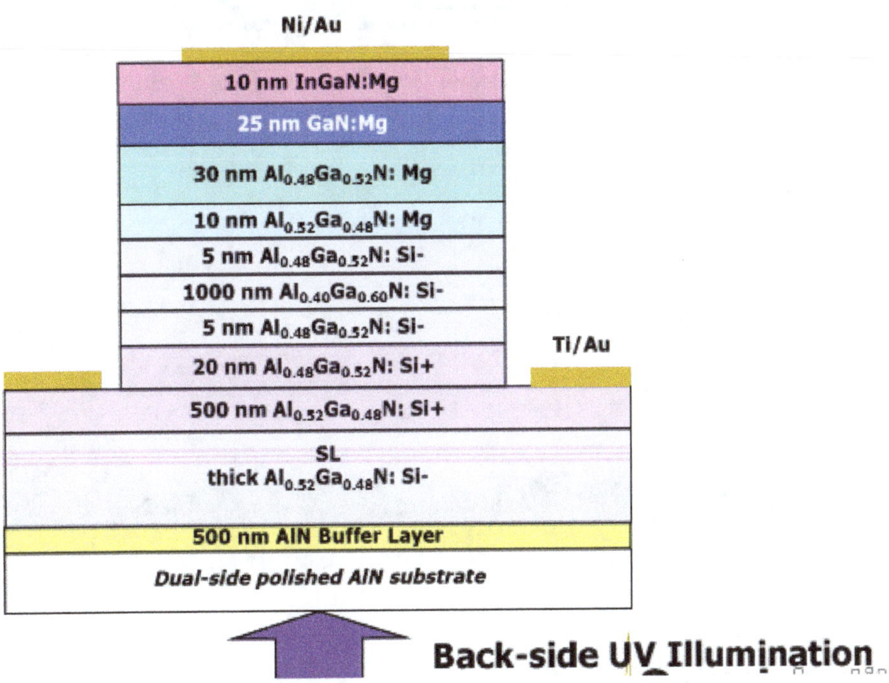

Figure 15. Device Structure Cross-section of prototype Back-Side Illuminated AlGaN UV APD [27]

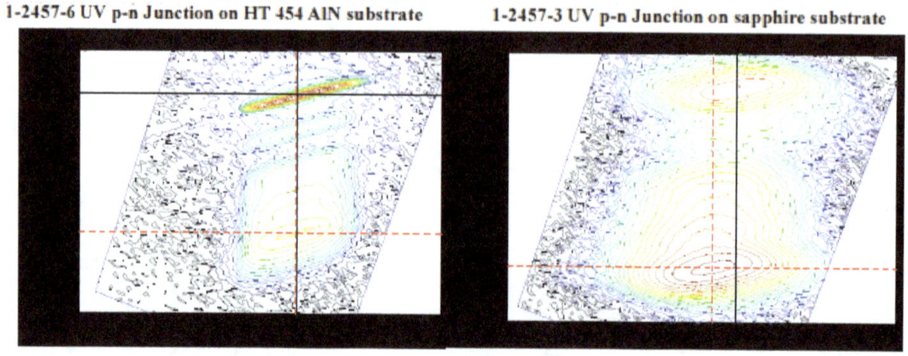

Figure 16. Reciprocal Space mapping of AlGaN p-n junctions on AlN and Sapphire Substrates [28]

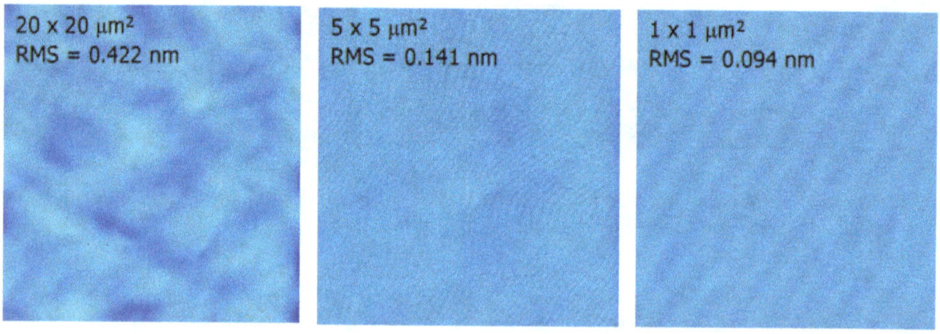

Figure 17. Microscopic surface morphology using AFM on GaN p-i-n structure grown on GaN/Sapphire template. No specific surface defects are observed [27]

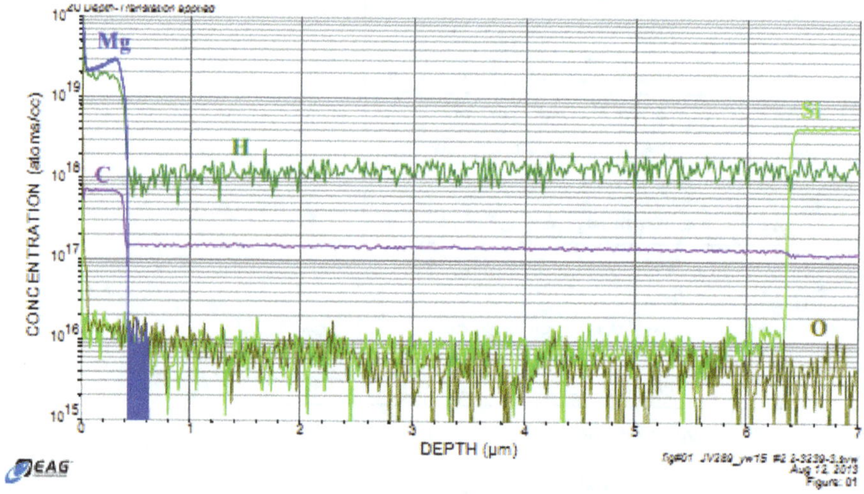

Figure 18. SIMS analysis of GaN p-i-n structure on GaN/Sapphire template, the data shows controlled Si and Mg doping for n- and p-type layers. The data shows low background doping concentration in GaN layer [28]

Figure 17 presents the microscopic surface morphology using AFM on GaN p-i-n structure grown on GaN/Sapphire template. No surface defects are observed. These results are encouraging to develop a low cost back-side illuminated UV APD detector array.

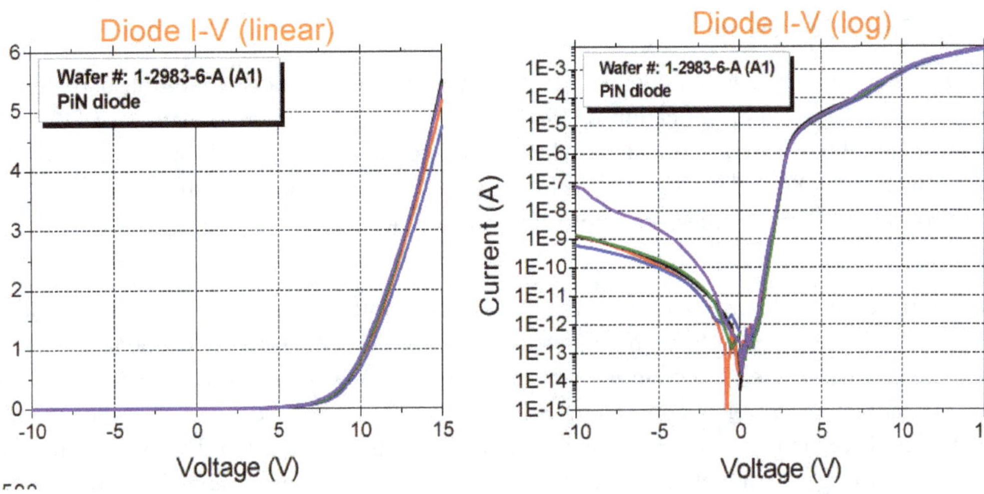

Figure 19. Current - Voltage characteristics of AlGaN UV APD Unpassivated test structure. Further reduction in the dark current will be expected with surface passivation [29].

Figure 18 presents the SIMS analysis of GaN p-i-n structure on GaN/Sapphire template, the data shows controlled Si and Mg doping for n- and p-type layers. The data shows low background doping concentration in GaN layer. The Mg doping is being increased for better p-type conductivity

Figure 19 presents the Current-Voltage characteristics of AlGaN UV APD with spectral response of 300 nm. Further reduction in the dark current will be expected with surface passivation. The future effort is underway to improve the growth characteristics Low-defect-density in substrates and high-quality epitaxial growth technologies are the keys to the successful implementation of a robust high-performance APDs for UV focal plane arrays [29-33].

6. Visible –Near Infrared (NIR) detector technologies

The Visible-Near Infrared band (0.4-1.7 μm) bridges the spectral gap between the visible and thermal bands in the electromagnetic spectrum. In this spectral band, the primary phenomenology of interest is the reflectance signature of the target, manifested as either its variations in brightness or spectral reflectance, or both.

Infrared imaging in the NIR /SWIR band offers several advantages: it can detect reflected light, offering more intuitive, visible-like images; is better suited for imaging in adverse environments and weather conditions, including fog, dust, and smoke; can also see in low light conditions, and use eye safe 1550 nm illumination; and can generate digital video outputs and thus offering a more dynamic range than traditional image intensifier night vision equipment. Under low light conditions, the sensitivity of the focal plane array is ultimately determined by the R_0A product of the photodiode.[34-36].

6.1. Si_{1-x} Ge_x (SiGe) detector arrays

Like the other two alloy semiconductors mentioned above, SiGe is another example of material that can be used for the fabrication of IR detectors. The key attractive feature of SiGe IR detectors is that they can be fabricated on large diameter Si substrates with size as large as 12-inch diameter using standard integrated circuit processing techniques. Furthermore, the SiGe detectors can be directly integrated onto low noise Si ROICs to yield low cost and highly uniform IR FPAs.

Some of the earlier attempts in developing SiGe IR detectors focused on their LWIR applications [34-36]. Renewed efforts are now developing these detectors for application in the NIR-SWIR band [36]. For the SiGe material to respond to the SWIR band, its cutoff wavelength is tuned by adjusting the SiGe alloy composition. Si and Ge have the same crystallographic structure and both materials can be alloyed with various Ge concentration. The lattice constant of Ge is 4.18% larger than that of Si, and for a Si_{1-x} Ge_x alloy the lattice constant does not exactly follow Vegard's law. The relative change of the lattice constant is given by [36]:

$$a_{Si1-x\ Gex} = 0.5431 + 0.01992x + 0.0002733x^2 (nm).$$

For a Si_{1-x} Ge_x layer with $x > 0$ on a Si substrate means that the layer is under compressive stress. A perfect epitaxial growth of such a strained heteroepitaxial layer can be achieved as long as its thickness does not exceed a critical thickness for stability. Beyond the critical thickness, the strain is relaxed through the formation of misfit dislocations which can cause an increase in the dark current.

Several approaches have been proposed to reduce the dark current in SiGe detector arrays by several orders of magnitude; these include Superlattice, Quantum dot and Buried junction designs [36-38]. Furthermore, some of these approaches have the potential of extending the wavelength of operation beyond 1.8-2.0 microns. The challenge is to take advantage of these innovative device designs and reduce the dark currents to 1-10 nA cm^{-2}. Figure 20 presents the SiGe /Ge detector array using buried junction approach to reduce the surface states and leakage current [36].

Figure 22 shows the Strained-Layer Superlattice (SLS) structure being evaluated for longer detector array response to 2 microns.. It consists of SiGe quantum wells and Si barrier layers, grown on p-type (001) Si substrates. Super lattices having differing Si barrier and Ge well thicknesses to control the strain are grown to optimize wavelength response and dark current.

The SiGe well thicknesses are kept below the critical layer thickness for dislocation formation. To complete the structure, the undoped superlattice is capped with a thin n+ Si cap layer to form the p-n junction. After growth the devices are patterned with a top contact, mesas are etched to provide isolation and the substrate contact is formed. The etched mesa can also be passivated to minimize surface recombination as indicated in Figure 22. The device shown in the figure 22 uses substrate illumination, as is needed for use in FPA arrays, and short wavelength response can be improved by thinning the Si substrate.

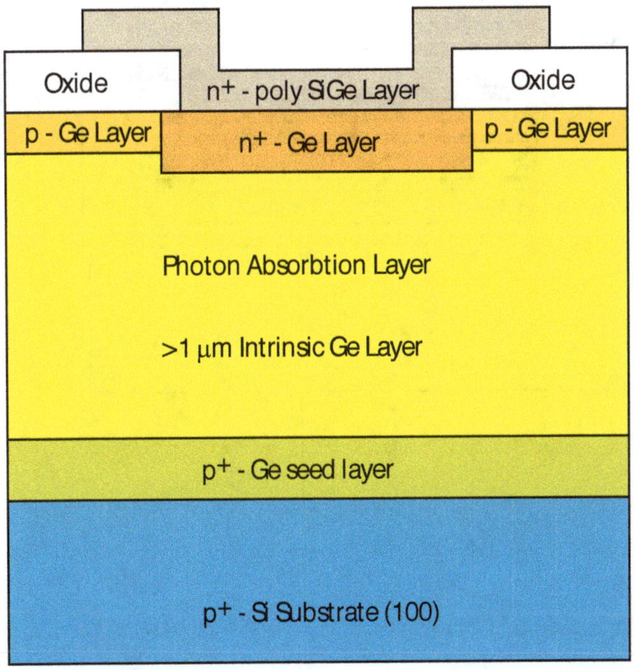

Figure 20. SiGe/Si based buried junction approach to be evaluated for reduced surface states and leakage current [36]

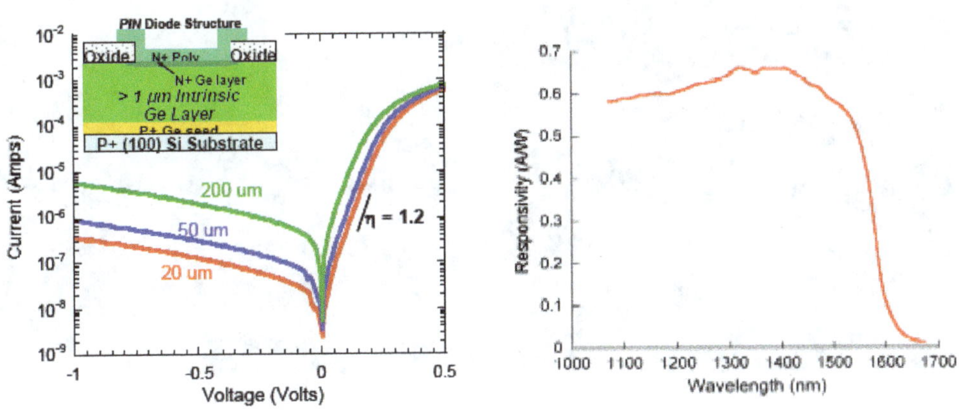

Figure 21. Measured room temperature I-V characteristics for large area diodes with 20, 50 and 200 micron unit cell. The inset shows the schematic device cross section. The spectral response data for SiGe detector is also presented. [38]

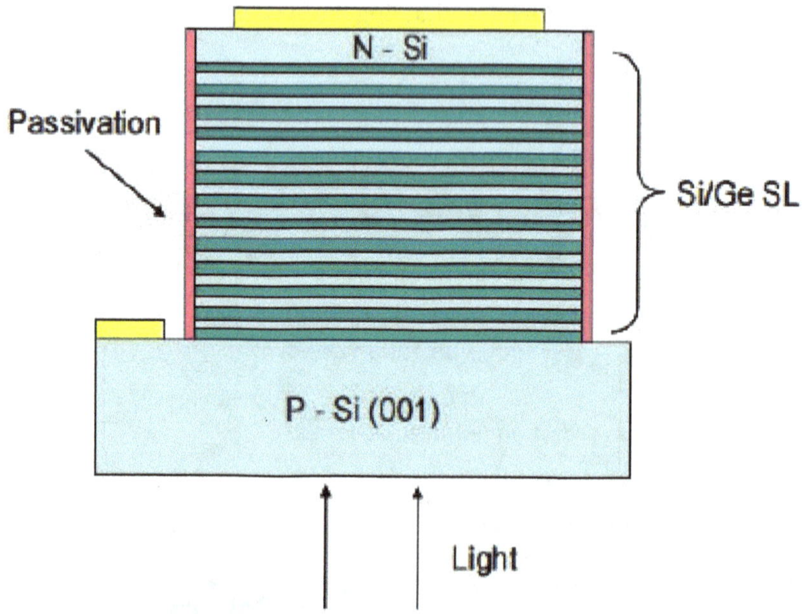

Figure 22. Schematic of detector array structure consisting of a SiGe /Si strained layer Superlattice grown on (001) silicon [38].

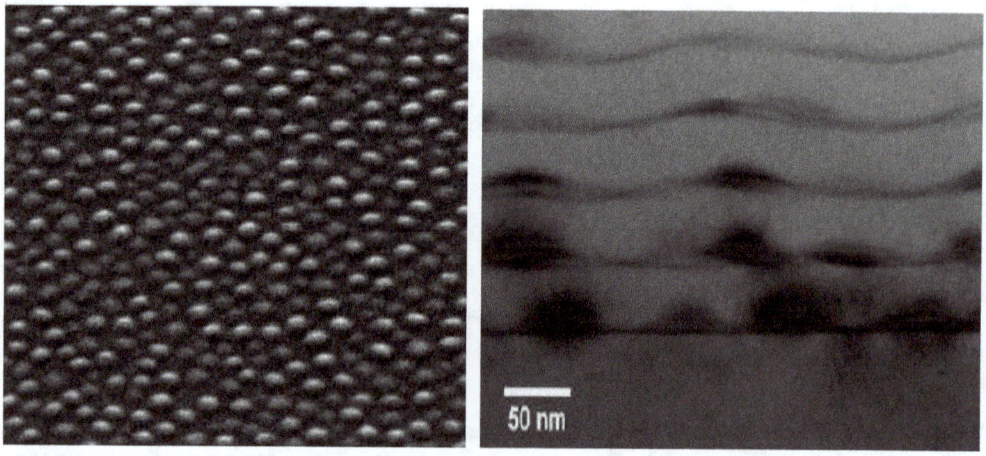

Figure 23. SEM image (45° tilt) of a Ge QD layer deposited on Si. The QDs are ~60 nm in diameter with a density of 1020 cm2. Also shown is a Cross-sectional TEM image of Ge/Si QDSL grown. Ge QDs appear with dark contrast compared to Si barriers. [38].

The strained-layer superlattice and quantum dot superlattice (QDSL) in the SiGe material system have the potential of developing Vis-NIR detector arrays with longer cutoff wavelength and potentially lower dark current. The advantage of quantum dots is the potential to exploit the optical properties of Ge while avoiding dislocation formation. Ge QDs grown on Si in Stranski-Krastanov mode can be deposited well beyond the critical thickness without dislocation nucleation [39].

Figure 23 shows an SEM image of an array of Ge nanodots grown by MOCVD. These dots are typically 50-75 nm in diameter with area coverage of ~20%. To increase optical absorption and sensitivity, MOCVD-based growth techniques is being developed for the deposition of Ge/Si quantum dot superlattices (QDSLs), where Ge QDs are alternated with thin (10-30 nm) Si barrier layers. A cross-sectional TEM image of QDSLs is shown in Figure 23.b.

7. SWIR detector technologies

The SWIR band (0.9-2.5 μm) bridges the spectral gap between the visible and thermal bands in the electromagnetic spectrum. In this spectral band, the primary phenomenology of interest is the reflectance signature of the target, manifested as either its variations in brightness or spectral reflectance, or both.

Infrared imaging in the SWIR band offers several advantages: it can detect reflected light, offering more intuitive, visible-like images; is better suited for imaging in adverse environments and weather conditions, including fog, dust, and smoke; can also see in low light conditions, and use eye safe 1550 nm illumination that is totally undetectable by regular night vision equipment; and can generate digital video outputs and thus offering a more dynamic range than traditional image intensifier night vision equipment. Under low light conditions, the sensitivity of the focal plane array is ultimately determined by the R_0A product of the photodiode.

7.1. In_x $Ga_{1-x}As$ detector array development

For SWIR imaging, InGaAs is one of the widely used detector materials due to its low dark current. The detector material can be prepared using any of the following techniques: Molecular beam epitaxy (MBE), metal-organic chemical vapor deposition (MOCVD), liquid phase epitaxy (LPE), hydride-transport vapor phase epitaxy (VPE), and atomic layer epitaxy (ALE). InGaAs layers are typically grown on lattice matched InP substrates using an alloy composition of x = 0.53 [40-42].

The spectral response typically covers 0.9-1.7μm at room temperature. By increasing the composition to x=0.82, InGaAs is able to extend its cutoff to 2.6 μm. However, the crystal defects due to epitaxy and the decreased shunt resistance, due to a smaller band gap, degrade performance at the longer cutoff wavelengths. [43].

The band gap [7-1] of the strained In_x Ga_{1-x} As: InP structure can be tailored by varying the alloy composition during crystal growth according to the equation:

$$E_g(eV)=\left(E_{g_{GaAs}}-\frac{\alpha_{GaAs}T^2}{T+\beta_{GaAs}}+\left(E_{g_{InAs}}-\frac{\alpha_{InAs}T^2}{T+\beta_{InAs}}-E_{g_{GaAs}}+\frac{\alpha_{GaAs}T^2}{T+\beta_{GaAs}}\right)x-0.475x\left(1-x\right)\right.$$

Where E_g is the band gap in (eV), α and β are fitting parameters, and x is the In: As ratio. The cut-off wavelength can be calculated from the expression $\lambda_{co}=hc/E_{gap}$

The response can be extended to include the visible wavelength range by removing the InP substrate. There has been an intensive effort to develop InGaAs arrays for Low Light Level (LLL) SWIR imaging [40-46]. An example is in astrophysical space based observatories that are very demanding on the detectors due to the very low IR flux levels. Such low flux levels represent the detection of few photons over long integration times and, therefore, require extremely low dark current photodiodes hybridized to a high performance ROIC stage. For such LLL applications there are challenges ahead to further lower noise, reduce pixel size, fabricate larger arrays, achieve higher operating temperatures, and reduce production cost.

The spectral response of InGaAs diodes at room temperature is in the 0.9 – 1.67 μm wavelength range which matches with the ambient night glow spectrum. Imaging under such low light conditions requires that the noise of the detector be extremely low. A significant portion of the noise is contributed by the dark current of the InGaAs detector and the readout noise. Dark current consists of unwanted thermally generated carriers that can cause the detector to produce a random varying output signal.

It is associated with interfacial, diffusional, G-R, and tunneling currents. The temperature dependence of the dark current is primarily due to the intrinsic carrier concentration which depends exponentially on the temperature. The dark current of the detector can be reduced through appropriate fabrication processes and device design. The impact of dark current noise as a function of read noise is shown in Figure 5.1, where the curves for different pixel pitch map the dark current noise into an equivalent read noise.

For a given read noise, the required dark current density increases as the pixel pitch is decreased. The challenge is to maintain a low dark current density as the pixel pitch is reduced. Simultaneously, the challenge for the read out circuit is to reduce the read noise. If the limitation is due to the detector and its noise level overwhelms the source signal, the solution may be to use an external illuminator or cool the detector. The choice of either solution will depend on a tradeoff between size, weight, and power requirements (SWaP).

As mentioned above, the dark current of the detector can be reduced through appropriate fabrication processes and device designs. By focusing on the growth conditions for the InGaAs absorption layer, heterointerfaces and the passivation layer, researchers have been able to demonstrate dark current density below 1.5 nA/cm2 at 77°C for 15 μm pitch arrays as shown Figure 24.

In scaling to small pixel pitch, further effort is continued to develop wafer processing parameters and methods that reduce surface related perimeter effects and enable small pixel pitch InGaAs detectors with dark current densities comparable to large (25 μm) pixel pitches detectors [5-6]. Figure 26 presents a plot of dark current density, measured at 20 °C, for eight different, 300 x 10 pixel test arrays distributed across a 3" wafer. The average dark current density at -100 mV was 2.95 nA/cm2.

Further effort is underway to demonstrate large format (>1Kx1K) and small pixel (<20μm) InGaAs focal plane arrays (FPAs) for a variety of low light level (LLL) imaging applications such as night vision. These applications demand extremely low detector dark current and Si read-out integration circuit (ROIC) noise [47].

Recent work [47] has demonstrated significant progress in InGaAs detector array development on a 4" wafer as shown in figure 27; and also reducing dark current density for 10-20μm pixel arrays, (3) developing sub-10μm pixel array technology and demonstrating the feasibility of making 5μm pixel arrays, and (4) reducing the capacitance of small pixels [47]. Figure 28 demonstrates recent results for spectral quantum efficiency (QE) as a function of wavelength measured on backside illuminated InGaAs photodiodes test array at different temperatures demonstrating Visible-Near IR response with InP substrate removed [47].

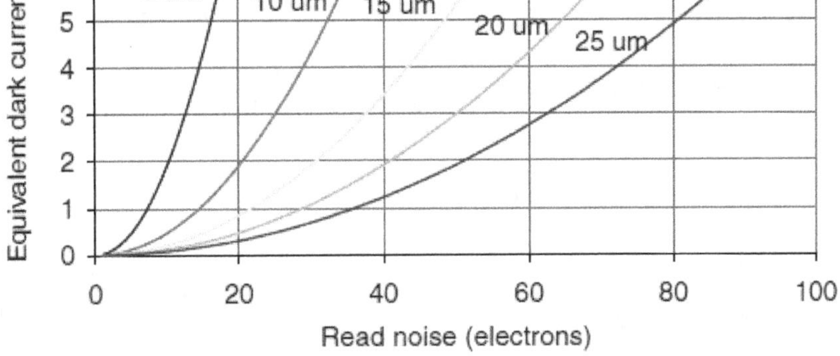

Figure 24. Dark current density versus read noise for different pixel pitches [44].

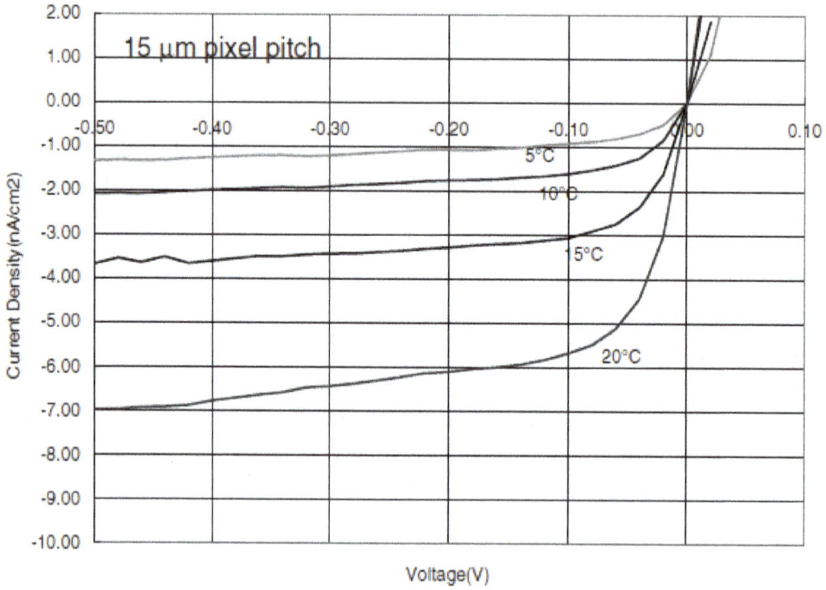

Figure 25. Dark current density at different temperatures using test structures on the wafer. Test arrays have 225 pixels (15 μm pitch) and the guard ring is not biased [45].

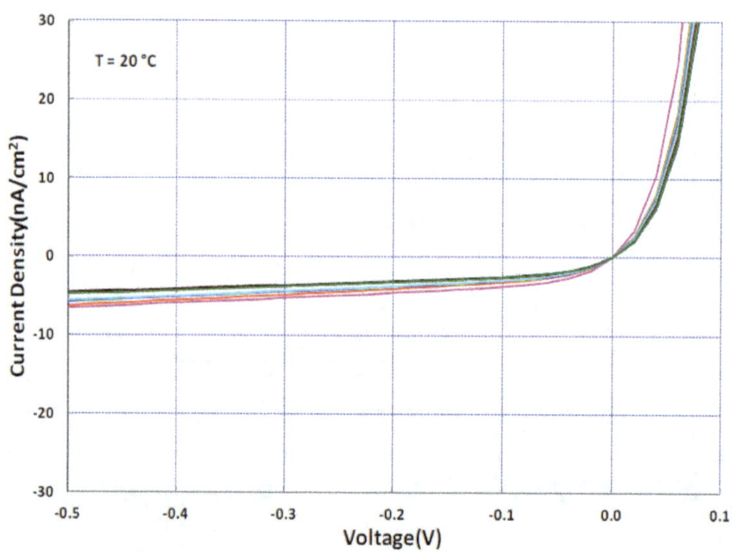

Figure 26. Experimental results for InGaAs test array demonstrating dark current density for eight separate 300 x 10, 15 μm pitch pixel test arrays measured across a wafer. The average dark current density for the test arrays at 100 mV reverse bias is 2.95 nA/cm2 at 20 °C [45].

Figure 27. Experimental 1280x1024/15μm arrays on 4" wafer surrounded by various test mini-arrays with pitch sizes of 5-20μm [47].

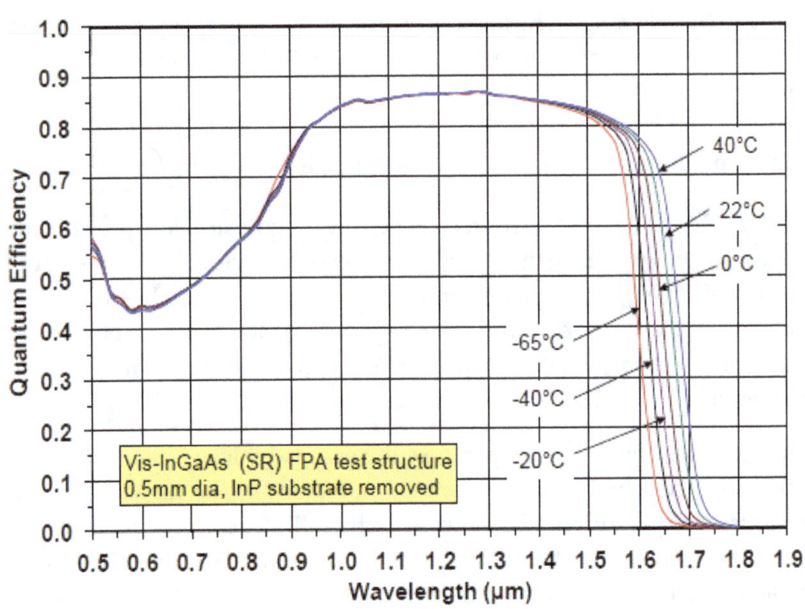

Figure 28. Spectral QE vs. wavelength at different temperatures measured for backside illuminated InGaAs photodiodes test array demonstrating Visible-Near IR response with InP substrate removed [47].

8. Nanostructured detector technology for MWIR and LWIR bands

EO/IR Sensors and imagers using nanostructure based materials are being developed for a variety of Defense Applications. In this section, we will present recent work under way for development of next generation carbon nanostructure based infrared detectors and arrays. We will discuss detector concepts that will provide next generation high performance, high frame rate, and uncooled nano-bolometer for MWIR and LWIR bands [52-55]. The critical technologies being developed include carbon nanostructure growth, characterization, optical and electronic properties that show the feasibility for IR detection. Experimental results on CNT nanostructures will be presented. Further discussion will be provided for the path forward to demonstrate enhanced IR sensitivity and larger arrays.

The microbolometer based on Si-MEMS device structure has been under development for over 20 years with support from DARPA and the US Army. Two most common Si-MEMS based structures utilize VOx and amorphous silicon based technologies. Several companies such as BAE systems and DRS Technologies are developing and producing 17 micron unit cell 640x 480 and larger arrays using VOx [48-51]. Similarly, L3Communications and other groups are developing and producing 640x480 with 17 micron unit cell using amorphous-Silicon technology [50-51].

We will discuss the use of carbon nanostructures for use as the high performance bolometric element of the MWIR and LWIR bands. As part of this effort, we are exploring development of smaller unit cell bolometer.i.e. 5-10 micron unit cell, with higher TCR and higher frequency response in the 1 to 10 KHz range. The feasibility of such an array can open up a larger number of defense and commercial applications. This section will discuss the efforts under way to explore these possibilities.

8.1. Design and modeling of CNT-based bolometer characteristics

To optimize bolometer design, we need to consider several physical key phenomena. From a fundamental point of view, we take the absorbing material to have an extremely large response to infrared radiation. The phonon modes of the material need to be able to easily couple to infrared radiation. Furthermore, once this coupling has been achieved, the absorbed radiation should greatly increase the population of the phonon modes thereby significantly increasing the lattice temperature [52-53].

On a macroscopic scale, this large temperature increase is typically described in terms of a large thermal resistance. The higher the thermal resistance of the material, in general, the higher the resulting temperature will be after absorbing IR. Of course, higher thermal resistance may also give rise to larger thermal noise. This mitigating factor must be balanced with signal response in order to optimize IR sensitivity, or the minimum detectable IR signal.

In summary, for DC operation, we work to maximize the thermal resistance R_{th} while achieving an acceptable noise, which maximizes the minimum signal that we can detect. In addition to DC operation, we try to maximize the thermal frequency response of the bolometer. This requires that we minimize the thermal capacitance C_{th} of the device and thereby minimize the thermal $R_{th}C_{th}$ time constant for the absorbing material.

Temperature Dependent Electrical Characteristics: In addition to considering the DC and transient thermal characteristics of the absorbing material, we need to optimize the electrical response as well. To achieve this, we associate the electrical response with the thermal response by considering the temperature dependent voltage-current characteristics of the electrical material.

A good response is obtained by utilizing a material that has an effectively large variation of electrical resistance with temperature. However, at the same time we want the material to have a temperature coefficient of resistance that is relatively independent of the absolute value of the resistance itself. Therefore, we look at materials that have an exponential relationship between electrical resistance and temperature. For such materials, the Temperature Coefficient of Resistance (TCR) is not strongly dependent on the absolute value of the electrical resistance of the material itself.

8.1.1. Calculating the thermal response of the CNT bolometer film

In this work we are assessing the possibility of designing a bolometer using carbon nanotubes as both the IR absorbing material and the electrical response material. Thus, our aim is to first determine the thermal response of the bolometer absorber that is composed of CNTs, and then determine how the electrical characteristics of the CNT material depend on its changes in temperature after IR absorption.

While there are numerous geometries of CNT based material we can consider, for this work we will focus on an absorbing material composed of a CNT film. The film is taken to consist of a random placement of CNTs that is two nanotubes thick.

Heat Flow Equation: To determine the temperature of the material in the presence of IR radiation, we start with the heat flow equation. This is a partial differential equation relating the time rate of change in temperature to the position and the rate of net heat that is absorbed by the material as a function of time and position.

$$Cv\frac{\partial T}{\partial t}=\kappa\nabla^2 T + H$$

In the equation C_v is the thermal capacitance (joules/degree-unit cm^3) and κ is the thermal diffusion coefficient in watts/degree-cm). *H* is the net power absorbed by the material in watts per unit volume. To solve this equation for the CNT bolometer, we have to first determine C_v and κ for the CNT and the CNT film.

Thermal Capacitance of CNT Absorber: To determine the heat capacity of a carbon nanotube, we first determine the internal vibrational energy of the CNT, and then take the derivative with respect to temperature. The internal energy is found by determining the energy of each vibrational mode, multiplying by the probability that the mode is populated using Bose-Einstein statistics, and then summing over all of the allowed modes.

The number of allowed modes will depend on the diameter and wrapping angle of the CNTs present, so we take a statistical sample. Multiplying the individual CNT heat capacity by the number of CNTs in the film provides a reasonable value for the heat capacity of the film. After following this procedure and inserting numerical values for physical constants, we arrive at the following average numerical value for thermal capacity of a CNT C_{vt}, where the length L is in microns and the diameter d is in nanometers

$C_{vt} \sim 1.4 \times 10^{-18} (L)(d)$

CNT Thermal Diffusion Coefficient: In addition to the thermal capacitance, we need to determine the thermal diffusion coefficient, and eventually the thermal resistance of a single CNT. Experiments on isolated CNTs have fit the coefficient of thermal diffusion to data obtaining the following expression [52-53]:

$$\kappa(L, T)=\left\{3.7x10^{-7}T+9.7x10^{-10}T^2+\frac{9.3}{T^2}\left[1+\frac{0.5}{L}\right]\right\}^{-1}$$

From κ we can obtain the thermal resistance R_T of a single CNT using the following definition:

$$R_T=\frac{4L}{\kappa\pi d^2}$$

Where L, d are the length and diameter of the CNT

Using average values for CNTs gives the following numerical value for the thermal resistance of a CNT

$R_T \sim 5 \times 10^8 (L/d^2)$

Where R_T is in units of degrees K/Watt, L is in units of microns and d is in units of nanometers. So a tube that is one micron long and one nanometer in diameter will have a thermal resistance R_T of approximately 5×10^8 K/W.

Net IR Radiation Power Absorbed: Now that we have the thermal diffusion and capacitance we are almost ready to begin solving the above heat flow equation to determine the temperature of the bolometer. However, before doing so, we need to determine H, the net IR power absorbed by the bolometer. We determine this power using the Stefan-Boltzmann Law of blackbody radiation, which relates the net power absorbed to the temperatures of the subject and the bolometer using the following expression:

$$H_{net}=\sigma A\varepsilon(T_{obj}^4-T_b^4)$$

Where H_{net}, σ, A, ε, T_{obj} and T_b are heat absorbed by the bolometer, the Stefan-Boltzmann constant, cross-sectional area, emissivity, object of interest temperature and bolometer absorber temperature, respectively. Figure 1 below shows the net IR power absorbed by the absorber as a function of bolometer temperature for radiating objects at 20°C and 36.5°C. Cooling the bolometer by 30°C below room temperature allows for significantly more power to be absorbed, which can give rise to a much stronger signal.

Calculating the Bolometer Temperature Distribution: Using the aforementioned expressions for C_{vt}, K and H, we expand on previous work and convert the heat flow equation above into a

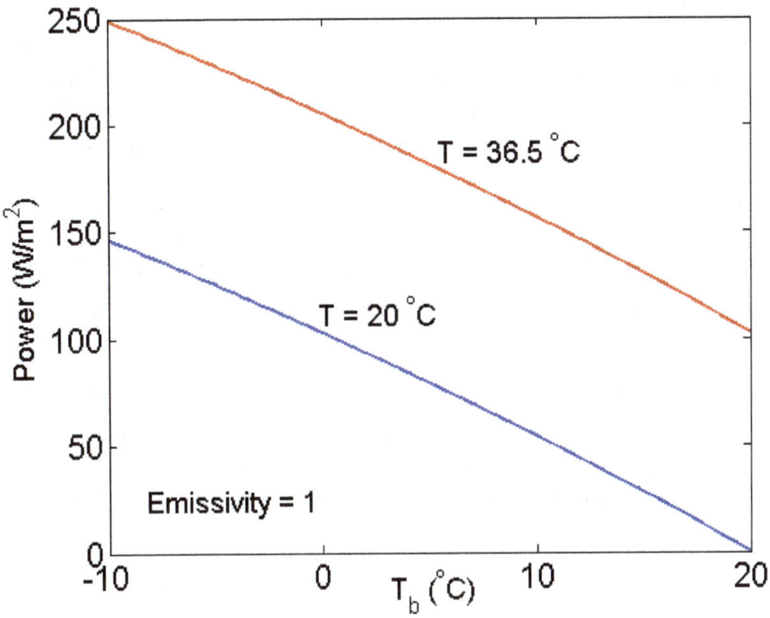

Figure 29. Net power received by bolometer as a function of bolometer temperature.[54]

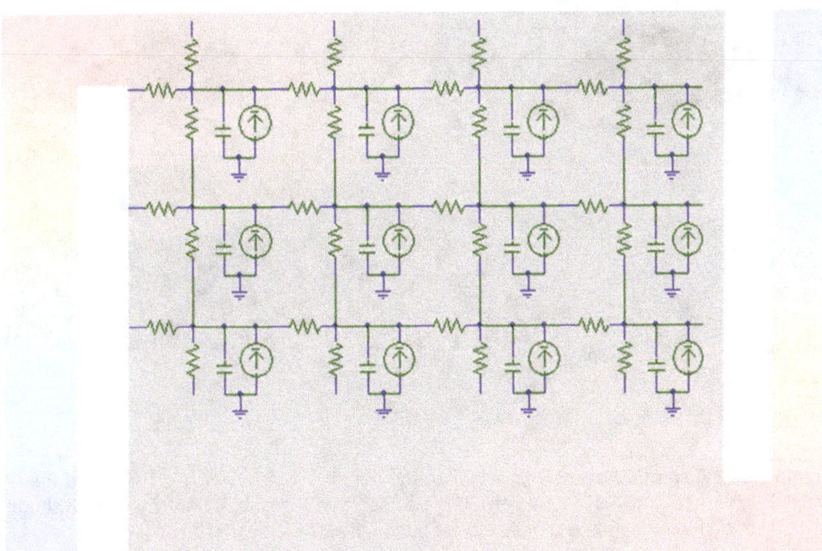

Figure 30. Illustration of thermal network superimposed on bolometer for calculating temperature map of CNT bolometer absorber.[55]

thermal network, illustrated in Figure 2 [6]. In actuality, there are thousands of nodes in the network for which we calculate the temperature for each.

In Figure 30, each resistor represents the thermal resistance of a CNT in series with the thermal resistance between adjacent CNTs. In addition, the capacitors represent the thermal capacity of a CNT, while the current sources represent the net IR radiation absorbed by each CNT. This thermal network contains thousands of nodes, and there is an equation relating the thermal resistance, capacitance and net power for each node. This system of equations is then solved for the temperature as a function of position and time throughout the bolometer absorber [54].

Results of these calculations for are shown in Figure 31 for different types of CNT networks. Here, we assumed that the net absorbed power is 1 nW, and the pixel is tightly packed with the CNTs. The entire pixel's temperature map is obtained with a 100×100 temperature resolution. For the tubes, we used two different thermal resistance values: 5×10^8 and 1×10^9 K/W.

As expected, higher the thermal resistance, higher the temperature difference from the ambient. We note that in general thermal resistance also rises with increasing temperature, resulting in further heating of hot spots compared to the case that this dependency is ignored. The temperature gradient of the contacts legs connecting the film to the readout IC (ROIC) is clearly shown.

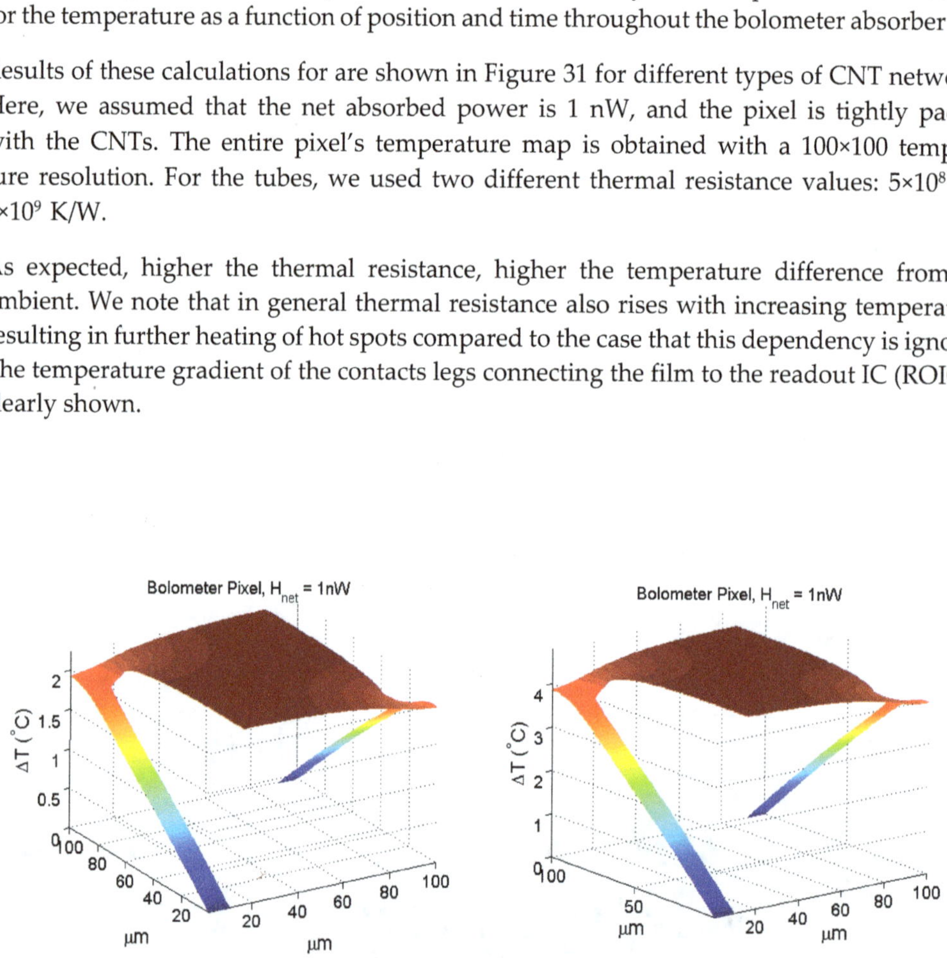

Figure 31. Temperature map of a bolometer pixel when the net absorbed power *H* is 1 nW, and the CNT thermal resistance is 5×10^8 K/W (left) and 1×10^9 K/W (left). We assumed that the pixel is tightly packed with the CNTs. The temperature gradient of the contacts legs connecting the film to the ROIC is clearly shown [54]

8.1.2. Calculating the electrical response of the CNT film

To read the temperature that the bolometer pixel reaches after an exposure to infrared radiation, one needs to measure the electrical resistance of the pixel. By comparing this resistance to a look-up table or using the a-priori knowledge of temperature coefficient of resistance, the pixel temperature can be determined. Therefore, in addition to having a large thermal resistance which translates into higher temperature rises, a large temperature coefficient of electrical resistance (TCR) is desirable to achieve a higher temperature resolution.

Here TCR is defined as the change in electrical resistance per degree Kelvin divided by the absolute electrical resistance measured at the quiescent point, as follows:

$$TCR = \frac{1}{R_e}\frac{dR_e}{dT}$$

Thus, the pixel electrical resistance after it reaches a temperature that is ΔT above its ambient becomes $R_e(T) = R_e(To)(1{+}TCR)$. Using this relationship, the pixel temperature is calculated.

To obtain a high temperature resolution, a large change in electrical resistance is needed upon heating. To achieve this, a substantial increase either in electron concentration or velocity (for a given electric field) is necessary. And to this end, materials with junctions where thermionic emission or tunneling are the electrical current bottlenecks offer a good solution. As the tunneling current exponentially rises with temperature, the effective change in their electrical resistance due to temperature becomes large compared to those observed in bulk materials where the change is proportional T^γ and γ is generally < 2.

Here, a film of CNTs is proposed as the bolometer pixel material, since it is likely to have large thermal resistance and TCR values simultaneously. Both of these favorable properties are partially owed to the junctions between the tubes. As the electrical current flows along the mat, it needs to jump from one tube to the next where they intersect.

At this intersection, the carriers see a potential barrier that they need to tunnel through which gives rise to exponential increase in current upon heating. Assuming that the electron transport across this barrier is governed by a Fowler-Nordheim-type tunneling or thermionic emission, the expected TCR values can be calculated using the following expression:

$$I = q\int_0^\infty T_t(E)v(E)DOS(E)f(E,\,T)dE$$

Where T_t, v, DOS and f, are the transmission coefficient, thermal velocity, density of states and distribution function, respectively for electrons in the CNT. We perform this calculation a function of barrier height and electric field. The results are shown in figure 4. The figure on the left shows a contour plot of the theoretical values of the TCR for a CNT film.

The TCR is plotted as a function of electric field between the tubes and the barrier height. Theoretical calculations predict an extremely large TCR, which can be attributed to the relatively large barrier height between adjacent CNTs. If a lower barrier height is assumed, on the order of 0.06eV, then a TCR of approximately 2.5% is obtained. It is also worth pointing out that such a large TCR comes at the price of extremely low output currents. The bolometer current densities, as a function of barrier height and electric field are shown in Figure 32, on the right.

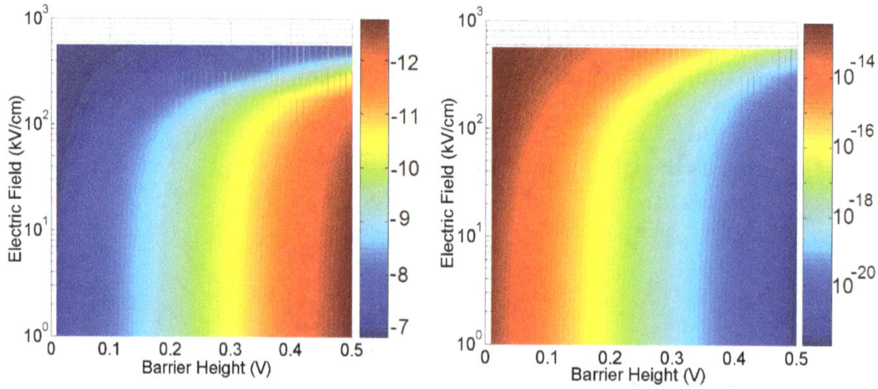

Figure 32. Left figure is a contour plot of TCR versus electric field and barrier height between CNTs of the film. The right figure shows the bolometer current also as a function of electric field and barrier height. The scales are the color bars on the right of each contour plot in units of %TCR and amperes, respectively.

8.1.3. CNT growth and charcterization

In this section, we will discuss growth and characterization of carbon nanotubes with single wall (SWCNT) and multiwall (MWCNT) for use as the high performance bolometric element for development of MWIR and LWIR sensitive detector elements.

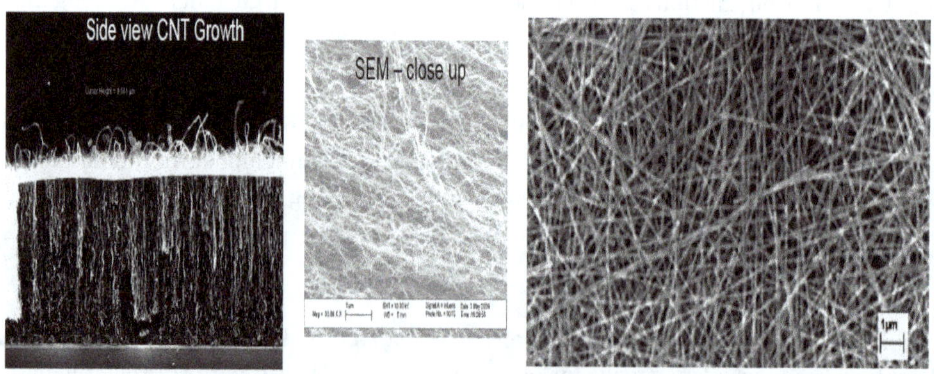

Figure 33. Growth of multiwall CNT forest with the ability to separate the form growth substrate with good length / diameter uniformity and the MWCNT released from the template.[55]

Figure 33 presents growth of dense oriented multi-walled CNT "forest like growth". The figure shows the CNT growth can be easily separated from the growth substrate. We have shown good length/diameter uniformity. Further work on the growth optimization is underway.

Figure 34 shows the prototype fixture to evaluate the CNT films for bolometric application. This fixture is being used for quick evaluation of both electrical and optical characteristics of the CNT samples. The figure also shows the preliminary results of reflectivity measurements for SWCNT and MWCNT samples with various sample treatments. We have carried out some preliminary measurements of TCR on CNT samples.

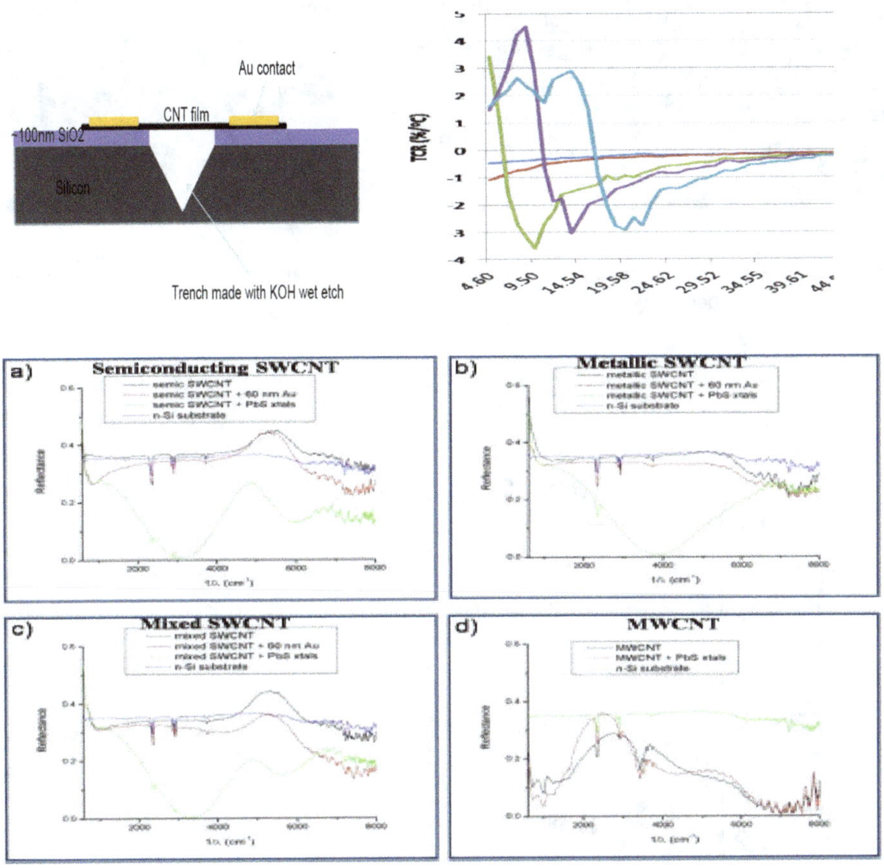

Figure 34. The composite figure shows prototype CNT film bolometer test fixture to evaluate the CNT film quality. Some preliminary data on reflectivity measurements for SWCNT and MWCNT are shown along with preliminary results on TCR measurements.[55]

Figure 35 shows scanning electron microscopy SEM. images of representative MWCNT films in the unsuspended left. and suspended.right. forms, respectively. Unlike their SWCNT counterparts, the MWCNT films contain substantial uncovered substrate areas. In addition, some minor deformation of recess is visible on suspended MWCNT films, which is similar to the SWCNT film case in the same thickness range. Figure 35.b includes a transmission electron microscopy.TEM. image of a representative individual MWCNT, which has a large hollow center of approximately 10–11 nm in diameter and contains approximately 40–50 CNT shells.

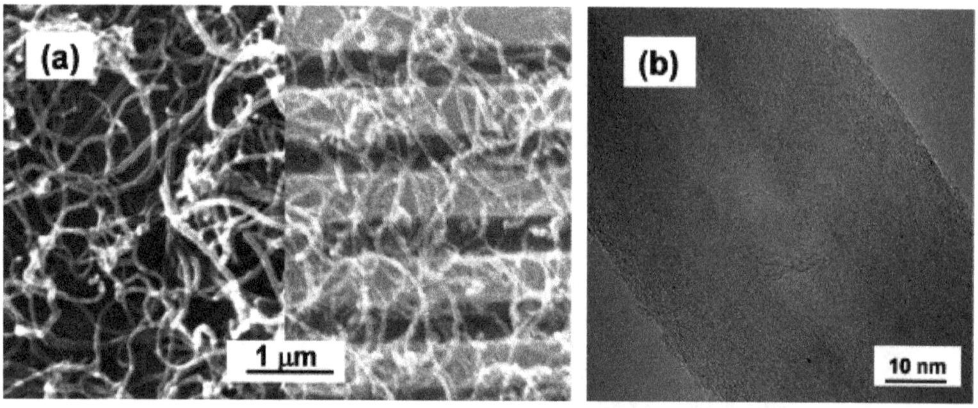

Figure 35. SEM images of unsuspended.left (a). and suspended.right (b). MWCNT films. A TEM image of a representative MWCNT. The shell number is estimated to be.40–50 for the MWCNTs [56]

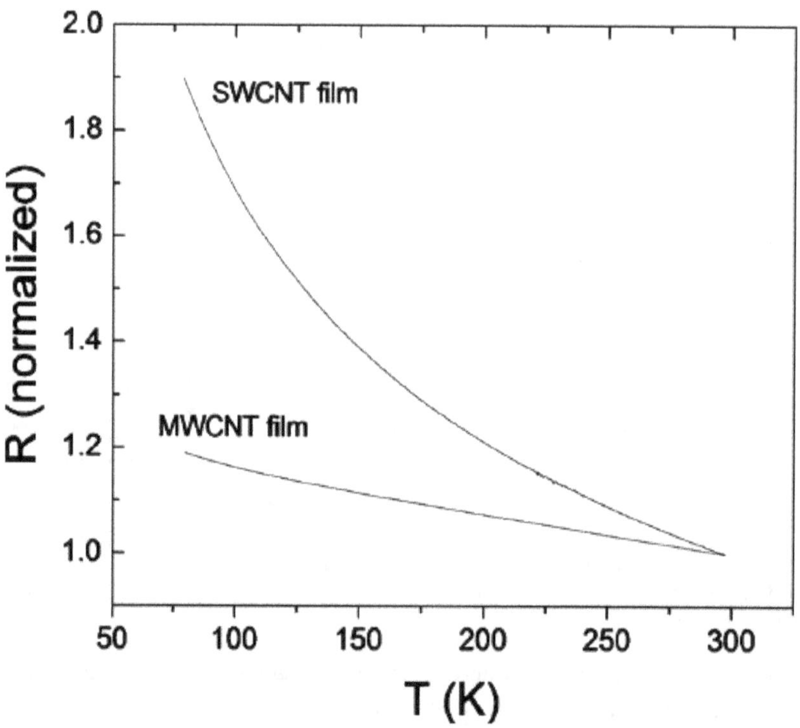

Figure 36. Resistance versus temperatures curves of SWCNT films and MWCNT films.[56]

All MWCNT films studied in this work [56] show semiconductive resistance-temperature.R-T. behaviors and a representative curves is depicted in Figure 36.. Nevertheless, the increase in the resistivity of MWCNT films is much less than that of SWCNT films with decreasing temperature, as shown in Figure 36. This is not unexpected considering a much smaller band gap in MWCNTs. The reduced temperature dependence also implies smaller TCR absolute value in MWCNTs. For example, the TCR absolute value at room temperature for MWCNT films is about 0.07%/K in contrast to.0.17%/K for SWCNT films. R-T curve after suspending the MWCNT film has been also measured.

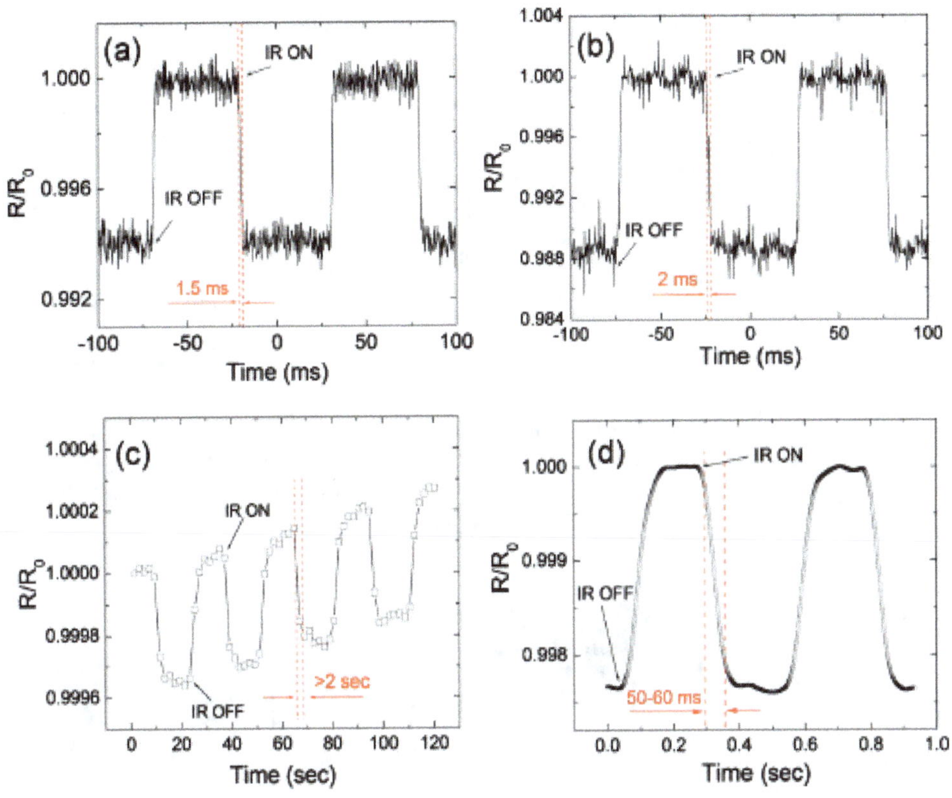

Figure 37. Photoresponse of unsuspended and suspended CNT films.(a) Unsuspended MWCNT film, f=10 Hz, in IR.3 mW/mm2 ;.(b). suspended MWCNT film, f=10 Hz, in IR.3 mW/mm2 ; (c) unsuspended SWCNT film, f=1/30 Hz, in IR.3.5 mW/mm2 ; and (d) suspended SWCNT film, f =2 Hz, in IR.3.5 mW/mm2 [56].

Figure 37 compares the Photoresponse R/R0 of MWCNT films in unsuspended (a) and suspended (b) cases, where R_0 is the sample resistance before IR radiation was turned on and the change in the resistance caused by IR radiation is defined as.R=R-R_0. For comparison, the results of their SWCNT counterparts are also included in Figure 37 (c).unsuspended and Fig. 37(d). suspended..

Two major differences are visible between MWCNT and SWCNT films, a significantly higher.R/R_0 and a much shorter response time in the cases of MWCNT. The.R/R_0 for MWCNT samples is typically in the range of a few percent, which is more than one order of magnitude higher than that of suspended SWCNT films and two orders of magnitude higher than the unsuspended SWCNT films at a comparable IR power. Considering a lower TCR absolute value in MWCNTs, the much enhanced Photoresponse of MWCNT films should be attributed to the naturally suspended inner CNT shells, which may provide an ideal configuration to enhance the bolometric effect by improving light absorption and reducing thermal link. Physical suspension of the films in both MWCNT.Fig. 37.(b) and SWCNT.Fig. 37(d.) cases results in a further improvement of.R/R_0 as compared to their unsuspended counterparts. The improvement is, however, much more pronounced in suspended cases [56].

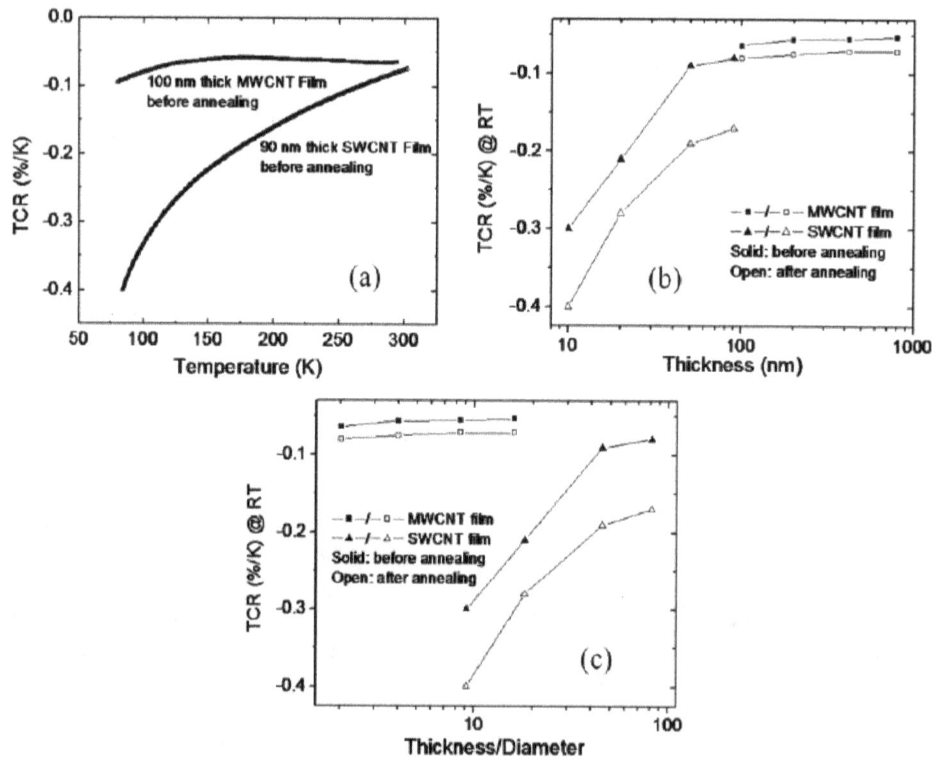

Figure 38. TCR as function of temperature for a 90 nm thick MWCNT film (a) and a 100 nm thick MWCNT film before annealing. (b) TCR versus thickness for MWCNT and SWCNT films with different thicknesses. (c) TCR versus thickness/diameter ratios for SWCNT and MWCNT films [57].

We have also shown the results of TCR as function of temperature in figure 38, for a 90 nm thick MWCNT film (a) and a 100 nm thick MWCNT film before annealing. (b) TCR versus thickness for MWCNT and SWCNT films with different thicknesses. (c) TCR versus thickness/diameter ratios for SWCNT and MWCNT films [57].

We have discussed recent efforts for modeling CNT based bolometer and the experimental work for development of next generation carbon nanostructure based infrared detectors and arrays. Our goal is to develop high performance, high frame rate, and uncooled nano-bolometer for MWIR and LWIR bands. We also discussed CNT growth system and its capability to grow samples of various orientations. We have also presented recent results on SWCNT and MWCNT samples that show promise for use of CNT for developing next generation high performance small pixel bolometer arrays.

9. Summary

In this chapter, we have discussed recent advances in nanostructured based detector technology, materials and devices for optical sensing applications. The chapter has presented an overview of recent work underway on a variety of semiconductors and advanced materials such as GaN, ZnO, Si/SiGe, InGaAs and CNT for optical sensing applications.

Optical sensing technology is critical for defense and commercial applications including optical communication. Advances in optoelectronics materials in the UV, Visible and Infrared, using nanostructures, and use of novel materials such as CNT have opened doors for new approaches to apply device design methodology that are expected to offer enhanced performance and low cost optical sensors in a wide range of applications.

We have covered the UV band (200-400 nm) and address some of the recent advances in nanostructures growth and characterization using GaN/AlGaN, ZnO/MgZnO based technologies and their applications. We have also discussed nanostructure based Si/SiGe technologies (400-1700 nm) that covers various bands of interest in visible-near infrared for detection and optical communication applications. The chapter has also discussed some of the theoretical and experimental results in these detector technologies.

Recent advancements in design and development of CNT based detection technologies have shown promise for optical sensor applications. We have presented theoretical and experimental results on these device and their potential applications in various bands of interest. It is anticipated that the current research and development presented in this chapter will enable a host of new integrated technologies for a variety of defense and commercial applications.

Although numerous research activities are ongoing in the area of Nanoscience and technology, we briefly made comments on such technologies to make readers aware of various research activities.

Acknowledgements

The authors gratefully acknowledge the contributions of the many distinguished scientists in the United States for development of nanotechnology based EO/IR detector technology for optical sensor applications.

Author details

Ashok K. Sood[1], Nibir K. Dhar[2], Dennis L. Polla[3], Madan Dubey[4] and
Priyalal Wijewarnasuriya[4]

1 Magnolia Optical Technologies Inc., Woburn, MA, USA

2 Defense Advanced Project Agency, Arlington, VA, USA

3 College of Science and Engineering, University of Minnesota, Minneapolis, MN, USA

4 Army Research Laboratory, Adelphi, MD, USA

References

[1] Raytheon Vision Systems Spectral Charts, Goleta, CA

[2] N. K. Dhar. " IR Material Research at the Army Research Laboratory" Invited Keynote Paper, Proceedings of SPIE, Volume 6542, 65420C (2007)

[3] J.P. Long, "UV detectors and focal plane array imagers based on AlGaN p-i-n photodiodes", Opto-Electronics Review 10(4), 251-260 (2002)

[4] M. Crawford, "Advances in AlGaN-based deep UV LEDs ", MRS Proceedings, Vol. 831, fall 2004.

[5] M.B. Reine, "Solar-blind AlGaN 256x256 p-i-n detectors and focal plane arrays", Proc. of SPIE, Vol. 6119 (2006)

[6] Ashok K. Sood, Robert A. Richwine, Yash R. Puri, Nibir K. Dhar, Dennis L. Polla, and Priyalal S. Wijewarnasuriya, " Multispectral EO/IR sensor model for evaluating UV, visible, SWIR, MWIR and LWIR system performance" Proceedings of SPIE 7300, 73000H (2009)

[7] Abdiel Rivera, John Zeller, Tariq Manzur, Ashok Sood and Mehdi Anwar, " MOCVD Growth and Characterization of ZnO Nanowire Arrays for UV Detectors" Proceedings of SPIE, Volume 8540, October 2012.

[8] Liang, S., Sheng, S., Liu, Y., Huo, Z., Lu, Y., and Shen, H., "ZnO Schottky ultraviolet photodetectors," J. Cryst. Growth 225, 110-113 (2001).

[9] Zhang, J., Que, W., Jia, Q., Ye, X., and Ding, Y., "Controllable hydrothermal synthesis of ZnO nanowires arrays on Al-doped ZnO seed layer and patterning of ZnO nanowires arrays via surface modification of substrate," Appl. Surf. Sci. 257(23), 10134-10140 (2011).

[10] Lee, C. H., Yi, G. C.., Zuev, Y. M., and Kim, P., "Thermoelectric power measurements of wide band gap semiconducting nanowires,"Appl. Phys. Lett. 94, 22106 (2009).

[11] Falyouni, F., Benmamas, L., Thiandoume, C., Barjon, J., Lusson, A., Galter, P., and Sallet, V., "Metal organic chemical vapor deposition growth and luminescence of ZnO micro- and nanowires," Journal Vac. Sci. Technol. B 87, 1662 (2009).

[12] Jeong, M. C., Oh, B.Y., Lee, W., and Myoung, J. M., "Comparative study on the growth characteristics of ZnO nanowires and thin films by metal-organic chemical vapor deposition (MOCVD)," Journal of. Crystal Growth 268, 149-154 (2004).

[13] Kim, S. W., Fujita, S., and Fujita, S., "ZnO nanowires with high aspect ratios grown by metal-organic chemical vapor deposition using gold nanoparticles," Appl. Phys. Lett. 86,153119 (2005).

[14] Lee, W., Jeong, M. C., and Myoung, J. M., "Catalyst-free growth of ZnO nanowires by metal-organic chemical vapor deposition (MOCVD) and thermal evaporation," Acta Mat. 52, 3949-3957 (2004).

[15] Liou, S. C., Hsiao, C. S., and Chen, S. Y., "Growth behavior and microstructure evolution of ZnO nanorods grown on Si in aqueous solution," Journal of. Crystal. Growth 274, 438 (2005).

[16] Dong, J. W., Osinski, A., Hertog, B., Dabiran, A. M., Chow, P. P., Heo, Y. W., Norton, D. P, and Pearton, S. J., "Development of MgZnO-ZnO-AlGaN heterostructures for ultraviolet light emitting applications," J. Electron. Mat. 34, 416-423 (2005).

[17] Rivera, A., Zeller, J., Sood, A.K., and Anwar, A. F. M., "A Comparison of ZnO Nanowires and Nanorods Grown Using MOCVD and Hydrothermal Processes," J. Electron. Mat. 42, 894-900 (2013).

[18] Ha, B., Ham, H., and Lee, C. J., "Photoluminescence of ZnO nanowires dependent on O_2 and Ar annealing," Phys. Chem. Solids 69, 2453-2456 (2008).

[19] Djurišić, A. B., Ng, A.M.C., and Chen, X.Y., "ZnO nanostructures for optoelectronics: Material properties and device applications," Progress Quantum Electronics 34, 191-259 (2010)

[20] Mehdi Anwar, Abdiel Rivera, Anaz Mazady, Hung Chou, John Zeller and Ashok K. Sood, "ZnO Solar Blind Detectors: from Material to System", Proceedings of SPIE Volume 8868, 8868B (2013).

[21] Zhong Lin Wang, Guang Zhu, Ya Yang, Sihong Wang and Caofeng Pan, " Progress in Nanogenerator for Portable Electronics" Materials Today Volume 15, Number 12 December 2012

[22] Wenzhuo Wu, Xiaonan Wen and Zhong Lin Wang, " Texel- Addressable Matrix of Vertical-Nanowire and Adaptive Tactile Imaging" Science, Volume 340, 24 May 2013.

[23] R.D. Dupuis, H.J. Ryou and D. Yoder. "High-performance GaN and AlGaN ultraviolet avalanche photodiodes grown by MOCVD on bulk III-N substrates", Proc. of SPIE, Vol. 6739 (2006)

[24] S.C. Shen, Y. Chang, J. B. Limb, J.H. Ryou, P. D. Yoder and R.D. Dupuis, " Performance of Deep UV GaN Avalanche Photodiodes Grown by MOCVD", IEEE Photonics Technology Letters, Volume 19, Number 21, November 2007.

[25] Y. Zhang, S.C. Shen, H. J. Kim, S. Choi, J. H. Ryou, R.D. Dupuis and B. Narayan, " Low-Noise GaN Ultraviolet p-i-n photodiodes on GaN Substrates" Applied Physics Letters, 94, 221109 (2009).

[26] Ashok K. Sood, Robert A. Richwine, Roger E. Welser, Yash R. Puri, Russell D. Dupuis, Mi-Hee Ji, Jemoh Kim, Theeradetch Detchprohm, Nibir K. Dhar and Roy L. Peters, " Development of III-N UVAPDs for Ultraviolet Sensor Applications" Proceedings of SPIE Volume 8868, 88680T (2013)

[27] Ashok K. Sood, Robert A. Richwine, Yash R. Puri, Russell D. Dupuis, Nibir K. Dhar and Raymond S. Balcerak "Development of GaN/AlGaN APD's for UV Imaging Applications" Proceedings of SPIE 7780, 77800E (2010)

[28] J.P. Long, "UV detectors and focal plane array imagers based on AlGaN p-i-n photodiodes", Opto-Electronics Review 10(4), 251-260 (2002)

[29] M. Crawford, "Advances in AlGaN-based deep UV LEDs ", MRS Proceedings, Vol. 831, fall 2004.

[30] M.B. Reine, "Solar-blind AlGaN 256x256 p-i-n detectors and focal plane arrays", Proc. of SPIE, Vol. 6119 (2006)

[31] Ashok K. Sood, Robert A. Richwine, Yash R. Puri, Nibir K. Dhar, Dennis L. Polla, and Priyalal S. Wijewarnasuriya, " Multispectral EO/IR sensor model for evaluating UV, visible, SWIR, MWIR and LWIR system performance" Proceedings of SPIE 7300, 73000H (2009)

[32] S.C. Shen, Y. Chang, J. B. Limb, J.H. Ryou, P. D. Yoder and R.D. Dupuis, " Performance of Deep UV GaN Avalanche Photodiodes Grown by MOCVD", IEEE Photonics Technology Letters, Volume 19, Number 21, November 2007.

[33] Y. Zhang, S.C. Shen, H. J. Kim, S. Choi, J. H. Ryou, R.D. Dupuis and B. Narayan, " Low-Noise GaN Ultraviolet p-i-n photodiodes on GaN Substrates" Applied Physics Letters, 94, 221109 (2009).

[34] B.-Y. Tsaur, C. K. Chen and S. A. Marino, "Long-wavelength $Ge_x Si_{1-x}$ /Si heterojunction infrared detectors and focal plane array," *Proc. SPIE 1540*, 580-595, 1991.

[35] H. Wada, M. Nagashima, K. Hayashi, J. Nakanishi, M. Kimata, N. Kumada and S. Ito, "512 x512 element GeSi/Si heterojunction infrared focal plane array," *Proc. SPIE* 3698, 584-595, 1999.

[36] A. K. Sood, R. A. Richwine, Y. R. Puri, N. DiLello, J. L. Hoyt, N. Dhar, R. S. Balcerak, and T. G. Bramhall, "Development of SiGe Arrays for Visible-Near IR Imaging Applications," *Proc. SPIE* 7780,77800F, 2010.

[37] V.T. Bublik, S.S. Gorelik, A.A. Zaitsev and A.Y. Polyakov, "Calculation on the Binding Energy of Ge-Si Solid Solution," *Phys. Status Solidi* 65, K79-84, 1974.

[38] A. K. Sood, R. A. Richwine, Y. R. Puri, N. DiLello, J. L. Hoyt, T. Akinwande, N. K. Dhar, R. S. Balcerak, and T. G. Bramhall, "Characterization of SiGe detector arrays for Visible-Near IR Imaging Sensor Applications," *Proc. SPIE* 8012,801240, 2011.

[39] Ashok K. Sood, Robert A. Richwine, Gopal Pethuraja, Yash R. Puri, Je-Ung Lee, Pradeep Haldar and Nibir K. Dhar, " Design and Development of Wafer-Level Short Wave Infrared Micro-Camera" Proceedings of The SPIE Volume 8704, 870439 (2013).

[40] R.J. van der A, R.W.M. Hoogeveen, H.J. Spruijt, and A.P.H. Goede, "Low noise InGaAs infrared (1.O-2.4tm) Focal Plane Arrays," *Proc. SPIE* 2957, 54-65, 1997.

[41] D. Acton, M. Jack, and T. Sessler, "Large format short-wave infrared (SWIR) focal plane array (FPA) with extremely low noise and high dynamic range," *Proc. SPIE* 7298, 72983E, 2009.

[42] B. M. Onat, W. Huang, N. Masaun, M. Lange, M. H. Ettenberg, and C. Dries, "Ultra low dark current InGaAs technology for focal plane arrays for low-light level visible-shortwave infrared imaging," *Proc. SPIE* 6542, 65420L, 2007.

[43] J. Boisvert, T. Isshiki, R. Sudharsanan, P. Yuan, and P. McDonald, "Performance of very low dark current SWIR PIN arrays," *Proc. SPIE* 6940, 69400L, 2008.

[44] M. MacDougal, J. Geske, J. Wang, and D. Follman, "Short-wavelength infrared imaging using low dark current InGaAs detector arrays and vertical-cavity surface-emitting laser illuminators," Optical Engineering 50(6), 061011, 2011.

[45] A. D. Hood. "Large InGaAs Focal Plane Arrays for SWIR Imaging" Proceedings of SPIE, Volume 8353, 83530A, 2012.

[46] H. Yuan, M. Meixell, J. Zhang, P. Bey, J. Kimchi, and L.C. Kilmer "Low Dark Current Small Pixel Large Format InGaAs 2-D Photodetector array Development" Proceedings of SPIE, Volume 8353, 835309, 2012.

[47] H. Yuan, G. Apgar, J. Kim, J. Laquindanum, V. Nalavade, P. Beer, J. Kimchi, and T. Wong, "FPA development: from InGaAs, InSb, to HgCdTe," *Proc. SPIE* 6940, 69403C, 2008.

[48] R. Blackwell, D. Lacroix, T. Bach, et. al " 17 micron microbolometer FPA Technology at BAE Systems" Proceedings of SPIE, Volume 7298,72980P (2009).

[49] C. Li, G. Skidmore, C. Howard, E. Clarke and J. Han, "Advancement in 17-micron pixel pitch uncooled focal plane arrays" Proceedings of SPIE, Volume 7298, 72980S (2009).

[50] T. Schimert, C. Hanson, J. Brady, et. al. " Advanced in small-pixel, large-format alpha-silicon bolometer arrays", Proceedings of SPIE, Volume 7298, 72980T (2009)

[51] C. Trouilleau, B. Fieque, S. Noblet, F. Giner et.al. "High Performance uncooled-amorphous silicon TEC less XGA IRFPA with 17 micron pixel pitch" Proceedings of SPIE Volume 7298, 72980Q, (2009).

[52] E. Pop, D. Mann, Q. Wang, K. Goodson, H. Dai, Thermal Conductance of an Individual Single-Wall Carbon Nanotube above Room Temperature Nano Letters 6, 96 (2006)

[53] A. Akturk, N. Goldsman, G. Metze, "Self-consistent modeling of heating and MOSFET performance in 3-D integrated circuits," IEEE Trans. on Elect. Dev. 52 (11): 2395-2403 (2005).

[54] Ashok K. Sood, E. James Egerton, Yash R. Puri, Gustavo Fernandes, Jimmy Xu, Akin Akturk, Neil Goldsman, Nibir K. Dhar, Madan Dubey, Priyalal S. Wijewarnasuriya and Bobby I Lineberry, " Design and Development of CNY based Micro-bolometer for IR Imaging Applications" Proceedings of SPIE, Volume 8353, 83533A, May 2012

[55] Gustavo Fernandes, Jin Ho Kim, Jimmy Xu, Ashok K. Sood, Nibir K. Dhar and Madan Dubey, " Unleashing Giant TCR from Phase –Changes in Carbon Nanotube Composites" Proceedings of SPIE Volume 8868, 88680S, September 2013

[56] Rongtao Lu, Jack J. Shi, F. Javier Baca and Judy Z. Wu, " High Performance Multiwall Carbon Nanotube Bolometer" Journal of Applied Physics, 108, 084305 (2010).

[57] Rongtao Lu, Rayyan Kamal and Judy Z. Wu, " A comparative study of 1/f noise and temperature coefficient of resistance in multiwall and single-wall carbon nanotube bolometers" nanotechnology, Volume 22, 265503 (2011).

Permissions

All chapters in this book were first published by InTech Open; hereby published with permission under the Creative Commons Attribution License or equivalent. Every chapter published in this book has been scrutinized by our experts. Their significance has been extensively debated. The topics covered herein carry significant findings which will fuel the growth of the discipline. They may even be implemented as practical applications or may be referred to as a beginning point for another development.

The contributors of this book come from diverse backgrounds, making this book a truly international effort. This book will bring forth new frontiers with its revolutionizing research information and detailed analysis of the nascent developments around the world.

We would like to thank all the contributing authors for lending their expertise to make the book truly unique. They have played a crucial role in the development of this book. Without their invaluable contributions this book wouldn't have been possible. They have made vital efforts to compile up to date information on the varied aspects of this subject to make this book a valuable addition to the collection of many professionals and students.

This book was conceptualized with the vision of imparting up-to-date information and advanced data in this field. To ensure the same, a matchless editorial board was set up. Every individual on the board went through rigorous rounds of assessment to prove their worth. After which they invested a large part of their time researching and compiling the most relevant data for our readers.

The editorial board has been involved in producing this book since its inception. They have spent rigorous hours researching and exploring the diverse topics which have resulted in the successful publishing of this book. They have passed on their knowledge of decades through this book. To expedite this challenging task, the publisher supported the team at every step. A small team of assistant editors was also appointed to further simplify the editing procedure and attain best results for the readers.

Apart from the editorial board, the designing team has also invested a significant amount of their time in understanding the subject and creating the most relevant covers. They scrutinized every image to scout for the most suitable representation of the subject and create an appropriate cover for the book.

The publishing team has been an ardent support to the editorial, designing and production team. Their endless efforts to recruit the best for this project, has resulted in the accomplishment of this book. They are a veteran in the field of academics and their pool of knowledge is as vast as their experience in printing. Their expertise and guidance has proved useful at every step. Their uncompromising quality standards have made this book an exceptional effort. Their encouragement from time to time has been an inspiration for everyone.

The publisher and the editorial board hope that this book will prove to be a valuable piece of knowledge for researchers, students, practitioners and scholars across the globe.

List of Contributors

Kaikai Xu and Qi Yu
State Key Laboratory of Electronic Thin Films and Integrated Devices, University of Electronic Science and Technology of China, Chengdu, China

Weifeng Sun
National ASIC System Engineering Research Center, Southeast University, Nanjing, China

Lukas W. Snyman
Department of Electrical Engineering, Tshwane University of Technology (TUT), Pretoria, South Africa

Kingsley A. Ogudo
Department of Electrical Engineering, Tshwane University of Technology (TUT), Pretoria, South Africa
Université Paris-Est, ESYCOM, ESIEE Paris, UPEM, Le Cnam, Cité Descartes, Noisy-Le-Grand-Cedex, France

Jean-Luc Polleux
Université Paris-Est, ESYCOM, ESIEE Paris, UPEM, Le Cnam, Cité Descartes, Noisy-Le-Grand-Cedex, France

Guannpyng Li
Integrated Nanosystems Research Facility (INRF), University of California, Irvine, California, USA

Oswaldo González and Marcos F. Guerra Medina
Department of Industrial Engineering, University of La Laguna, Spain

Inocencio R. Martín
Department of Physics, University of La Laguna, Spain

S.H. Mousavi
INM -Leibniz Institute for New Materials, Germany
Physics Department, Shahrood University of Technology, Shahrood, Iran

P.W. de Oliveira
INM -Leibniz Institute for New Materials, Germany

Vjaceslavs Bobrovs, Jurgis Porins, Sandis Spolitis and Girts Ivanovs
Riga Technical University, Institute of Telecommunications, Latvia

H. Haratizadeh
Physics Department, Shahrood University of Technology, Shahrood, Iran

S.A. Jafari Mohammdi
Aryasolar Chemical Company, Tehran, Iran

Farzaneh Fadakar Masouleh
Physics Department, University of Guilan, Rasht, Guilan, Iran

Narottam Das
Department of Electrical and Computer Engineering, Curtin University, Perth, WA, Australia
Department of Electrical and Computer Engineering, Curtin University, Miri, Sarawak, Malaysia

Alejandro García Juárez, Luis Arturo García Delgado, Ana Lilia Leal Cruz and Alicia Vera Marquina
University of Sonora, Department of Physics Research, México

Ignacio Enrique Zaldívar Huerta
National Institute of Astrophysics, Optics and Electronics Department of Electronics, México

Antonio Baylón Fuentes
Inst. FEMTO-ST, Université de Franche-Comté, Besançon, France

María del Rocío Gómez Colín
University of Sonora, Department of Physics, México

Y. Ben-Ezra and B.I. Lembrikov
Department of Electrical Engineering, Holon Institute of Technology, Holon, Israel

Ashok K. Sood
Magnolia Optical Technologies Inc., Woburn, MA, USA

Nibir K. Dhar
Defense Advanced Project Agency, Arlington, VA, USA

Dennis L. Polla
College of Science and Engineering, University of Minnesota, Minneapolis, MN, USA

Madan Dubey and Priyalal Wijewarnasuriya
Army Research Laboratory, Adelphi, MD, USA

Index

www.ingramcontent.com/pod-product-compliance
Lightning Source LLC
LaVergne TN
LVHW080617210926
848485LV00003B/175